CHEMICAL KNOWLEDGE IN THE EARLY MODERN WORLD

EDITED BY

Matthew Daniel Eddy, Seymour H. Mauskopf, and William R. Newman

OSIRIS | 29

A Research Journal Devoted to the
History of Science and Its Cultural Influences

Osiris

Series editor, 2013–2018

ANDREA RUSNOCK, *University of Rhode Island*

Volumes 28 to 32 in this series are designed to connect the history of science to broader cultural developments, and to place scientific ideas, institutions, practices, and practitioners within international and global contexts. Some volumes address new themes in the history of science and explore new categories of analysis, while others assess the "state of the field" in various established and emerging areas of the history of science.

Series editor, 2002–2012

KATHRYN OLESKO, *Georgetown University*

Cover Illustration:

Detail from *The Iatrochemist* by Balthasar van den Bossche, 17th–18th century. Courtesy of the Chemical Heritage Foundation Collection.

OSIRIS 2014   SECOND SERIES   VOLUME 29

Acknowledgments

This volume was conceived at a Philadelphia workshop funded by the Chemical Heritage Foundation. It was then refined at a second workshop funded by the Riksbankens Jubileumsfond and hosted by the Office for History of Science at the Department of History of Science and Ideas, Uppsala University. The editors would like to thank all three institutions for their generous support.

An Introduction to *Chemical Knowledge in the Early Modern World*

by Matthew Daniel Eddy,* Seymour H. Mauskopf,†
and William R. Newman‡

THE SCOPE OF EARLY MODERN CHEMICAL KNOWLEDGE

The essays in this volume collectively cover the development of chemistry in the "early modern world," that is to say, from the fifteenth century through the eighteenth century. Until comparatively recently, this period was of less interest to historians of chemistry than the succeeding era of the emergence of "modern" chemistry, with its familiar chemical elements, compounds, and equations. But recent research, exemplified by the essays of this volume, has shown how exciting and complex this era in the history of chemistry was in its own right. And its backdrop of early modern European and world history was critically significant for the development of the modern world. The beginning of this period witnessed the high water mark of the Renaissance, the inception of global "outreach" of sea voyages and explorations by Europeans, the Protestant Reformation, and the beginning of bureaucratic national monarchies and smaller political entities. Its conclusion was marked by those revolutionary sequels to the Age of Enlightenment that also ushered in the modern world: the French and Industrial Revolutions.

Our "early modern" centuries, in turn, divide up into two fairly distinct research epochs for the history of early modern chemistry. The first is late fifteenth- through seventeenth-century "chymistry."[1] The second is the chemistry of the eighteenth century. About half of the essays present research dealing primarily with the first epoch. The rest of the essays treat aspects of eighteenth-century chemistry, except for the final essay (Bensaude-Vincent), which offers a general commentary on the entire early modern period.

* Matthew Daniel Eddy, Department of Philosophy, 50/51 Old Elvet, Durham University, Durham DH1 3HN, U.K.; m.d.eddy@durham.ac.uk.
† Seymour H. Mauskopf, Department of History, Duke University, Box 90719, Durham, NC 27708-0719; shmaus@duke.edu.
‡ William R. Newman, Department of History and Philosophy of Science, Goodbody Hall 130, Indiana University, 1011 East Third Street, Bloomington, IN 47405; wnewman@indiana.edu.

[1] This use of "chymistry" for the early modern field of alchemy-chemistry is now widely accepted in the scholarly world and has been acknowledged by the *Oxford English Dictionary* (see the entry in the electronic *OED* for "Chemistry"). See also William R. Newman and Lawrence M. Principe, "Alchemy vs. Chemistry: The Etymological Origins of a Historiographic Mistake," *Early Sci. Med.* 3 (1998): 32–65.

CHEMISTRY AND HISTORY

The past century has witnessed a number of significant changes in the ways in which scholars have written about the history of chemistry. Long before the history of science developed as a professional field, the history of chemistry was pursued vigorously. The nineteenth century was replete with major historical writings and editions;[2] Hermann Kopp's *Geschichte der Chemie*[3] served as a standard reference work well into the twentieth century. Even a cursory glance at these works would reveal that they are different kinds of histories from those being written about chemical knowledge today. What brought about this change? In this section we would like to note the changes in the history of alchemy and chemistry that occurred over the previous century with a view to showing the historiographical background to the themes covered by the essays in this volume.

Most of the historians of chemistry before the mid-twentieth century had been trained as scientists. Perhaps the most recognized of these was the English physical chemist James Riddick Partington (1886–1965). Best known in the history of science for his four-volume *History of Chemistry*,[4] he also did research on important aspects of eighteenth-century chemistry, such as the evolution of the phlogiston theory.[5] Indeed, a number of British chemists, including Partington's collaborator in the phlogiston study, Douglas McKie, contributed to the development of what might be seen today as the "standard view" of eighteenth-century chemistry, that is to say, a narrative that centered around the phlogiston theory, the development of pneumatic chemistry in Britain, and the Chemical Revolution (against the phlogiston theory), associated with Antoine-Laurent Lavoisier (1743–94) and his French disciples.[6]

As in other pre-twentieth-century scholarly endeavors, research in the history of chemistry was primarily a male domain. By the twentieth century, however, women scholars entered the field,[7] and the publications of one of them, Hélène Metzger (1889–1944), were to become transformative.[8] Metzger treated the history of chemistry as a species of intellectual history very much a part of the milieu of French historical and philosophical studies being carried out by her contemporaries, such as Gaston Bachelard, Émile Meyerson, and Alexandre Koyré. What they did—particularly Metzger and Koyré—was to historicize their subject matter. A specialist of seventeenth- and eighteenth-century French chemistry, Metzger attempted to get into the mindsets of her seventeenth- and eighteenth-century protagonists with as

---

[2] E.g., Thomas Thomson, *The History of Chemistry*, 2 vols. (London, 1830–1); Albert Ladenburg, *Vorträge über die Entwicklungsgeschichte der Chemie in den letzten hundert Jahren* (Brunswick, 1869); Jean Baptiste Dumas and Eduard Grimaux, *Oeuvres de Lavoisier*, 6 vols. (Paris, 1862–93); Marcellin Berthelot, *Les origines de l'alchimie* (Paris, 1885).

[3] Kopp, *Geschichte der Chemie* (Brunswick, 1843–7).

[4] Partington, *A History of Chemistry* (London, 1961–70).

[5] James R. Partington and Douglas McKie, "Historical Studies on the Phlogiston Theory," *Ann. Sci.* 2 (1937): 361–404; 3 (1938): 1–58, 337–71; 5 (1939): 113–49.

[6] McKie himself published important studies on Joseph Black and Joseph Priestley and wrote extensively on Lavoisier. He was also important in establishing the history of science as an academic discipline in the United Kingdom.

[7] I. Freund, *The Study of Chemical Composition: An Account of Its Method and Historical Development* (Cambridge, 1904).

[8] Metzger's major works in the history of chemistry were *La genèse de la science des cristaux* (Paris, 1918); *Les doctrines chimiques en France du début du XVII*[e] *à la fin du XVIII*[e] *siècle* (Paris, 1923); *Newton, Stahl, Boerhaave et la doctrine chimique* (Paris, 1930).

little reference as possible to whether they were ultimately "right."[9] This approach gained influence, and by the 1960s it was being discussed by Thomas Kuhn, who, early on in *The Structure of Scientific Revolutions*, cites Metzger (along with Meyerson and Koyré) as having shown him "what it was like to think scientifically in a period when the canons of scientific thought were very different from those current today."[10] Metzger's perspective on eighteenth-century chemistry was, consequently, different from the above-mentioned "standard view." In some ways, it prefigured the perspectives of the contributors to this volume.[11]

By the mid-twentieth century, the history of science emerged as a distinct academic discipline in the United States, the United Kingdom, and the rest of Europe. In the United States, one of the principal "enablers" was the chemist and administrator (president of Harvard University) James Bryant Conant. He produced one of the first recognizable syntheses of the standard view as a chapter titled "The Overthrow of the Phlogiston Theory: The Chemical Revolution of 1775–1789" in his own *Harvard Case Histories in Experimental Science*.[12] Framed as a duel between the phlogistonist, Joseph Priestley, and the antiphlogistonist revolutionary, Lavoisier, Conant's standard view was adopted by Thomas Kuhn in *The Structure of Scientific Revolutions*.

By the time Kuhn's work appeared (1962), historians and historically trained scientists were writing about the history of science. This influenced the history of chemistry, and scholars began to look at wider cultural influences at play in early material theories. The most prominent historian in the United States was Henry Guerlac, who, the year prior to the publication of Kuhn's book, had published a major study of the origins of Lavoisier's Chemical Revolution, *Lavoisier—the Crucial Year: The Background and Origin of His First Experiments on Combustion in 1772*.[13] During the 1960s and 1970s, Guerlac and his students developed something of a research industry on Lavoisier and the Chemical Revolution, in which the historical context, the "background" in Guerlac's words, and the details of the life, career, and scientific development of Lavoisier were fleshed out.[14]

During the 1960s and 1970s, there were a few attempts to provide an alternative to the standard "overthrow of the phlogiston theory" view of eighteenth-century chemistry, and there were a number of social studies of chemistry that looked at national

[9] For recent scholarship, see G. Freudenthal, ed., *Études sur Hélène Metzger* (Leiden, 1990).

[10] T. S. Kuhn, *The Structure of Scientific Revolutions*, 3rd ed. (Chicago, 1996), viii.

[11] One should note that historians of alchemy were also employing a value-neutral approach independently of Metzger. For the valuable scholarly corpus of Julius Ruska, see http://juliusruska.digi library.de/digital_library.html (accessed 16 November 2013). Ruska's student Paul Kraus was one of the giants of twentieth-century Islamic scholarship. See his remarkable *Jābir ibn Ḥayyān: Contribution à l'histoire des idées scientifiques dans l'Islam*, Mémoires présentés à l'Institut d'Egypte, vols. 44–5 (Cairo, 1942–3). Moreover, the historical perspective of historians like Metzger (and Koyré) was not quite that of twenty-first-century historians, as will be detailed in this volume. For example, despite her pioneering attempt to empathize with her premodern chemical protagonists, she did give privilege to certain "modern" components of their thought, such as Nicolas Lémery's seemingly mechanical corpuscularian explanations of chemical reactions. And her approach was very much in terms of the history of ideas; chemical laboratory practice, not to mention chemical artisanal and industrial interests, did not figure in her studies.

[12] James Bryant Conant and Leonard K. Nash, eds., *Harvard Case Histories in Experimental Science*, 2 vols. (Cambridge, Mass., 1950), 1:67–115, case 2.

[13] Guerlac, *Lavoisier—the Crucial Year: The Background and Origin of His First Experiments on Combustion in 1772* (Ithaca, N.Y., 1961).

[14] These included Rhoda Rappaport, Jerry B. Gough, and Carlton Perrin. Mention should also be made of Maurice Dumas, *Lavoisier, théoricien et expérimentateur* (Paris, 1955).

or social factors.[15] But while some books like Archibald Clow and Nan L. Clow's *Chemical Revolution* focused on the artisanal, medical, and industrial components of chemistry,[16] the standard view continued to hold sway until the mid-1980s. Moreover, it remained firmly intellectualist in that chemical ideas were largely treated as disembodied entities operating relatively free of any social or cultural constraints. To a certain extent this tradition was a reflection of the general tenor of the history of science in the United States during its first decades as an academic field, when the intellectualist exemplar of Alexandre Koyré was dominant.

THE REHABILITATION OF ALCHEMY

From the 1980s forward, a number of significant changes occurred in the historiographical methods used to investigate early modern chemical knowledge. Perhaps the biggest changes occurred at first in the history of alchemy. Whereas the standard Chemical Revolution view provided a shared focal point for eighteenth-century historians, frameworks that addressed the nature and scope of alchemical knowledge for the preceding two centuries were more diffuse. A good part of the reason for this was that these centuries were classified under the temporal framework of the Scientific Revolution. Historical study of this revolution, centering on the development of astronomy and physics (and, to a degree, experimental anatomy), was a central component in the emergence of history of science as an academic discipline.[17] But the chemical sciences were at best an outlying domain in the master narrative of the Scientific Revolution before the 1970s.[18] Moreover, alchemy, today recognized as a vital component of the "chymical" sciences of these centuries, was derided and dismissed. This sentiment can be seen, for example, in A. R. Hall's *Scientific Revolution* (1954):

> The most remarkable feature of all alchemical writings is that their authors prove themselves utterly incapable of distinguishing true from false, a genuine observation (according to our modern knowledge) from the product of their own extravagant imaginations. . . . The theoretical contribution of alchemy to science was very small.[19]

[15] The phlogiston historiography was challenged by Robert Siegfried and Betty Jo Dobbs, "Composition, a Neglected Aspect of the Chemical Revolution," *Ann. Sci.* 24 (1968): 275–93; and Arnold Thackray, *Atoms and Powers: An Essay on Newtonian Matter Theory and the Development of Chemistry* (Cambridge, Mass., 1970). For socio-institutional histories, see Arthur Donovan, *Philosophical Chemistry in the Scottish Enlightenment: The Doctrines and Discoveries of William Cullen and Joseph Black* (Edinburgh, 1975); and Karl Hufbauer, *The Formation of the German Chemical Community, 1720–1795* (Berkeley and Los Angeles, 1982).

[16] Archibald Clow and Nan L. Clow, *The Chemical Revolution: A Contribution to Social Technology* (London, 1952). This book dealt with the industrial role of chemistry in the first Industrial Revolution. Curiously, one of the earliest articles of Henry Guerlac, who became the leading scholar of an intellectualist standard Lavoisian model of the Chemical Revolution, "Some French Antecedents of the Chemical Revolution," *Chymia* 5 (1959): 73–112, actually details the mid-eighteenth-century transfer of metallurgical chemical technology from Germany to France through the translation of technical treatises that contained, among other things, the phlogiston theory.

[17] Two works of the 1950s were emblematic: A. R. Hall, *The Scientific Revolution 1500–1800: The Formation of the Modern Scientific Attitude* (Boston, 1954); and Thomas S. Kuhn, *The Copernican Revolution: Planetary Astronomy in the Development of Western Thought* (Cambridge, Mass., 1957).

[18] The pervasive disparagement of alchemy by early historians of the Scientific Revolution and its surprising echoes in more modern surveys of seventeenth-century science have been discussed at length in William R. Newman, *Atoms and Alchemy: Chymistry and the Experimental Origins of the Scientific Revolution* (Chicago, 2006), 6–12.

[19] Hall, *Scientific Revolution* (cit. n. 17), 307.

In Hall's book, Paracelsus, soon to become a locus of research on early modern "chemical philosophy," was also treated negatively, if more ambivalently than alchemy generally:

> His was not in any sense a modern mind. He believed in the philosopher's stone. He believed in the alchemical theory of transmutation, and in others yet more wonderful. . . . He had in full measure the faculty for self-deception characteristic of the Hermetic tradition.[20]

As noted above, historical research in the last decades of the twentieth century on science in the early modern period moved away from the progressivist and presentist perspective exhibited so prominently in Hall's *Scientific Revolution* toward serious and sympathetic study and assessment of those domains dismissed by Hall as not "modern": the so-called Hermetic tradition[21] and Paracelsus.[22] In many ways, the work of Allen Debus on Paracelsus and his followers[23] established the contours of the historical narratives for this period just as Guerlac and his students had done for the eighteenth century.

The alchemical scholarship of the past quarter century has significantly widened and enriched the purview established by these major researchers. Although Debus was alive to the importance of the worldviews of Paracelsus and the Paracelsians to early modern chemical thought and considered very seriously their medical aspects and activities, he was less interested in exploring the details of the chrysopoetic (gold-making) traditions of the early modern period. To some degree, Debus still presented a "sanitized" picture of early modern chemistry in which the content that did not conform to modern chemistry was minimized or pushed aside.[24]

It was left to the next generation of scholars to delve more deeply into the material and conceptual logic of alchemy. An early and sympathetic treatment of alchemy in the context of the Scientific Revolution was provided by Betty Jo Teeter Dobbs and her studies on Newton;[25] subsequent scholarship has resulted in a comprehensive

[20] Ibid., 309–10. Hall did give Paracelsus the credit of being an "iconoclast" and of being the originator of medical chemistry, which, as it developed away from Paracelsus's own "incoherent, obscure, megalomanic writings," did point the way to rational chemistry.

[21] The term "Hermetic tradition," popularized by Yates and widely used by historians in the decades after her groundbreaking work, has lost credence among more recent scholars. See Frances Yates, *Giordano Bruno and Hermetic Tradition* (London, 1964).

[22] Walter Pagel, *Paracelsus: An Introduction to Philosophical Medicine in the Era of the Renaissance* (Basel, 1958).

[23] Allen G. Debus, *The English Paracelsians* (London, 1965) and many subsequent books.

[24] This had been noted in the 1960s and 1970s in reviews of his books; e.g., "It is rash to assert that in England 'the occult aspects of his work were rejected while the new [chemical] remedies were eagerly adopted, provided they proved their worth'"; C. H. Josten, review of *The English Paracelsians*, by Allen G. Debus, *Brit. J. Hist. Sci.* 3 (1967): 296. Consequently, he did not focus on the alchemical aspects of Paracelsianism. Charles Webster, reviewing Debus's *The Chemical Philosophy: Paracelsian Science and Medicine in the Sixteenth and Seventeenth Centuries*, 2 vols. (New York, 1977), took Debus to task on this point: "Debus has disregarded the extensive surviving body of Renaissance alchemical literature which was circulating in manuscript form, and he has not taken account of the fact that the educated elite possessed a ready knowledge of Latin, so that the absence of alchemical or natural philosophical works was no barrier to their acquaintance with the various traditions of Renaissance philosophy"; Webster, Essay Review, *Isis* 70 (1979): 588–92, on 590.

[25] Betty Jo Teeter Dobbs, *The Foundations of Newton's Alchemy: or, "The Hunting of the Greene Lyon"* (Cambridge, 1975). This project was initiated in the late 1960s. Unfortunately, in this book Dobbs adopted the approach of Carl Jung; the Jungian analysis of alchemy has come under withering criticism in more recent scholarship.

reassessment of the very meaning of "alchemy." Far from being a peripheral figure sequestered from the public eye, the "alchemist," as shown in the research of Pamela Smith and others, was an artisanal "expert" who played a core role in regulating and disseminating natural knowledge across Europe.[26] Likewise, the work of William R. Newman and Lawrence M. Principe revealed that, instead of being based on abstruse or even nonsensical theories, alchemy was a serious experimental enterprise that was coextensive with sixteenth- and seventeenth-century "chymistry."

Over the past two decades, historians have come to recognize that chymistry was an early modern field that incorporated a wide range of productive chemical and medical technologies as well as a long-standing belief in the transmutability of metals and other materials.[27] It is now much easier to see why it was no accident that scientific luminaries such as Isaac Newton, Robert Boyle, Gottfried Wilhelm Leibniz, and John Locke involved themselves deeply in the chrysopoetic side of chymistry—if anything, this subject was the idée fixe of the Age of Gold. Furthermore, it has emerged in the last few years that even late medieval alchemy seriously challenged the widely held antiatomist matter theory of Thomist and Scotist writers, creating a dialectic that became even more apparent in the sixteenth century after the Society of Jesus adopted Thomas Aquinas as their master in theology. Centuries before Lavoisier, alchemists were already employing analytical processes to arrive at a sort of "chymical atomism" that regarded a range of material substances to be operationally indivisible and capable of retrieval from seemingly "perfect" mixtures.[28] Needless to say, a discipline that challenged contemporaneous views in such a fundamental way could hardly fail to have further repercussions as well. Accordingly, as our historical knowledge of chymistry as an early science has broadened, it has come increasingly to include textual, literary, and religious themes alongside a deepening appreciation of alchemical experimentation.[29]

CHEMICAL REVOLUTIONS

The previous volume of *Osiris* devoted to the history of chemistry appeared in 1988 and was titled *The Chemical Revolution: Essays in Reinterpretation*.[30] Published at a time when the history of chemistry was becoming more introspective, the subtitle suggests a sense of disquiet with the standard view of the Chemical Revolution as simply the "overthrow of the phlogiston theory."[31] Indeed, a number of

---

[26] See Pamela H. Smith, *The Business of Alchemy: Science and Culture in the Holy Roman Empire* (Princeton, N.J., 1994).

[27] See Newman and Principe, "Alchemy vs. Chemistry," and their "Some Problems with the Historiography of Alchemy," in *Secrets of Nature: Astrology and Alchemy in Early Modern Europe*, ed. William R. Newman and Anthony Grafton (Cambridge, Mass., 2001), 385–431.

[28] The significance of the European alchemical tradition for important features of the Scientific Revolution, such as the corpuscular philosophy of matter, has been stressed particularly in the publications of William R. Newman and Lawrence M. Principe. For the role of late medieval alchemy in reframing atomism, see Newman, *Atoms and Alchemy: Chymistry and the Experimental Origins of the Scientific Revolution* (Chicago, 2006).

[29] A work of major importance that relates early modern chemistry to humanistic and literary contexts is Owen Hannaway, *The Chemist and the Word: The Didactic Origins of Chemistry* (Baltimore, 1975). As the subtitle indicated, Hannaway's narrative traced "the invention of chemistry as a discipline" back to the work of the humanist Andreas Libavius, who initiated a didactic chemical textbook tradition with his own textbook, the *Alchemia* (Frankfurt am Main, 1597).

[30] Arthur Donovan, ed., *The Chemical Revolution: Essays in Reinterpretation*, vol. 4 of *Osiris* (1988).

[31] Arthur Donovan, "Introduction," in ibid., 6

essays broadened the connections of this episode with other (and earlier) features of eighteenth-century chemistry.[32] But the primacy of the Chemical Revolution as the telos of eighteenth-century chemistry remained unchallenged. However, in the following year (the bicentenary of the publication of Lavoisier's *Traité élémentaire de chimie* and, of course, the start of the French Revolution), Frederic L. Holmes launched a challenge to precisely this perspective in *Eighteenth-Century Chemistry as an Investigative Enterprise*. There he stated,

> Historians of science have found it difficult to view eighteenth century chemistry as anything other than the stage on which the drama of the chemical revolution was performed. So strong has the disposition been to identify the advent of the modern science with the chemical system established by Lavoisier between 1772 and 1789 that all earlier activity has been treated most often as a prologue to these climactic events.[33]

In many respects, Holmes's view extended a rising sentiment, expressed by other historians of chemistry like Rhoda Rappaport and Rachel Laudan,[34] that the Chemical Revolution narrative downplayed the overarching material models that united chemical theories and practices in the laboratory and in the field. Over the course of the book, Holmes outlined a much broader view of eighteenth-century chemistry, focusing on the larger traditions of early modern chemistry pursued at the Paris Académie Royale des Sciences (as well as German and Swedish chemical developments). Regarding chemical research, Holmes's purview extended far beyond the traditional triune foci of phlogiston theory, pneumatic chemistry, and the Lavoisian Chemical Revolution. In particular, he focused on salts and plant materials, which, he showed, had their own independent, progressive research traditions, producing important conceptual and methodological developments. This led him to give serious and sympathetic attention to chemists such as Wilhelm Homberg, Nicolas and Louis Lémery, and Etienne-François Geoffroy, who had not figured prominently in discourses on eighteenth-century chemistry since Metzger.[35]

Holmes also foregrounded historical research on other aspects of eighteenth-century chemistry that had lain somewhat submerged because of the primacy of the standard view. One was chemical laboratory instruments and techniques.[36] He pointed out that, although the development of the pneumatic trough for the collection

---

[32] Notably, J. B. Gough, "Lavoisier and the Fulfillment of the Stahlian Revolution," and John G. McEvoy, "Continuity and Discontinuity in the Chemical Revolution," both in Donovan, *The Chemical Revolution* (cit. n. 30), 15–33 and 195–213, respectively.

[33] Frederic Lawrence Holmes, *Eighteenth-Century Chemistry as an Investigative Enterprise* (Berkeley, Calif., 1989), 3. Significantly, Holmes lauded Metzger as "the most conspicuous exception to the historiographic patterns that have dominated the treatment of eighteenth-century chemistry" (8).

[34] See the collection of Rhoda Rappaport's early essays in *Studies in Eighteenth Century Geology* (Aldershot, 2011). Rachel Laudan also wrote a number of notable essays during the 1980s and 1990s, laying the foundation for *From Mineralogy to Geology: The Foundations of a Science, 1650–1830* (Chicago, 1994).

[35] Geoffroy had received attention in connection with his *Table des differents rapports observés entre differentes substances* and the subsequent development of chemical affinity theory. Holmes also highlighted hitherto neglected German chemists such as Johann Pott, Andreas Marggraf, as well as Georg Ernst Stahl and his student, Caspar Neumann.

[36] In this connection, he cited the important but neglected study of Jon Eklund, *The Incomplete Chemist: Being an Essay on the Eighteenth-Century Chemist in His Laboratory, with a Dictionary of Obsolete Chemical Terms of the Period*, Smithsonian Studies in History and Technology no. 33 (Washington, D.C., 1975). More recently, the topic of chemical experimental practice has been addressed in a book of essays, a number of which deal with the early modern period; Frederic L. Holmes

of "airs" had been highlighted in the standard view, other kinds of substances had not figured prominently. Holmes gave particular attention to an important analytical laboratory technique called the "wet" way, or analysis through liquid (usually watery or humid) agents or mediums. Another feature that Holmes reintroduced was "chemistry and industry," where he tried to connect eighteenth-century chemical technology and industry "to a story of eighteenth century chemical science that has been less adequately told."[37]

In a concluding reconsideration of the Chemical Revolution, Holmes laid down a position that is echoed strongly—and developed—in this volume: "If my portrayal of earlier eighteenth century chemistry is valid, then the chemical revolution cannot have overturned the science of chemistry as a whole, or have established a science for the first time."[38] In recent decades this assessment has been taken in many directions, and the concept of a singular Chemical Revolution has been transformed into a more pluralistic notion, one that is not confined to the late eighteenth century and which is more properly construed as a series of chemical revolutions that drew strongly from theories, practices, and instruments that grew out of the "chymical" tradition that emerged in the late seventeenth century. The chemists who drove this change were not only polite, financially independent savants like Lavoisier, they were also professionals and artisans who used chemical knowledge on a daily basis in both iterative and innovative ways.

In recent years the themes of Holmes's work have been extended so that we now have a much better idea of what eighteenth-century chemists were actually doing, that is to say, what they were reading, analyzing, and synthesizing. Key to this extension was the notion that advances in what might be seen as "pure" chemistry were intimately tied to "practical" concerns of mining, industry, and medicine. In other words, chemistry was a technoscience, a hybrid of science and technology that engendered a host of instrumental, managerial, and experimental revolutions during the eighteenth and nineteenth centuries. As shown in the influential works of Ursula Klein and Bernadette Bensaude-Vincent, many of the key players in this climate, like their forebears in alchemy, were artisans and professionals, a large number of whom worked in mines, apothecary shops, and factories.[39]

OVERARCHING THEMES

It is clear that, by the 1990s, the historiographical perspectives and purviews of early modern alchemy and chemistry were undergoing profound transformations. Moreover, the changes mirror broader thematic developments taking place in the history of science, as well as the history of technology and medicine, such as an increasing

---

and Trevor H. Levere, eds., *Instruments and Experimentation in the History of Chemistry* (Cambridge, Mass., 2000).

[37] Holmes, *Eighteenth-Century Chemistry* (cit. n. 33), 102.

[38] Ibid., 107. The passage continued: "It must instead have transformed certain extensive areas of a science whose scope exceeded these areas. Lavoisier himself recognized, within contexts conducive to such recognition, that what he had transformed represented only large parts within a larger whole."

[39] Ursula Klein and Wolfgang Lefèvre, *Materials in Eighteenth-Century Science: A Historical Ontology* (Cambridge, Mass., 2007); Bernadette Bensaude-Vincent, *A History of Chemistry* (Cambridge, Mass., 1996). Klein and Bensaude-Vincent also coedited a number of important works, including Ursula Klein and Emma Spary, eds., *Materials and Expertise in Early Modern Science* (Chicago, 2010); Bernadette Bensaude-Vincent and Christine Blondel, eds., *Science and Spectacle in the European Enlightenment* (Aldershot, 2008).

emphasis on experimental and artisanal practice and a broadening of what constituted "science" in the early modern period. This has entailed the abandonment of normative viewpoints emanating from presentism or progressivism that derided and dismissed out-of-date science, much less scientific activity, such as alchemy, that fell on the wrong side of constructed demarcations between science and pseudoscience. The essays in this volume both testify to these transformations and expand on them.

Thus, many of the essays demonstrate interest in how chemical knowledge was gained, lost, preserved, and circulated. Chemical knowledge itself is treated as a set of skills and routines that required specific kinds of artifacts such as instruments and substances that gained and lost meaning over time. In a word, knowledge is something you *do*. This perspective has generated an intense interest in the history of material culture, with scholars asking which substances, specifically, were used in experiments and what did experimentalists actually do to manipulate them. As can be seen by the materials examined in this volume, historians are now fascinated with the expanding array of substances and compounds that were bought and sold in the service of early modern chemical knowledge. Thus, as will be shown, the materials (and the instruments used to study and manipulate them) are as much objects of society, culture, and commerce as of nature.[40]

Theoretical issues concerning the nature and hierarchy of material substances range from the challenges of identifying mysterious substances derived from venerable texts in the earlier era to delineating the natures of the tangible materials used in the laboratory and in commerce in the eighteenth century. The role of instruments for studying and manipulating chemical substances figures prominently in some of this volume's essays on eighteenth-century chemistry. Of particular note are the dynamic interactions between the artisanal craftsmen of these instruments and the laboratory chemists who commission and employ them.

An aspect of material culture specific to the history of early modern chemistry is alchemy, especially chrysopoeia. We have already spoken about the "rehabilitation" of alchemy as a rational investigative enterprise for the sixteenth and seventeenth centuries. Many essays concerned with this era focus on or deal with alchemy. What is more surprising is that, as some of the essays demonstrate, alchemical investigations were pursued by major chemists throughout much of the eighteenth century. But changing institutional and cultural contexts, namely, the ascendancy and proliferation of public scientific societies and state institutions requiring the deployment of chemical knowledge, had an impact on the relationship between alchemical and chemical activities and interests. The most striking aspects of this were the increasing "privatization" of alchemy, the concern to promote a positively scientific and utilitarian public image of chemistry, and the tensions within individual chemists over how to deal with (and conceal) their persistent alchemical interest. This had precedence in the work and activities of Robert Boyle; it is only in recent decades that we have come to appreciate how much of an alchemical "adept" he was. But in the eighteenth century, these changes—and tensions—became more marked.

One particularly important aspect of materials is their consideration as objects of commerce. This is the focus of our third theme: the artisanal, industrial, and commercial aspects of the early modern chemical enterprise, both for the pre-eighteenth-century era and the eighteenth century itself. Chemists were often craftsmen and

---

[40] Klein and Lefèvre, *Materials in Eighteenth-Century Science* (cit. n. 39).

tradesmen by occupation (e.g., apothecaries) and as such were as much involved in commercial as in natural philosophical activities both on their own and in the service of the state. One of our contributors goes so far as to consider eighteenth-century French chemistry as prefiguring (and paralleling), to a degree, contemporary tech-noscience.[41] The commercial aspect of the chemical enterprise has also begun to at-tract scholarly attention to broader contexts such as (*a*) the emergence of bureau-cratic nation-states (with mercantilist objectives) and national scientific institutions; (*b*) worldwide exploration and colonization and the appearance of new, commercially valuable materials; and (*c*) the incorporation of chymistry/chemistry in educational institutions, both at the university level in the medical faculty and in the formal (and informal) instruction of craftsmen.[42] Indeed, the most striking changes from the ear-lier to the later era are the role of the bureaucratic nation-state and its science-related institutions and of the university as a locus of chemical pedagogy. Both sets of in-stitutions served as patrons and facilitators of interactions between chemists (now legitimated as "experts" by these institutions) and craftsmen.

In a number of essays, the issue of chemical pedagogy, academic and artisanal, naturally leads to the consideration of the construction and delineation of an "ex-pert," and who was recognized or certified as the authoritative possessor and imparter of chemical knowledge, both natural knowledge (and practice) and artisanal knowl-edge. The term "expert"—like "scientist"—was not in use before the nineteenth cen-tury, at least in English, but we can recognize progenitors of "experts" (and "scien-tists") in many of our fifteenth- to eighteenth-century actors.

THE ESSAYS

In his essay, John A. Norris examines the life and works of the Lutheran preacher Johann Mathesius, author of the famous *Sarepta, oder Bergpostill* (1562). The col-lection was composed while Mathesius was pastor of Joachimstal (now Jáchymov), then a boomtown in the rapidly developing mining area of the Erzgebirge (Krušné hory). By considering Mathesius's seemingly novel concept of "gur," a putative Ur-substance out of which metals were thought to grow, Norris provides a sensitive study of the relationship between miners' beliefs and traditional alchemical ideas of metal-logenesis in sixteenth-century Germany. As Norris convincingly shows, gur was not just an empirical discovery of unlearned miners, but a fusion of mineralogical obser-vation and the already old theory of the alchemical principles of mercury and sulfur. Norris's study has important implications for the relationship of early modern arti-sanal and learned culture more generally, and particularly for the evolving study of the subterranean world that one witnesses in Central Europe during this period.

Whereas the word gur seems to have been a new term in Mathesius's day, Jenni-fer M. Rampling's essay examines a very different linguistic phenomenon—the fact that alchemists typically used the same terms over long periods of time to describe

---

[41] Bernadette Bensaude-Vincent, "Concluding Remarks: A View of the Past through the Lens of the Present," in this volume.

[42] For craftsmen, see Jonathan Simon, *Chemistry, Pharmacy and the Revolution in France, 1777–1809* (Aldershot, 2005); Klein and Spary, *Materials and Expertise* (cit. n. 39). For the incorporation of chemistry into university settings, see Matthew Daniel Eddy, *The Language of Mineralogy: John Walker, Chemistry and the Edinburgh Medical School, 1750–1800* (Farnham, 2008); and John C. Powers, *Inventing Chemistry: Herman Boerhaave and the Reform of the Chemical Arts* (Chicago, 2012).

very different concrete referents and practices in the material world. The fortuna of George Ripley's fifteenth-century corpus is a particularly apt vehicle for studying this feature because of his marked authority in the world of early modern alchemy: his was a "name to conjure with." Hence, when Ripley's work was read and used in the sixteenth and seventeenth centuries, there was a need to bring him up to date by interpreting his alchemical terms to fit the most current techniques and materials available to alchemists. Ripley's practice was heavily based on "sericon," an obscure term that seems to have originally meant minium or red lead oxide, but when his authority was appropriated by George Starkey, the American alchemist and friend of Robert Boyle whose popular works circulated under the name "Eirenaeus Philalethes," Starkey argued that the key material behind Ripley's alchemy was crude antimony or stibnite. Rampling points to the new interest among historians of science in the material culture of alchemy and the complex interaction between text and practice.

While Bruce T. Moran shares a focus on language with Rampling, his concern is not the ongoing transformation of referents while the terms remain unchanged. Moran's focus lies rather in the interaction of learned chymistry, represented by the Saxon pedagogue Andreas Libavius, and various less polished chymical authors, such as the little-known Italian physician Joseph Michael. Libavius expended huge efforts in decoding medieval alchemists such as pseudo-Raymond Lull and pseudo-Arnold of Villanova, so he considered himself an expert in the art of deciphering *Decknamen*. Michael, in the view of Libavius, had focused too exclusively on a single meaning for the elixir described in the alchemical works of Roger Bacon, and in the process Michael had reduced the scope of chymistry to a single, monolithic practice. Combining humanist concepts of art as a multifarious endeavor, Libavius argued that chymistry should actually engage in a host of different technological and medical pursuits. Libavius's emphasis on the multiple utility of the discipline lay behind his important role in establishing the pedagogical foundations of the discipline in the form of chymical textbooks.

Anna Marie Roos also takes us into the world of chymical expertise, but in a rather different way from the expertise that Andreas Libavius employed in deciphering pseudo-Lull or that George Starkey demonstrated by arriving at the "true" meaning behind George Ripley's alchemy. Roos's paper focuses on Robert Plot, "first keeper of the Ashmolean, secretary of the Royal Society, and Oxford's first professor of chymistry." Despite these various public roles based on his scientific expertise, Plot was also seriously involved in the more secret pursuit of the philosophers' stone, as Roos's examination of his manuscripts reveals. Plot was not just a private aspirant to chrysopoetic and medical secrets, however; like the alchemical employees of many a Continental prince, he entered into alchemical contracts with various parties in order to finance his research. Roos has unearthed several examples of these fascinating documents that reveal the close interaction between commerce and "the searching out of secrets" in the minds of early modern chymists. It is no accident that Plot's legal arrangements with his backers remind us of the consulting agreements between university chemists and industry today.

Ku-ming (Kevin) Chang's contribution carries us well into the eighteenth century and fills an important gap in our previous understanding of the relationship between Georg Ernst Stahl's chemistry and contemporaneous work being done at the Académie Royale des Sciences. It has long been known that the mid-eighteenth century witnessed a strong French reception of Stahl's phlogiston theory, but was this the sudden discovery of a previously isolated German figure's work, or the culmination of a much longer interest? Chang shows convincingly that the latter was the case, and that significant elements of Stahl's theory had already been incorporated into the famous *Table des differents rapports* of Etienne-François Geoffroy, published in 1718. More than this, Geoffroy continued to insert features of Stahlian chemistry throughout his career, as did many other chemists at the Académie. It would appear that Stahl's influence was even more pervasive than previously thought, and that there was a hitherto little noted network of communications between German chemists and their French counterparts operating throughout the first half of the eighteenth century.

Whereas Chang deals with the transmission of ideas and practices between Germany and France, William R. Newman's paper considers the influence of Robert Boyle on the German chymists Johann Joachim Becher and Stahl. Since the publication of a seminal paper of the 1950s by Thomas Kuhn, it has repeatedly been claimed by prominent historians of science that Boyle, for all his fame as a mechanical philosopher, exercised little real influence on the history of chemistry. Newman gives a close analysis of Becher's corpus as it evolved from the 1650s through the 1660s and shows that in all probability the German polymath's main source for his hierarchical theory of matter was Boyle's work, especially *The Sceptical Chymist*. This matter theory was subsequently adopted by Stahl and—in modified form—by French Stahlians such as Pierre-Joseph Macquer. Newman also argues that Boyle's corpuscular theory and indeed "chymical atomism" more generally often went hand in hand with a belief in chrysopoeia; the phenomenon is not surprising when one understands the theoretical and practical needs that chymical atomism served.

The essay by Bernard Joly addresses some of the same characters adduced by Chang, especially Etienne-François Geoffroy, about whose biography Joly has extracted interesting new information. Joly goes into considerable depth to show that Geoffroy's famous debate with Louis Lémery about the supposed resynthesis of iron in plant ashes had ramifications extending far beyond the specifics of this experiment. Allying himself with the mechanistic physics of the seventeenth century, Lémery advertised the fact that the iron experiment derived from J. J. Becher's work in order to link Geoffroy to chrysopoetic attempts that had fallen into public disrepute at the Académie Royale des Sciences. As Joly points out, Geoffroy did in fact owe a strong debt to Becher, along with the numerous French *Cours de chimie*, which provided much of the empirical data that would be formulated in Geoffroy's famous *Table des differents rapports*. Joly makes the important point that Geoffroy's *Table* was a synopsis of previous knowledge rather than a revolutionary new advance. As in Chang's essay, the picture of Geoffroy that begins to emerge is one of a figure far more connected to existing chymical traditions than one might expect from reading other scholarship in the history of chemistry.

Lawrence M. Principe's paper extracts a wealth of new archival and manuscript data to enrich and expand upon the same general conclusion drawn by Joly. Not only was Geoffroy employing alchemical sources in his famous debate with Louis Lémery and in his 1722 attack on chrysopoetic frauds, "Des supercheries concernant la pierre philosophale," but Geoffroy was actively pursuing alchemical goals himself. Under the tutelage of Wilhelm Homberg, himself an avid student of alchemical authors such as Eirenaeus Philalethes (George Starkey), Geoffroy published on the old alchemical desideratum of potable gold. Principe presents considerable additional evidence to show that other mainstream French chemists after Geoffroy, such as Pierre-Joseph

Macquer and Guillaume-François Rouelle, were also interested in chrysopoeia; the latter may even have kept a private laboratory for his alchemical project. This new information leads Principe to the surprising conclusion that alchemy did not die at the hands of Geoffroy and his intellectual heirs as often asserted—it merely "went underground." This is a startling observation indeed about a man whose supposed rejection of transmutation and alchemical matter theory has been claimed elsewhere as a radical step in the development of chemistry, dissociating it from alchemy.

Perhaps the chemistry teacher par excellence at the beginning of the eighteenth century was Professor Hermann Boerhaave of Leiden University. Focusing on Boerhaave's lectures, John C. Powers's chapter argues that, rather than being a dry and pedestrian concern, pedagogy played an important role in reshaping the experimental techniques of chemistry. He shows that Boerhaave actively employed new, innovative instruments such as Daniel Gabriel Fahrenheit's thermometers. Powers also reveals how such instruments became a central part of chemical theory and practice through their use in the classroom. In making this point, Powers identifies a flexible way of thinking about new instruments that emerged long before the experimental innovations traditionally associated with the Chemical Revolution.

From the seventeenth century onward there was a significant shift in what leading chemists like Boerhaave counted as the basic building blocks of matter. Hjalmar Fors looks at how this ontological transformation unfolded in the chemical mineralogy practiced in northern European mines during the eighteenth century. He shows how influential chemists such as Georg Brandt and Axel Fredrik Cronstedt combined the artisanal practice of assaying with the more scholarly tradition of natural history to create a form of classification that treated individual metals as foundational units of matter. He argues that this move was motivated by the pragmatic epistemology of what might be called "mining knowledge" and that it laid the foundation for the concept of elementary substance advocated by Lavoisier.

During the twentieth century, one of the most neglected aspects of the history of early modern chemistry was pharmacy, especially in French historiography of chemistry. This neglect elided the important contributions of chemically trained apothecaries and physicians. Jonathan Simon addresses this negligence by arguing for the importance of pharmacy as an ensemble of essential practices that affected how French chemists and the reading public viewed the discipline of "chimie." Using widely read publications like the French pharmacopoeia and the textbooks of Nicolas Lémery and Antoine Baumé, Simon argues that the reading public played a crucial role in shaping how pharmacists conceptualized and communicated chemical knowledge. By reflecting on the changing nature of the audience and content of chemistry, he helps us understand the continued presence of pharmacy in chemistry.

Unlike the recent rediscovery of pharmaceutical chemistry, the close relationship between eighteenth-century chemistry and industry has long been recognized by historians of science, technology, and society. This relationship, however, is often portrayed as a unidirectional flow of knowledge in which the findings and theories of experts trickled down into industry. Ursula Klein's chapter turns this historiographical model on its head by arguing that there were many kinds of chemical "experts" and that many of them gained their expertise on the floors of factories. In making this argument, she sets aside the traditional association often made between chemistry as pure science and technology as an applied science. Concentrating on the many different kinds of chemists operating in the large Royal Prussian Porcelain Manufactory, she avers that chemistry was a "technoscience" and that chemical expertise was not confined to elite settings like the academy.

Christine Lehman also examines the prominence of mid-eighteenth-century chemical experts, many of whom operated as agents of government *and* industry. Whereas Klein explains how a range of actors developed different kinds of chemical expertise in one setting, Lehman's chapter uses the impressive career of the French chemist Pierre-Joseph Macquer to show how the many roles of one person generated useful, and important, chemical knowledge. She points out that Macquer was a physician, teacher, academician, and inspector and that these roles allowed him to showcase his expertise in settings such as a classroom, garden, study, and even the Château de Saint-Germain-en-Laye. Following him on his public inspections and even on secret missions for the crown, Lehman presents a geography of chemical knowledge in which material facts and theories traveled side by side on the roads of France and the Low Countries.

As intimated above, the ideas of Antoine-Laurent Lavoisier played a central role in the standard model used to frame the Chemical Revolution. This tradition tended to portray him as a wealthy chemist whose prescient theories of combustion and composition triumphed over the supposedly antiquated chemistry of artisanal settings. Focusing on Lavoisier's instruments and experimental techniques, Marco Beretta's chapter problematizes this view by showing that Lavoisier relied on the technical input of the instrument makers and artisans who were employed in his laboratory. Beretta argues that these artisans, rather than being unseen technicians, were Lavoisier's collaborators. In making this argument, Beretta reconstructs for the first time the network of instrument makers who worked in Lavoisier's laboratory, showing that, instead of being a solitary genius, Lavoisier was intimately dependent upon a chorus of collaborators for his experimental successes.

Whereas Beretta's essay raises the importance of the material culture of instrumentation, Matthew James Crawford's chapter underscores the imperial origins of the material substances used by chemists. He points out that there were many people living in Asian and American colonies who paid very close attention to the composition of pharmaceuticals and industrial materials. More specifically, he tells the story of how cinchona bark became a chemical commodity in Spain's South American colonies. Focusing on the career of a "botanist-chemist" who acted as an agent of the Spanish Crown, Crawford explains how chemistry was an imperial science and highlights the informal, but vital, pathways through which chemical techniques circulated in early modern colonial settings.

As the many illustrations included with the essays of this volume clearly show, we are currently living at a time when early modern historians pay much more attention to the role of visual culture. Of course, the world of alchemical imagery has long attracted the attention of historians of art and science alike. Yet aside from research on the depiction of instruments, the visual culture of eighteenth-century chemistry has remained relatively unexplored. Matthew Daniel Eddy's chapter addresses this lacuna by focusing on the diagrammatic pictures used by Scotland's Joseph Black to teach hundreds of students from the 1750s to the 1790s. Treating Black's affinity diagrams as a collective system, Eddy approaches them via a visual anthropology that allows him to reveal how Black skillfully appropriated preexisting visualizations to teach his students chemical attraction and repulsion, that is, two core chemical forces that formed the theoretical basis for late Enlightenment chemistry.

In the concluding historiographical reflection, Bernadette Bensaude-Vincent focuses on the current interest among historians of chemistry in the material, artisanal, and commercial aspects of early modern chemistry, interests certainly well represented in this volume. She sees this upsurge of interest as related to the ascendant "technoscientific" nature of much of present-day science and industry, notably chemical "technosciences." In these technoscientific enterprises, scientific research, application, and commercial development are almost seamlessly melded. Bensaude-Vincent herself perceives and explores deep parallels between modern technoscience and eighteenth-century French chemical practices and activities. She also notes differences—often cultural—such as the radically altered valuation of chemical activities (highly regarded in the eighteenth century; highly problematic today) and the transformed perception of temporal change and destiny (optimistic in the eighteenth century; problematic today).

Our intent in this volume is to bring to an audience wider than the cognoscenti of the history of early modern chemistry the historical visions of pre-nineteenth-century chemistry that have begun to emerge. These portrayals of chemistry (and "chymistry") are much more complex and even amorphous than they were before the reconstructions of the past quarter century. In part, this reflects the liberation of the historiography of chemistry from the clearly defined shadow cast by the more mathematical sciences. Although some attempts were formerly made to shoehorn chemistry into the highly mathematical story of physics and astronomy extending from Copernicus to Newton and their heirs, the fact is that Lavoisier himself required nothing more sophisticated than arithmetic to quantify his experimental results. Neither eighteenth-century chemistry nor its forebears in the realm of medieval alchemy and early modern chymistry can be seen as waves emanating from a central event such as the Cartesian geometrizing of nature or the development of calculus by Leibniz and Newton. Instead, mining technology, chrysopoeia, chymical medicine, pigment making, refining of salts, and the trade in distilled spirits all played a part, along with other pursuits, in the development of chemistry. To an earlier generation of scholars, this meant that early modern chemistry was theoretically underdeveloped, dominated by ad hoc "rules of thumb," and hence largely unworthy of study by professional historians of science. With the current emphasis on material and visual culture, the integration of scientific practice with theory, and the role of expert knowledge in the developing industry and commerce of early modern Europe, the table has been turned. The present volume underscores the diversity and richness of early modern chemical traditions, sometimes undeniably bewildering in their approaches and goals. To anyone familiar with chemistry of the past century, this should not be too surprising, for it, too, is an enormous and polymorphous entity.

THE AGE OF CHYMISTRY (1450–1700)

Transmuting Sericon:
Alchemy as "Practical Exegesis" in Early Modern England

by Jennifer M. Rampling*

ABSTRACT

An influential strand of English alchemy was the pursuit of the "vegetable stone," a medicinal elixir popularized by George Ripley (d. ca. 1490), made from a metallic substance, "sericon." Yet the identity of sericon was not fixed, undergoing radical reinterpretation between the fifteenth and seventeenth centuries as Ripley's lead-based practice was eclipsed by new methods, notably the antimonial approach of George Starkey (1628–65). Tracing "sericonian" alchemy over 250 years, I show how alchemists fed their practical findings back into textual accounts, creating a "feedback loop" in which the authority of past adepts was maintained by exegetical manipulations—a process that I term "practical exegesis."

INTRODUCTION

What is "sericon"? In fifteenth-century England, the term was used in alchemical writing to denote a metallic body: the calx or ash of "adrop." Adrop was also known as the "Green Lion," and sericon was thus the "Blood of the Green Lion." Together, these expressions provided a group of related cover names, or *Decknamen*, used by alchemical authors to disguise the true identity of their materials and processes. In isolation, such names do not seem particularly helpful to the modern historian. Yet early modern alchemists also experienced perplexity and frustration when faced by these exegetical barriers to praxis. One solution was to import knowledge from other sources—including their own practical observation—in order to decipher the obscure instructions of their sources. Over time, the meaning of particular *Decknamen* changed, allowing historians a glimpse of the methods by which alchemical practitioners tested, amended, and adapted their earlier authorities in light of their own experience.

The identity of sericon is significant because this term was used to denote the *prima materia* of the "vegetable stone," a powerful alchemical elixir thought to heal sickness and prolong human life. The vegetable stone was one of the core pursuits of

* Princeton University, Department of History, 129 Dickinson Hall, Princeton, NJ 08544-1017; rampling@princeton.edu.

This research was funded by a Wellcome Trust postdoctoral research fellowship (090614/Z/09/Z) and Darwin Trust of Edinburgh Martin Pollock doctoral scholarship. My thanks also to interlocutors at the *Osiris* workshop (Uppsala), AD HOC reading group (London), Early Modern Medicine reading group (Cambridge), and History Department, Princeton University, for their valuable suggestions.

English chemical medicine, and the identity of sericon was therefore key to success in this pursuit. Yet sericon's nature was not fixed. Both its identity and the alchemical practice it represented underwent radical reinterpretation between the fifteenth and seventeenth centuries. One outcome of that reinterpretation was that, over time, the original practice came to be eclipsed by other methods and materials. This eclipse is one reason why we find few traces in modern scholarship of either sericon or the medicinal vegetable stone. Indeed, pre-Paracelsian chemical medicine remains relatively little studied in an English context, in comparison to other alchemical pursuits, particularly chrysopoeia (gold making).[1]

In this chapter, I shall use sericon as a case study for examining a problem faced by historians of science: the difficulty of isolating and tracking changes in chemical ideas, practices, and nomenclature over long periods, while avoiding anachronism. Alchemical receipts, like other genres of "secret," enjoyed a certain social mobility and exchange value in early modern Europe. Copied, excerpted, swapped, and annotated, procedural instructions could be "tweaked" to reflect the taste, experience, and wider reading of their copyists.[2] Yet such adjustments, a common feature of recipe collections, are also encountered in more substantive alchemical treatises—reflecting the particular character of these writings, in which key terms are frequently encoded or concealed, demanding an element of translation. The resulting alterations influenced the character of early modern practice yet easily pass unnoticed in the present day: since many alchemical texts were printed relatively late in their "life cycle," the versions most readily available to scholars may capture a moment long after their original meaning changed. In such cases, modern readings of late medieval alchemical writing may unwittingly be shaped by the views of sixteenth- and seventeenth-century glossators.[3]

Identifying these shifts in meaning brings other benefits. In recent years several scholars, notably William Newman and Lawrence Principe, have unpacked alchemical allegories to show that even seemingly fanciful accounts may disguise workable chemical procedures.[4] Alchemical writers employed well-established *Decknamen* to confer authority on their own writings by signaling a relationship with venerable predecessors, while also retaining a certain flexibility of interpretation.[5] By teasing

[1] Notable exceptions include Charles Webster, "Alchemical and Paracelsian Medicine," in *Health, Medicine and Mortality in the Sixteenth Century*, ed. Charles Webster (Cambridge, 1979), 301–34; Michela Pereira, "Mater Medicinarum: English Physicians and the Alchemical Elixir in the Fifteenth Century," in *Medicine from the Black Death to the French Disease*, ed. Roger French, Jon Arrizabalaga, Andrew Cunningham, and Luis Garcia-Ballester (Aldershot, 1998), 26–52; Linda Ehrsam Voigts, "The Master of the King's Stillatories," in *The Lancastrian Court: Proceedings of the 2001 Harlaxton Symposium*, ed. Jenny Stratford (Donington, 2003), 233–52.

[2] I borrow the term from Elaine Leong, "Tweaking as Creating: Recipes and Knowledge Production in Early Modern England" (paper presented at "Alchemy and Medicine from Antiquity to the Enlightenment," Centre for Research in the Arts, Social Sciences and Humanities, Cambridge, 22–24 September 2011).

[3] Through intertextual relationships, originally distinct works may also come to influence one another over time. See Anke Timmermann, *Verse and Transmutation: A Corpus of Middle English Alchemical Poetry* (Leiden, 2013).

[4] William R. Newman, *Gehennical Fire: The Lives of George Starkey, an American Alchemist in the Scientific Revolution* (Cambridge, Mass., 1994), chap. 4; Lawrence M. Principe, *The Secrets of Alchemy* (Chicago, 2012). Alchemical allegories, of course, encapsulated meanings beyond the merely practical: see, inter alia, Barbara Obrist, *Les débuts de l'imagerie alchimique: XIVe–XVe siècles* (Paris, 1982).

[5] See William R. Newman, "Decknamen or Pseudochemical Language? Eirenaeus Philalethes and Carl Jung," *Rev. Hist. Sci.* 49 (1996): 159–88, on 164–5.

apart these interpretative layers, we may also catch glimpses of how alchemy was conceived and practiced at different historical moments—of the ways in which practice produced chemical knowledge, while being shaped and reported according to the conventions of an established textual tradition. Interventions in the transmission of a text may also mark points at which early modern practitioners diverged from their sources and sought to identify solutions compatible with their own experience, while still retaining the authority of their predecessors.

In the process, they faced similar exegetical challenges to those encountered by modern scholars attempting to decipher the various levels of meaning encountered in this demanding body of literature. Alchemical books and manuscripts are often encrusted with annotation: evidence of readers' attempts to trace references and distill workable practices from their obscure and sometimes contradictory contents. Each generation of practitioners grappled with the challenge of understanding those that came before: rendering authoritative pronouncements down into coherent and, importantly, *replicable* procedures. Their findings fed back, in turn, into the textual tradition, as earlier sources were glossed to accommodate practical knowledge, and new observations framed using the language of established tropes. In the process, meanings shifted, and terms were reinterpreted in light of textual exegesis and practical experience—a feedback loop that I term "practical exegesis." Yet the retention of authoritative language often serves to mask these practical interventions. To trace an alchemical process over time, it is therefore necessary to understand not only the nature of the original practice but also subsequent stages in its reception.

"WHICH BY MASTERS IS CALLED SERICON"

Late fifteenth-century England provides a fertile site for examining changes in alchemical ideas and practices. From the midcentury, English alchemy was enriched by an influx of treatises pseudonymously attributed to "Raymond": the Majorcan philosopher Ramon Llull (commonly anglicized as Raymond Lull; ca. 1232–ca. 1316).[6] Unknown to their readers, the core works of pseudo-Lullian alchemy had in fact been composed at different times and by different people, who sometimes proposed incompatible goals and methods. Identifying and reconciling points of conflict were therefore inevitable and necessary tasks for commentators such as George Ripley (d. ca. 1490), canon-regular of Bridlington Priory in Yorkshire, and perhaps the best-known expositor of pseudo-Lullian alchemy.[7]

Ripley's engagement with "Raymond" is particularly evident in his *Medulla alchimiae*, or "Marrow of Alchemy" (1476).[8] This Latin treatise is addressed to a high-ranking prelate, often identified (although not in the earliest manuscript copies) as

[6] On pseudo-Lull, see particularly Michela Pereira, *The Alchemical Corpus Attributed to Raymond Lull* (London, 1989); Pereira, *L'oro dei filosofi: saggio sulle idee di un alchimista del Trecento* (Spoleto, 1992).

[7] On Ripley, see Jennifer M. Rampling, "Establishing the Canon: George Ripley and His Alchemical Sources," *Ambix* 55 (2008): 189–208; Rampling, "The Catalogue of the Ripley Corpus: Alchemical Writings Attributed to George Ripley (d. ca. 1490)," *Ambix* 57 (2010): 125–201 (hereafter *CRC*); Rampling, *The Making of English Alchemy* (Chicago, forthcoming).

[8] On the authenticity of the *Medulla*, see Jennifer M. Rampling, "The Alchemy of George Ripley, 1470–1700" (PhD diss., Univ. of Cambridge, 2009), chap. 1; *CRC* (cit. n. 7), 130–1. The earliest extant witness of the original Latin text is Cambridge, Trinity College Library, MS R.14.58, Pt. 3 (hereafter *Medulla*), 1r–6r.

George Neville, the Archbishop of York.[9] In his preface, Ripley promises to strip away the layers of obfuscation from his medieval authorities, revealing the practical kernel—or marrow—of their art. This he proposes to do in three chapters, treating the "mineral," "vegetable," and "animal" stones, respectively: three elixirs, each with a distinct mode of manufacture and a different application.[10]

Of these, the vegetable stone is a medicinal elixir that, unlike the transmutational mineral stone, does not contain toxic corrosives. Instead, this stone was thought to employ a solvent derived from wine: either distilled alcohol or tartar. One of Ripley's authorities, the pseudo-Lullian *Liber de secretis naturae, seu de quinta essentia*, prescribes highly rectified spirit of wine as a "resolutive menstruum" for dissolving metallic bodies.[11] This, however, Ripley knows to be infeasible:

> Some assert that this fire is a water drawn from wine, according to the common way, and should be rectified, being distilled as many times as possible . . . yet, when water of this kind (which fools call the pure spirit), even if rectified a hundred times, is put upon the calx of whatever body, however well prepared, nevertheless we see it will be found weak and entirely insufficient for the act of dissolving our body.[12]

To complicate matters further, in another work, the *Repertorium*, Raymond stated that the solvent should be drawn from a metallic body. Ripley fastens onto the contradiction:

> If, as Raymond says, the resolutive menstruum springs from wine or the tartar thereof, how is what the same philosopher says to be understood: "Our water is a metalline water, because it is produced from a metalline kind"?[13]

By drawing attention to these problems in his source text, Ripley demonstrates his own knowledge of both textual authorities and material substances. His aim is to show that he understands Raymond's true meaning: the resolutive menstruum is not spirit of wine (which must therefore be a cover name) but a solvent possessed of both vegetable and mineral qualities, drawn from an imperfect metallic body using a solvent that is nevertheless derived from wine. In light of this exegetical juggling, Ripley proposes an alternative recipe for the vegetable stone:

> Take the sharpest humidity of grapes, distilled, and in it dissolve the body, well calcined into red (which by masters is called *sericon*) into crystalline, clear and heavy water. Of which water let a gum be made, which tastes like alum, which by Raymond is called *vitriolum azoqueus*.[14]

---

[9] Although Ripley does not name his ecclesiastical patron, Neville (who died in 1476) is a likely candidate, as archbishop of Ripley's diocese of York.

[10] The trope of the animal, vegetable and mineral stone appears in the pseudo-Aristotelian *Secretum secretorum*, and recurs in the pseudo-Lullian *Epistola accurtationis*, one of the *Medulla*'s main sources. See Rampling, *Making of English Alchemy* (cit. n. 7), passim.

[11] Ibid., chap. 2. On *De quinta essentia*, see also Michela Pereira, "'*Vegetare seu transmutare*': The Vegetable Soul and Pseudo-Lullian Alchemy, in *Arbor Scientiae: der Baum des Wissens von Ramon Lull: Akten des Internationalen Kongresses aus Anlaß des 40-jährigen Jubiläums des Raimundus-Lullus-Instituts der Universität Freiburg i. Br.*, ed. Fernando Domínguez Reboiras, Pere Villalba Varneda, and Peter Walter (Turnhout, 2002), 93–119.

[12] *Medulla* (cit. n. 8), 5r. For the original Latin text of this and subsequent passages, see Rampling, *Making of English Alchemy* (cit. n. 7), chap. 2. All translations are mine unless otherwise stated.

[13] *Medulla* (cit. n. 8), 5r.

[14] Ibid., 5v.

In Ripley's process, the "sharpest humidity of grapes" apparently refers to distilled wine vinegar, in which the metallic body is gradually dissolved. Yet what of sericon?

Like many *Decknamen*, the term has authoritative provenance. Sericon is one of the substances mentioned by the philosopher "Mundus" in the *Turba philosophorum*, an early thirteenth-century Latin translation of an Arabic text probably composed around 900 CE.[15] While its early etymology is obscure, the term seems to have originally signified a red pigment.[16] One contender is minium, or red lead: an orange-red powder made by calcining litharge (modern lead oxide). An inexpensive substitute for the costlier vermilion, minium-based pigments were familiar to every medieval scriptorium, supplying manuscripts with their rubricated capitals.

The identification of sericon with red lead is supported by its various appearances in English alchemical recipes of the late fifteenth century, where it is usually related to "adrop" (another *Deckname* for lead). Sometimes the terms are used interchangeably,[17] but more often sericon is described as a product of adrop— "Sericon, which is otherwise called burned Adrop."[18] In alchemical writing the term was used to denote minium, or calcined lead, a definition still in use in 1612, in Martin Ruland's famous *Lexicon alchemiae*.[19]

In choosing sericon as his starting matter, Ripley may have been influenced by various factors. Just as his study of Lullian texts highlighted logical contradictions, so his personal knowledge of chemical procedures ruled out a literal reading of Raymond's implausible "spirit of wine." On the other hand, both logic and experience could be satisfied if Raymond's resolutive menstruum were explained in terms of lead and vinegar. Lead salts dissolved in strong vinegar were well known to produce "sugar of lead," a sweet-tasting, crystalline, and toxic compound that, barring impurities, would correspond to modern lead acetate.

Ripley's recipe provides a reasonably detailed overview of this process: once dissolved in the distilled vinegar, sericon forms a gum (lead acetate), which produces a white vapor (acetone) when distilled. Upon condensation, this liquid is found to have several interesting properties, including a sharp taste and a bad smell that earn it the name of *menstruum foetens* ("stinking menstruum"). It is also extremely volatile, and if the practitioner wishes to proceed to the elixir, he must do so within an hour of its distillation. When added to its calx, the menstruum boils without the addition

---

[15] Julius Ruska, ed., *Turba Philosophorum: Ein Beitrag zur Geschichte der Alchemie* (Berlin, 1931), 169: "Oportet igitur, ut plumbum in nigredinem convertatur; tunc decem praedicta in auri fermento apparebunt cum sericon, quod est compositio, quod et decem nuncupatur nominibus." Ruska gives the Arabic name as *sīrīqūn*, 30.

[16] Isidore of Seville describes *Syricum* as a red pigment used to add the capital letters to books, which he explicitly differentiates from *sericum*, silk: W. M. Lindsay, ed., *Isidori Hispalensis Episcopi Etymologiarum Sive Originum Libri XX*, Book XIX (Oxford, 1911). Alchemical texts in Byzantine Greek also use σηρικόν to denote a red pigment, although this usage may simply refer back to the Latin *Syricum*. See also Dietlinde Goltz, *Studien zur Geschichte der Mineralnamen in Pharmazie, Chemie und Medizin von den Anfängen bis Paracelsus* (Wiesbaden, 1972), 190–1.

[17] E.g., "Take Adrop otherwise callid sericon." British Library, MS Sloane 3579 (ca. 1475–1500), 6r.

[18] "Recipe Sericon quod aliter vocatur Adrop combustum." Bodleian Library, MS Ashmole 1450, Pt. VII (ca. 1560–1600), 82.

[19] "Sericon, id est, minium." Martin Ruland Jr., *Lexicon Alchemiae sive Dictionarivm Alchemisticvm, Cum obscuriorum Verborum, & Rerum. Hermeticarum, tum Theophrast-Paracelsicarum Phrasium . . .* (Frankfurt am Main, 1612), 431. Ruland defines minium as either "mercurius Saturni praecipitatus vel crocus minii" (mercury of Saturn precipitate, or saffron of minium) or "gemeine rote Farb/vnd gebrandt Bley" (ordinary red pigment and burnt lead), 336.

of any extraneous heat. For this reason, only enough liquid should be added to just cover the calx.[20]

We might wonder to what extent accounts like this, deeply embedded within an existing textual tradition, reflect their authors' own experimentation.[21] Throughout the *Medulla*, Ripley provides numerous accounts of processes and observations, suggesting that his careful study of pseudo-Lullian doctrines has indeed been put to the test. His dismissal of rectified spirit of wine as "entirely insufficient for the act of dissolving our body" smacks of personal experience, as do his warnings concerning the volatility of his own vegetable menstruum.

Availability of materials offered another constraint. The Englishman reports that he could not use another menstruum recommended by Raymond, a tartar "blacker than the tartar of the Catalonian grape," because "this thing is rare in these parts."[22] However, another authority, Guido de Montanor, "has discovered another unctuous humidity, sprung from wine," which provides an adequate substitute. Although Ripley discusses his options primarily in terms of authority rather than logistics, his adoption of a lead- rather than a tartar-based process may reflect local knowledge (Yorkshire was a lead-mining area).[23]

Yet Ripley's *practicae* are just as likely as his *theoricae* to be shaped by written accounts. For instance, in likening his gum to Raymond's *vitriolum azoqueus* and his condensed vapor to *menstruum foetens*, Ripley links his observations to substances reported in another pseudo-Lullian authority, the *Testamentum*[24]—thereby generating a Lullian endorsement for a process that he admits has not been taken directly from Raymond. Ripley has also varied the usual recipe for sugar of lead, by employing red lead rather than the more readily available litharge. In altering the chemistry, Ripley may have acted on information gleaned from his own experience or that of others. However, his usage may also reflect the authority of the *Turba philosophorum*, a text he mentions earlier in the *Medulla*. Ripley's description of the body, "which by masters is called sericon," sounds like a reference to the gathered adepts of the *Turba*'s "crowd of philosophers," who endorsed a substance of that name: a substance that by Ripley's time seems to have become identified with minium. The alchemical significance of the color red—associated with both blood and the culminating *rubedo* stage of the philosophers' stone—may also have counted in its favor.[25]

We might regard Ripley's adaptation as a kind of "practical exegesis," in which

---

[20] *Medulla* (cit. n. 8), 5v.

[21] For instance, secrets literature is rife with meticulously described yet suspect procedures, some of ancient provenance; see Pamela H. Smith, "What Is a Secret? Secrets and Craft Knowledge in Early Modern Europe," in *Secrets and Knowledge in Medicine and Science, 1500–1800*, ed. Elaine Leong and Alisha Rankin (Burlington, Vt., 2011), 47–66.

[22] *Medulla* (cit. n. 8), 5r. This substance, called *nigrus nigrius nigro*, is mentioned in the pseudo-Lullian *Epistola accurtationis*.

[23] Bridlington Priory owned a mine from which lead was exported to the continent; Colin George Flynn, "The Decline and End of the Lead Mining Industry in the Northern Pennines, 1865–1914: A Socio-Economic Comparison between Wensleydale, Swaledale and Teesdale" (PhD diss., Durham Univ., 1999).

[24] Michela Pereira and Barbara Spaggiari, eds., *Il Testamentum alchemico attribuito a Raimondo Lullo: Edizione del testo latino e catalano dal manoscritto Oxford, Corpus Christi College, 255* (Florence, 1999), 310, 318.

[25] On red's alchemical significance, see Leah DeVun, *Prophecy, Alchemy, and the End of Time* (New York, 2009), 116–27; Pamela H. Smith, "Vermilion, Mercury, Blood, and Lizards: Matter and Meaning in Metalworking," in *Materials and Expertise in Early Modern Europe: Between Market and Laboratory*, ed. Ursula Klein and Emma C. Spary (Chicago, 2010), 29–49, on 41–5.

specific processes and products (Raymond's resolutive menstruum and vegetable stone) have been reinterpreted to accommodate such considerations as availability of materials and compatibility with the practitioner's own empirical observations. Just as Ripley manipulated conflicting textual sources to obtain consensus, so he modified recipes to fit practical findings and practices to fit established tropes. Thus, while Ripley's practices and theoretical arguments have their origins in recognizable fourteenth- and fifteenth-century exemplars, his *Medulla* may be reduced neither to a compilation of earlier treatises nor to a straightforward recipe collection. In its consistent elaboration of pseudo-Lullian doctrines, supported by source criticism and applied to material pursuits, it offers both a commentary on a preexisting tradition and a serious practical engagement with the challenges posed by a confusing and—unknown to Ripley—pseudepigraphic corpus. Between the cracks of Ripley's familiar sources, we catch glimpses of flexibility and innovation in the staging of his own empirical work: a source of knowledge that would feed back into his writings and those of his readers.

THE FIXING OF SERICON

The sixteenth century saw an increase in the number of works attributed both to Raymond and to his increasingly well-known expositor, Ripley, as their respective corpora expanded to include new glosses and commentaries. The "sericonian" method, based on the dissolution of lead salts in wine- and tartar-based solvents, became a staple of English alchemical treatises, recipes, and patronage suits.

As an approach to alchemical practice, this method offered several benefits to practitioners and their patrons. The ingredients were both cheap and readily available. Besides providing manuscript illuminators with an alternative to vermilion, minium offered alchemists an inexpensive prime matter—an "imperfect" metallic body rather than the precious metals praised in pseudo-Lullian treatises. For instance, the *Pupilla alchimiae*, a macaronic treatise probably composed in the last quarter of the fifteenth century, describes a recognizably sericonian process using vinegar and "red lede," which is readily obtainable "at the appotecaries redy *prepar*at and of litell prise."[26]

It was also adaptable. While the substitution of vegetable solvents like vinegar for mineral acids raised the possibility of medicinal applications, sericon could also be used for transmutation. Ripley describes how his vegetable stone could be combined with the "mineral stone" (made using vitriol and saltpeter, the main ingredients of *aqua fortis*) to produce a compound water, or *aqua composita*.[27] The inclusion of mineral corrosives restricted the use of the compound water, which was therefore intended only for the healing of metals, rather than people.

These benefits were not, however, unique to lead—leaving open the door to a range of alternative readings and practices. For this reason, practitioners seeking to replicate Ripley's processes sometimes struggled with the precise nature of sericon. In a collection of recipes related to the vegetable stone, the Elizabethan merchant Clement Draper glossed his own transcription of the *Medulla* with alternative readings of

---

[26] British Library, MS Sloane 3747, 48v. When transcribing texts, I have retained original spelling and capitalization, using italics to denote the expansion of contractions or abbreviations. Information necessary to convey the sense of a word or passage (including the translation of symbols) is included within square brackets.

[27] *Medulla* (cit. n. 8), 3r.

sericon: ceruse of lead, or else a "vittrioll made of most sharpe moysture of grapes to be verdigreace for yt is made of vineger and tarter."[28] While Draper's first suggestion takes account of sericon's long-standing association with lead, his second proposes verdigris, a copper acetate that—like red lead—was widely used as a pigment. These readings may offer us a glimpse of Draper's exchanges with practitioners within London's King's Bench prison, where Draper spent his confinement compiling extensive alchemical notebooks and testing their contents.[29]

Another Elizabethan alchemist, Thomas Potter, compiled a compendium around 1579–80: now British Library MSS Sloane 3580A and B. Potter also seems to have been in contact with other alchemical adherents, to judge by his success in accessing and collating copies of alchemical treatises, including Ripley's earlier poem, the *Compound of Alchemy* (1471).[30] Ripley there identified "Venus" (the standard *Deckname* for copper) with the "green lion," as Potter later observed in his own annotations to the *Compound*: "The lyon greene, is mercury of venus, which muste be calcyned with [gold]. & [silver]."[31] This identification may have influenced Potter's speculations on the nature of sericon in other Ripleian works.[32] For instance, he added a marginal note to the *Whole Work of the Composition of the Stone Philosophical*: one of the most overtly practical texts attributed to Ripley, which opens by directing the adept to "First take 30 pound weight of sericon."[33] Potter here glossed sericon as "a minium powder of metal," while leaving the choice of metal open: "copper. &c."[34]

If Potter's diffident labeling suggests the privileging of text rather than experience, no such hesitation is apparent in a treatise of the French diplomat and translator, Blaise de Vigenère (1523–96). De Vigenère, inventor of a famous cipher, apparently deciphered Ripley's sericon without difficulty—although he disagreed with it on practical grounds, opting for litharge instead:

> Some, as *Riply*, and others, have taken the *minium* of lead, but it is . . . of an uneasie resolution, as also ceruse & calcined lead. For my part I have found litharge, which is nothing else but lead . . . poure thereon distilled boiling vinegar, stirring it strongly with a staffe, and sodainly the vinegar will charge itself, with the dissolution of litharge.[35]

However, by far the most frequent rereading of the alchemists' red lead was "antimony," often used to denote stibnite (natural sulfide of antimony) rather than metallic

[28] British Library, MS Sloane 1423, 29r.

[29] On Draper's alchemy and note-taking practices, see Deborah E. Harkness, *The Jewel House: Elizabethan London and the Scientific Revolution* (New Haven, Conn., 2007), chap. 5.

[30] On Potter's editing strategies, see George R. Keiser, "Preserving the Heritage: Middle English Verse Treatises in Early Modern Manuscripts," in Stanton J. Linden, ed., *Mystical Metal of Gold: Essays on Alchemy and Renaissance Culture* (New York, 2007), 189–214; Jennifer M. Rampling, "Depicting the Medieval Alchemical Cosmos: George Ripley's Wheel of Inferior Astronomy," *Early Sci. Med.* 18 (2013): 45–86, on 75–6.

[31] British Library, MS Sloane 3580A, 142r; cf. a similar note on 144v. On Ripley's use of copper, see Rampling, "Establishing the Canon" (cit. n. 7), 205–6.

[32] Potter highlighted Ripley's use of sericon in the *Medulla*, although without recording an interpretation: Sloane 3580A (cit. n. 31), 143v.

[33] On the *Whole Work*, see *CRC* (cit. n. 7), 197.

[34] Sloane 3580A (cit. n. 31), 214v.

[35] Blaise de Vigenère, *A Discourse of Fire and Salt, Discovering many secret Mysteries, as well Philosophicall, as Theologicall* (London, 1649), 70; English translation of de Vigenère's posthumously printed *Traicté du Feu et du Sel* (Paris, 1618).

antimony itself. Indeed, sixteenth-century practitioners were generally more interested in antimony than their medieval forebears and hence alert to possible references in their source texts.[36] The recommendation of antimonial compounds as medicinal purgatives in Paracelsian treatises, and their increasing use by medical practitioners throughout the sixteenth century, culminating in the notorious "antimony wars," provides one likely context for the increasing substitution of antimony in sericonian recipes.[37] Another is the value of antimony in metallurgy, as a means of purging gold or silver of impurities. From the late sixteenth century, the influential writings of Alexander von Suchten (1520–75) and the fictitious "Basil Valentine"[38] also helped secure antimony's place as a key alchemical ingredient.

The reinterpretation of red lead as antimony is assisted by similarities between both sets of procedures. Medicinal preparations of antimony often employed the vegetable solvents of wine, vinegar, and tartar, creating a clear, practical bridge between antimonial and lead-based practice.[39] Writing in 1350–1, John of Rupescissa had described the preparation of a "quintessence of antimony" using spirit of wine, which he explicitly differentiated from sugar of lead—maintaining that the antimonial product was sweeter and better than ceruse dissolved in vinegar.[40] This procedural correspondence may have been deepened further by the perceived relationship between various "leaden" metals that were not always sharply distinguished in early chemistry, including antimony, marchasite, and lead: a correspondence noted by both Rupescissa and his later reader, Paracelsus.[41] A "red antimony" also existed, in the form of kermesite—making "red lead" a plausible cover name for the brownish-red antimonial ore. Whatever the deciding factor, the substitution of antimony for red lead helped shape the reception of two other widely circulated works that later came to be associated with Ripley: the *Pupilla alchimae* and the *Whole Work*.[42]

Like the *Medulla*, the *Pupilla* describes two solvents, made using different ingredients: a "red lion" for medicine and transmutation and a "green lion" for transmutation only. In the earliest known copy, MS Sloane 3747, the red lion is explicitly identified

[36] The major exception is the *Liber de consideratione quintae essentiae* of John of Rupescissa (ca. 1351–2), which describes the distillation of antimony with quintessence of wine: *Ioannis de Rupescissa qui vixit ante CCCXX annos, de consideratione Quintae Essentiae rerum omnium, opus sanè egregium* (Basel, 1597), 88.

[37] On the "antimony wars" (1566–1666), see Didier Kahn, *Alchimie et Paracelsisme en France à la fin de la Renaissance (1567–1625)* (Geneva, 2007). Although mentioned in fifteenth-century English texts, including Ripley's *Compound of Alchemy* (where Ripley rejects it), antimony often appears in the context of recipes derived from Rupescissa. Thus Sloane 3747 (cit. n. 26), 94r, describes a "quintassence of antemony" excerpted from John's *De consideratione*.

[38] For instance, Alexander von Suchten, *Liber unus de Secretis Antimonii, das ist von der grossen Heimligkeit des Antimonii* (Strasbourg, 1570); *Antimonii Mysteria Gemina . . .* (Leipzig, 1604); Basil Valentine, *Triumph-Wagen Antimonii . . . An Tag geben, durch Johann Thölden . . .* (Leipzig, 1604).

[39] The archetypical antimonial product, emetic tartar (antimony potassium tartrate), is made using tartaric acid. On antimonial preparations, see R. Ian McCallum, *Antimony in Medical History* (Durham, 1999), 99–102.

[40] *Ioannis de Rupescissa* (cit. n. 36), 90.

[41] For instance, "Gold and the Marcasite, Antimony and Lead, the which in their framing and Constelation, may be compared to each other mutually, but are neverthelesse Separated in Virtue." English translation of Paracelsus's *Archidoxa . . . Zehen Bücher* (Basel, 1570), in *Paracelsus, his Archidoxis comprised in ten books: disclosing the genuine way of making quintessences, arcanums, magisteries, elixirs, &c . . .*, trans. J. H. (London, 1660), Book VI, 82–3. The precise identity of Rupescissa's antimony, "a leaden marchasite" (*Marchasita plumbea*), is unclear. However, he clearly differentiates it from lead compounds: *Ioannis de Rupescissa* (cit. n. 36), 90.

[42] I question Ripley's authorship of the *Pupilla* in CRC (cit. n. 7), 130 and 187, although this cannot be definitively excluded.

with red lead, and the green lion with Roman vitriol.[43] The writer praises red lead as the subtlest of lead compounds, being easily soluble in a "vegitable mercury," or "water":

> Take redde lede which is verey spongeouse rather then eny other lede which is neither so spongeouse ne so sotill for in it anon the water will entre and make his dissolucion as nedefull is to be.[44]

This straightforward identification is taken from the earliest witness of the *Pupilla*: a fifteenth-century copy lacking the *practica*.[45] Other copies date from the mid-sixteenth century onward, by which time the "red lion" seems to have acquired a gloss immediately after the reference to red lead:"[id est] antimonye prepared."[46] This note, an apparent interpolation by an unknown reader, was subsequently preserved in other copies of the *Pupilla*, including the only printed edition of the work (in which the "green lion," vitriol, has furthermore been glossed as "sublimed mercury").[47] By transforming "red lead" into the *Deckname* for an antimonial compound, this reading essentially altered the underlying alchemy of the treatise. Indeed, were it not for the solitary witness of MS Sloane 3747, we would have no clue that the red lion ever signified anything other than antimony.

An identical amendment was made to the pseudo-Ripleian *Whole Work*, in which the prescription of "30 pounds of sericon" seems to have originally referred to minium.[48] However, some readers clearly viewed it as antimony, including the mathematician John Dee, who glossed his own early seventeenth-century transcription as "red leade, [antimony]."[49] By 1683, the antimonial reading was sufficiently entrenched to be silently incorporated into the version published by the London bookseller William Cooper: "30 pound weight of Sericon or Antimony."[50]

In these examples, substantive changes in interpretation and practice—specifically, the substitution of an antimonial practice for a lead-based one—have been inferred from relatively minor alterations to manuscript copies. Yet even minor changes to a recipe could have serious practical implications for the outcome of a chemical operation.[51] Over several centuries, textual accounts became uncoupled from their

---

[43] "The redde lyon and the grene lyon that is red lede and vitriol romayn." Sloane 3747 (cit. n. 26), 49r.

[44] Ibid., 49v.

[45] Ibid., 47r–50v.

[46] Bodleian Library, MS Ashmole 1480, pt. 1 (sixteenth century), 91v, and other manuscripts (*CRC* [cit. n. 7], 188–90).

[47] "Sunt enim plumbum rubeum, sive minera plumbi, h. e. Antimonium minerale praeparatum; & Victriolum Romanum, id est [mercur]ius sublimatus." George Ripley, *Opera omnia chemica*, ed. Ludwig Combach (Kassel, 1649), 301.

[48] This reading is accepted in several early copies, including that of Thomas Potter (see above) and British Library, MS Sloane 1095 (1550–1600), 75r, where sericon is glossed: "Plumbum cum comburitur coloris rubei efficitur quod a magistris Sericon appellatur" (Lead once it is burned is made of a red color, which by masters is called sericon).

[49] Bodleian Library, MS Ashmole 1486, Pt. 5, 1. On Dee and sericon, see Jennifer M. Rampling, "John Dee and the Alchemists: Practising and Promoting English Alchemy in the Holy Roman Empire," *Stud. Hist. Philos. Sci.* 43 (2012): 498–508.

[50] *The Bosome-Book of Sir George Ripley* (London, 1683), included in *Collectanea chymica a Collection of Ten Several Treatises in Chymistry, Concerning The Liquor Alkahest, the Mercury of Philosophers, and other Curiosities worthy the Perusal* (London, 1684), 101. See also the annotations to the *Whole Work* in British Library MS Sloane 319, 4r–v; and the complete reframing of the *Whole Work* in antimonialist terms in the *Liber Secretissimus Georgii Riplei*, in MS Sloane 689, 20r–31r.

[51] See Lawrence M. Principe, "'Chemical Translation' and the Role of Impurities in Alchemy: Examples from Basil Valentine's *Triump-Wagen*," *Ambix* 34 (1987): 21–30.

original traditions and paired with new techniques—transformations that could be unwittingly mediated by practitioners whose intention was simply to clarify or supplement their authority. Thus, although the sericonian paradigm of menstrua and vegetable stones continued to circulate, it faced increasing competition from other sixteenth- and seventeenth-century practices, as sericonian works were adopted and adapted by proponents of different methods.

THE TRANSMUTATION OF PRACTICE

As recent scholarship has emphasized, much of the vigor of early modern alchemy stemmed from its dual identity as *scientia* and *ars*. This identity demanded, besides practical skill, the ability to construe texts.[52] The resulting negotiation between authority and experience evokes the quid pro quo of physicians seeking to substitute local flora and fauna for ancient (and often unidentifiable) *materia medica*. Yet alchemical readers brought an additional preoccupation to their texts, with the assumption that literal readings alone were insufficient to divine intended meaning. While these deliberately abstruse writings might be susceptible to the techniques of medieval scriptural exegesis or humanist philology,[53] practical observation offered another tool by which readings might be assessed and revised. Such revisions are not, however, always explicit and can sometimes be inferred only as their cumulative effects on succeeding literature become apparent.

The transformations wrought by practical exegesis affected not only ingredients and processes but even the ends of alchemy. The implications of these shifts are most striking when we compare treatises that draw upon a common authority yet still advocate different "alchemies": the medicinal sericonian approach, for instance, versus an antimonial method aimed at transmutation. Ambiguity in a source might produce multiple readings, as in the case of the *Vision*, a short, allegorical poem translated from Latin into English during the 1570s by the gentleman alchemist Samuel Norton, who attributed it to Ripley.[54]

The *Vision* describes the poet's vision of a red toad, which expires after consuming the "juice of grapes":

> When busie at my booke I was upon a certeine night,
> This Vision here exprest appear'd unto my dimmed sight,
> A *Toade* full rudde I saw did drinke the juce of grapes so fast,
> Till over charged with the broth, his bowells all to brast.[55]

---

[52] See, inter alia, William R. Newman and Lawrence M. Principe, *Alchemy Tried in the Fire: Starkey, Boyle, and the Fate of Helmontian Chymistry* (Chicago, 2002); Bruce T. Moran, *Andreas Libavius and the Transformation of Alchemy: Separating Chemical Cultures with Polemical Fire* (Sagamore Beach, Mass., 2007); Moran, "Eloquence in the Marketplace: Erudition and Pragmatic Humanism in the Restoration of Chymia," in this volume.

[53] See Peter J. Forshaw, "Vitriolic Reactions: Orthodox Responses to the Alchemical Exegesis of Genesis," in *The Word and the World: Biblical Exegesis and Early Modern Science*, ed. Kevin Killeen and Peter J. Forshaw (Basingstoke, 2007), 111–36.

[54] On Norton's role as "discoverer," translator, and editor of Ripleian works, including the *Vision* and *Whole Work*, see *CRC* (cit. n. 7), 132–3. On Norton, see Scott Mandelbrote, "Norton, Samuel (1548–1621)," in *Oxford Dictionary of National Biography* (Oxford, 2004; online ed., 2008), http://www.oxforddnb.com/view/article/20357 (accessed 23 January 2008).

[55] "The Vision of Sir George Ripley," in *Theatrum Chemicum Britannicum*, ed. Elias Ashmole (London, 1652), 374. On the various versions of the poem, see *CRC* (cit. n. 7), 195–6.

The remainder of the poem describes the death agonies and putrefaction of the unfortunate toad. Norton interpreted this allegory in the first chapter of his own *Key of Alchemie*, a work dedicated to Queen Elizabeth I that Norton hoped would elicit both a license to practice and financial investment.[56] He offers a sericonian reading, in which the juice of grapes is distilled vinegar, which "commeth of the vine, & hath vertue ingressive." This virtue causes the body of the toad to lose its original form: "Heere Riplies toad drinks so fast, that his Bowells be all burst."[57]

Key to the poem is the identity of the red toad. Norton explains, "By this toad [Ripley] meaneth red Ledd that is Adrop or Minium or Saturne, or Capricorne or Rupescissus Antimonie."[58] This string of *Decknamen* might give us pause for thought, particularly given the inclusion of John of Rupescissa's antimony—which, as we have seen, did not indicate minium, at least in its original sense. Helpfully, Norton priced up his materials in a shopping list appended to the *Key*. This starts with "Red lead or minium in waight 280" (at four pence the pound), then, "Item for the first solution of the same . . . 280 gallons of distilled vinegar" (at ten pence the gallon).[59]

The interpretative cycle has come full circle: while red lead was elsewhere taken as a cover name for antimony, Norton here reads Rupescissa's antimony as lead, preserving the authority rather than the substance of the medieval treatise. The reference to "capricorn" (which Norton also equates with minium elsewhere in the *Key*) alludes to another work, by Lucas Rodargirus.[60] Yet despite these textual appropriations, Norton claims to have extracted a known and replicable chemical process from the *Vision*, as attested by his list of materials and prices, detailed drawings of furnaces, and technical exposition of the obscure instructions of his authorities.

Some seventy-five years after Norton's *Key*, another "Exposition" of the *Vision* was written by George Starkey (1628–65). Starkey is well known both for his role in tutoring the young Robert Boyle in chemistry and for the success of alchemical works written under his pseudonym.[61] Writing as Eirenaeus Philalethes, Starkey produced a series of commentaries during the early 1650s on those Ripleian works that had already been printed in English, including the *Vision*.[62]

Starkey may not have realized as he set out his own line-by-line commentary that the *Vision* was based on Norton's translation of a Latin original. Reconstructing the

[56] The *Key* survives only in manuscript: Bodleian Library, MS Ashmole 1421 (transcribed ca. 1611 by Thomas Robson) and Getty Research Institute, MS 18, Vol. 10, Pt. 2 (hereafter Getty). There is no record of whether the Queen ever received Norton's proposal.

[57] Getty (cit. n. 56), 32. This passage is omitted in Ashmole 1421 (cit. n. 56).

[58] Getty (cit. n. 56), 31–2.

[59] Ibid., 156.

[60] "Lead also is by Rodagirius [*sic?*] named Capricornus, & being burnt or Calcined they Call that Minium," Getty (cit. n. 56), 23. The reference is to Rodargirus, "Pisces Zodiaci inferioris vel de Solutione Philosophica," printed in *Theatrum chemicum, præcipuos selectorum auctorum tractatus de chemiæ et lapidis philosophici antiquitate, veritate, iure, præstantia et operationibus . . .* , comp. Lazarus Zetzner (Strasbourg, 1622), V:723–65.

[61] For Starkey, see Newman, *Gehennical Fire* (cit. n. 4).

[62] "Sir George Ripley's Epistle, to King Edward unfolded" was printed (without Starkey's consent) in *Chymical, medicinal, and chyrurgical addresses: made to Samuel Hartlib, Esquire* (London, 1655), while a commentary on the *Compound*'s "Recapitulation" appeared in *A breviary of alchemy; or a commentary upon Sir George Ripley's recapitulation: being a paraphrastical epitome of his twelve gates . . .* (London, 1678). The Ripley commentaries, including "The Vision of Sr George Ripley, Canon of Bridlington, Unfolded," were collected in *Ripley Reviv'd: or An exposition upon Sir George Ripley's hermetico-poetical works . . .* (London, 1677–8).

authentic text in its proper fifteenth-century context was not, however, Starkey's goal. His exegesis marks an approach that is strikingly different both from that of Norton and from those of authorities like Ripley who advocated the use of red lead for a medicinal vegetable stone.

In Starkey's interpretation, the "ruddy toad" is not lead but gold.[63] The "juice of grapes" is a *Deckname* for the "philosophical mercury" already described in his earlier work, the *Introitus apertus ad occlusum Regis palatium* (The Open Entrance into the Closed Palace of the King).[64] In describing the toad's repast, Starkey here invokes the same processes reiterated elsewhere in his Ripley commentaries.[65] As William Newman has shown, these describe the reduction of antimony ore with iron to produce the "star regulus" of antimony, which is circulated with quicksilver to create a "philosophical mercury" capable of dissolving gold.[66]

Often throughout the commentaries Starkey constructs his own riddles, disguising his process with obscure language and imagery that may be decoded by those well versed in alchemical literature and practice. In the case of the *Vision*, the riddle comes ready-made: the death of Ripley's toad is recast as the ingression of philosophical mercury into the body of gold.[67]

In providing commentaries on the famous poem, both Norton and Starkey demonstrate their ability to decipher perplexing alchemical authorities and consequently their own fitness to propagate the art. Yet their approaches to authority differ. Based on his close reading of Raymond's *Epistola accurtationis*, Ripley's *Medulla*, and other sericonian authorities, Norton offers Elizabeth I his interpretation of the cover names applied to successive stages of the vegetable stone's manufacture. From lead is made first adrop (minium), then sericon or the green lion (the gum made from minium), menstruum (an oil drawn from the gum), and finally the blood of the green lion (a menstruum made from the oil).[68] The result is a complex and nuanced reading of the various terms employed in sericonian practice—Norton's only innovation being to recast sericon as the gum (sugar of lead) rather than minium itself. Several decades later, in 1599, Norton would record these steps in a handsome "tabula," printed in 1630: a tree that sprouts, appropriately enough, from the body of the red toad (*Bufo rubea*), which signifies lead (*Saturnus plumbum*; fig. 1).[69]

In contrast, although Starkey frames his own exposition in relation to Ripley's

---

[63] "Vision of Sr George Ripley" (cit. n. 62), 2.

[64] *Introitus apertus ad occlusum Regis palatium* (Amsterdam, 1667), 4–5.

[65] "Vision of Sr George Ripley" (cit. n. 62), 7.

[66] Newman deciphers one such passage from Starkey's account of the first gate of Ripley's *Compound* in *Gehennical Fire* (cit. n. 4), 115–69, particularly 125–33.

[67] "Vision of Sr George Ripley" (cit. n. 62), 8.

[68] "When they meant to hid the materiall of the vegetable stone; they then termed their Lead, Lead of Philosophers; & being Calcined, which they afore termed Minium they againe Called Adrop; & the gumme which to ye vegetable worck proceedeth of that matter; they Call Sericon; The oile which proceedeth of that Gumme Menstrue. Moreover they termed the Gumme, the green Lion of the philosophers; & this Menstrue is Called the bloud of the green Lion." Getty (cit. n. 56), 23.

[69] Samuel Norton, *Libri tres tabulorum arboris philosophicalis*: British Library, MS Sloane 3667, 24r–89v, 11r, 12r, 15r (incomplete; includes "tabulae"); Bodleian Library, MS Ashmole 1478, pt. 6, 42r–96v, dated 20 May 1599 (104v). The York physician Edmund Deane (1572–ca. 1640) later published the *Libri tres* as a series of eight tracts beginning with *Mercurius Redivivus, seu Modus conficiendi Lapidem Philosophicum* . . . (Frankfurt am Main, 1630). On visual representations of alchemical toads, see also Joachim Telle, *Buchsignete und Alchemie im XVI. und XVII. Jahrhundert: Studien zur frühneuzeitlichen Sinnbildkunst* (Hürtgenwald, 2004).

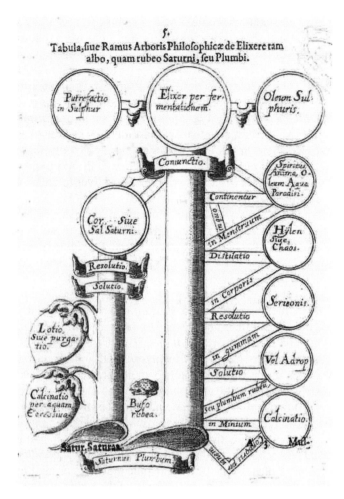

Figure 1. *"Tabula" from Samuel Norton,* Saturnus Saturatus dissolutus, et coelo restitutus, seu modus componendi lapidem philosophicum . . . , *ed. Edmund Deane (Frankfurt am Main, 1630), 5.*

poem, declaring his intention to "unfold *Ripley*'s Knots,"[70] for the actual chemistry he turns to continental sources, including the antimonial practice of Alexander von Suchten. This contrasts with Norton's goal of a medicinal panacea—the vegetable stone described by his sericonian sources. Yet Suchten's work is primarily a metallurgical practice, and Starkey reads it as such: meticulously repurposing the *Vision*'s dying toad as a harbinger of transmutation.

Since both the sericonian and antimonial approaches require the dissolution of a metallic body in a solvent, the toad's fatal thirst provides an apt allegory for either reading. While Norton's interpretation is doubtless more faithful to the original sense of the poem, pragmatism surely underlies his decision to relate his own practice to that of a revered English authority when seeking royal patronage. Norton and Starkey swathe their methods in the familiar cloak of authority, yet in each case the underlying practice proceeds from a recognizable and replicable starting point, using techniques that could be taught, deciphered, and (at least up to a point) re-created.

[70] "The Author's Preface to his Expositions upon Sir George Ripley's Compound of Alchymy, &c.," *Ripley Reviv'd* (cit. n. 62), sig. [*5]v.

CONCLUSION: LEAD INTO ANTIMONY?

Starkey's antimonial chemistry departs from the sericonian method beloved of fifteenth-century alchemists. Curiously, this shift occurs without any diminution in the authority of sericon's most famous advocate, Ripley. In the preface to *Ripley Reviv'd*, Eirenaeus Philalethes praises the Canon above all other authorities: "*Ripley* to me seems to carry the Garland."[71] Later, he compares his own mineral process to those "pitiful Sophisters" who "dote on many Stones, Vegetable, Animal, and Mineral."[72] Yet the approach rejected by Starkey is integral to Ripley's *Medulla*, and a staple of many sixteenth-century commentaries, including Norton's *Key*. Ripley's authority remains, but the alchemy is no longer his, as the structures and metaphors of his familiar works are appropriated to serve new practices.

Later commentaries also reveal a shift toward chrysopoetic alchemy. The multipurpose vegetable stone, effective as both medicine and transmuting agent, plays no part in these accounts and is sometimes explicitly excluded, for instance, by the pseudonymous "Hortolanus Junior," whose *Golden Age* was printed by Starkey's publisher, William Cooper.[73] This book essentially compiles statements relating to the antimonial alchemy of Eirenaeus Philalethes, while mocking the "superfluity of *Menstruums*" described in erroneous tracts: some made from "philosophical wine" and others from aqua fortis, not to mention "Mineral *Menstruums* Compounded of Vegetable, and Mineral *Menstruums* mixed together."[74]

This attack rejects an entire tradition of alchemical medicine: both the straightforward use of spirit of wine and the manufacture of vegetable-and-mineral compound waters. Alchemists pursuing multiple menstrua are dismissed as "Slipp-slop-Sawse makers." Yet Ripley, who took exactly this approach in the *Medulla*, retains his authority by virtue of his new relationship with Eirenaeus Philalethes; becoming, instead, one of the volume's major authorities.

Throughout these metamorphoses, authoritative instructions offered more than empty vehicles for reinterpretation. Just as Ripley labored to reconcile conflicting aspects of his pseudo-Lullian sources, so the marginalia encountered in later manuscripts speak of attempts by his own readers to derive replicable procedures from his writings. Identifying a work's principal ingredients might lead to reinterpretation: changes that in turn passed back into the textual tradition through the exegete's own writings. Through successive rereadings, often over the course of centuries, the life of authoritative texts was extended through practical exegesis, as practitioners wrestled authority and experience into an approximation of uniformity.

While an awareness of the prehistory of seventeenth-century works is necessary if we are to avoid unwittingly anachronistic readings, these early modern practices

---

[71] Ibid., sig. *3v.

[72] "Sir George Ripley's Epistle" (cit. n. 62), 23.

[73] *The Golden Age: or, the Reign of Saturn Reviewed, Tending to set forth a True and Natural Way, to prepare and fix common Mercury into Silver and Gold . . .* (London, 1698). On Cooper, see Lauren Kassell, "Secrets Revealed: Alchemical Books in Early-Modern England," *Hist. Sci.* 49 (2011): 61–87.

[74] *Golden Age* (cit. n. 73), 3–4. One likely target is the Lithuanian chemist and scholar Johann Seger Weidenfeld, who compiled sericonian sources in his *De secretis adeptorum sive de usu spiritus vini Lulliani libri IV. Opus practicum per concordantias philosophorum inter se discrepantium . . .* (London, 1684; rev. ed., Hamburg, 1685), published in English as *Four books of Johannes Segerus Weidenfeld concerning the secrets of the adepts, or, of the use of Lully's spirit of wine . . .* (London, 1685).

of interpolation, substitution, and reconciliation themselves provide telling evidence for the attempts of later readers to construe their texts. As they read, practitioners were alert to clues that might help them integrate each text into a preexisting fabric of knowledge. Indeed, even modern scholars—myself included—may find themselves engaged in a similar kind of exegetical endeavor as they seek to weave together the encoded and temporally distant strands of this fabric.[75] By re-creating experimental practices in modern laboratory settings, we may also (perhaps unwittingly) replicate our actors' historical practices: grounded in close reading of past texts, yet inevitably bringing contemporary chemical knowledge to bear.[76]

The results of these diachronic interactions stand as a warning to compilers of alchemical lexicons, reminding us that the "materiality" of early modern substances does not always collapse into concrete forms. In the late fifteenth century, sericon suggested red lead, the major ingredient in an influential branch of alchemico-medical practice. Over the next century and a half, it transformed into antimony. Time and exegesis accomplished what no alchemist could achieve alone: the transmutation of one metal into another.

[75] On re-creating Norton's practice, see Rampling, *Making of English Alchemy* (cit. n. 7), chap. 6. On Starkey's experiments, see Principe, *Secrets of Alchemy* (cit. n. 4), 164–66; and "Chymical Products," in *The Chymistry of Isaac Newton*, ed. William R. Newman, http://webapp1.dlib.indiana .edu/newton/reference/chemProd.do (accessed 25 November 2012).

[76] That such contemporary knowledge may both inform and confound the original sense of historical texts is suggested by Principe, "'Chemical Translation'" (cit. n. 51).

Auß Quecksilber und Schwefel Rein:
Johann Mathesius (1504–65) and *Sulfur-Mercurius* in the Silver Mines of Joachimstal

by John A. Norris*

ABSTRACT

The *Sarepta, oder Bergpostill* (1562) by Johann Mathesius is a book of sermons on mining and mineral subjects in which the composition and generation of metals in ore veins are discussed in terms of the *sulfur-mercurius* theory. Gur was an embodiment of *mercurius* or of *sulfur* and *mercurius*. *Sulfur* was evident in the sulfurous odor of the mines, in the supposed effects of subterranean heat, and in the deposition of mineral sulfur during the roasting of the ores. The toxic smoke given off during smelting was considered to be an additional manifestation of *mercurius*. Mathesius's sermons offer a glimpse of the ways miners' understanding of ores overlapped with alchemists' theories.

INTRODUCTION

The idea that metals are comprised of sulfurous and mercurial compositional principles is well known to historians of alchemy and early chemistry. This theory, which originated in medieval Arabic science, postulates that the differences between the seven principal metals (gold, silver, copper, mercury, tin, iron, and lead) arise from variations in the proportions and purity of their *sulfur* and *mercurius*.[1] For example, lead, generally considered to be the compositionally crudest of the metals, was said to be composed of less purified forms of these principles. Finer metals like copper and silver, on the contrary, exemplified higher degrees of purity. Variations in the relative proportions of the two principles accounted for the qualitative differences between each metal. This theory was directly applicable to the natural generation of metals, in which impurities would be gradually driven out as the protometallic mass continued to be refined in the Earth's subterranean heat. Common occurrences of two or more metals in a given ore deposit could thus be understood to result from uneven or incomplete processing of the *sulfur* and *mercurius* matter. The possibility of in-

---

* Vodní 1A, 602 00 Brno, Czech Republic; norrisjohn1@gmail.com.

I would like to thank Laura Gibbs for her help with the two Latin epigrams from Mathesius's text. Much of the research presented here was done while benefiting from a Mellon Travel Fellowship at the History of Science Collections, University of Oklahoma, during May and June 2012.

[1] The convention of designating the sulfur and mercury principles as *sulfur* and *mercurius* has been adopted after the usage in Ivo Purš and Vladimír Karpenko, eds., *Alchemie a Rudolf II* (Prague, 2011). In Mathesius's text, as in most early modern German writings on this subject, the *mercurius* principle is usually called *Quecksilber*, with distinctions between it and metallic mercury (also called *Quecksilber*) based on context.

---

creasing the purity of these compositional principles by laboratory methods seemed plausible under this theory and its subsequent variants, thus inspiring efforts to transmute metals at least from the Middle Ages onward into the eighteenth century. In this way, the *sulfur-mercurius* theory grounded the idea of metallic transmutation in the natural generation of metals, and this connection between laboratory synthesis and nature is present in many printed alchemical texts of the sixteenth and seventeenth centuries.

Despite this direct involvement with the generation of metals, there has been little consideration of how naturally occurring metallic ores were understood within the framework of this theory. As an idea that entered Western thought in alchemical texts and practices, the manner of its adoption into early modern European mining has not been traced with certainty. In a recent article, Warren Dym has called attention to this issue, also pointing out that some historians of the mid-twentieth century incorrectly denied that the *sulfur-mercurius* theory became involved in the sphere of European mining at all.[2] Such a claim is incompatible with references to *sulfur-mercurius* in early mining literature. According to the *Sarepta, oder Bergpostill* (1562), a collection of sermons on mining and mineral subjects by Johann Mathesius (1504–65), the Lutheran pastor in the mining town of Joachimstal (now Jáchymov in the Czech Republic), the generation and composition of metals and ores were commonly interpreted in terms of *sulfur-mercurius* within the mining culture to which he belonged.[3] Consideration of Mathesius's text informs us about the ways in which the *sulfur-mercurius* theory was applied to observations within the mines and how at least some miners viewed the composition and generation of the ore veins in which they toiled.

These interpretations, as recorded in the *Sarepta*, are based largely on the occurrence of gur, a substance widely believed to be protometallic matter corresponding to *mercurius* but sometimes presented as being a mixture of both principles. Gur has recently been discussed by Ana Maria Alfonso-Goldfarb and Marcia H. H. Ferraz, who cite Mathesius but focus on the seventeenth and eighteenth centuries.[4] The present discussion involves a more detailed examination of Mathesius's descriptions than has yet appeared in the current literature. It also provides an opportunity to further consider the identity of gur and its connection to *sulfur-mercurius* in the opinions of those who encountered it. Mathesius also cited the presumed effects of subterranean heat as an indication of the presence of *sulfur*, while the toxic smoke from the roasting and smelting of ores was associated with *mercurius*. This correlation between smelting smoke and the known toxicity of mercury as evincing the presence of a *mercurius* principle in metallic ores does not seem to have been discussed in previous literature about the *sulfur-mercurius* theory.

Sixteenth-century miners would generally not have been motivated to support this theory by the plausibility of performing metallic transmutations in the laboratory. It is therefore of interest to seek an understanding of how this "alchemical" theory

---

[2] Warren Alexander Dym, "Alchemy and Mining: Metallogenesis and Prospecting in Early Mining Books," *Ambix* 55 (2008): 232–54, on 234–5.

[3] The beliefs of miners regarding the origin of ores and metals is sure to have varied at different times and places and even within single communities. I assume the validity of Mathesius's claims without demanding that these opinions were universally shared among those involved in mining.

[4] Ana Maria Alfonso-Goldfarb and Marcia H. H. Ferraz, "Gur, Ghur, Guhr or Bur? The Quest for a Metalliferous Prime Matter in Early Modern Times," *Brit. J. Hist. Sci.* 46 (2011): 1–15.

made sense to the miners of whom Mathesius wrote. It will be shown below that Joachimstal in Mathesius's day was a place where the generation of metals seemed evident and ongoing, and where *sulfur* and *mercurius* seemed physically present in the *realia* of mining and smelting.

The early period of mining in the northwestern Bohemian town of Joachimstal is especially favorable for exploring this issue. Large-scale mining began there during the early sixteenth century, a time in which a literature of metallogenic theory was developing. Furthermore, observations concerning the apparent generation, regeneration, and compositional relations between metals and their ores were made when the mine tunnels were still relatively new, lending them a greater sense of validity than those involving places that had been mined for hundreds of years. Most importantly, such observations, interpreted in terms of *sulfur-mercurius*, were contemporaneously recorded by Mathesius in the *Sarepta*.

MATHESIUS IN JOACHIMSTAL

At the beginning of the sixteenth century, there were only sparsely populated, thick forests in the part of the Krušné hory (the Czech side of the Erzgebirge) where Joachimstal was eventually founded. Prospecting in this area seems to have begun around 1511. Initial efforts uncovered encouraging silver finds around the village of Konradsgrün in 1512, but the scale of operations remained very small. In 1516, Count Stefan Schlick (1487–1526), whose family owned domains nearby, bought up the area of Konradsgrün, organized further prospecting and mining activity, and gave the valley the name Sankt Joachims Thal, the valley of St. Joachim. At the beginning of that year, there were only about 100 people living in Konradsgrün, but as news spread of the rich ore quickly uncovered under Schlick's direction, workers from declining mining towns in Bohemia and Germany, as well as from the Tyrol and Switzerland, headed to northwestern Bohemia. By the end of the year, the number of inhabitants in the newly renamed town of Joachimstal had grown to more than 1,000. The population continued growing as the number of rich mines multiplied. By 1520 there were nearly 5,000 inhabitants, increasing to 17,000 by 1530, and cresting at just over 18,000 in 1534, making Joachimstal the second most populous Bohemian town after Prague (50,000 in 1534).[5] During the 1530s, the period of peak production, the Joachimstal mines produced around 10,000 kilograms of silver annually.[6]

It was during this period of rapid growth that Johann Mathesius came to Joachimstal. He was born in 1505 in the Saxon town of Rochlitz, the son of a successful mine owner. After attending the Latin school in Nuremberg, he studied for several years at the University of Ingolstadt. Necessity forced him to interrupt his studies in 1525, and he worked for several years as a private librarian, and later as a private tutor. Around 1529, a stipend from his hometown allowed him to spend one year studying at the University of Wittenberg, where he attended the lectures of Martin Luther, thereby solidifying his interest in the Reformation. Following another period of private tutoring, he was appointed rector of the new Latin school in Joachimstal in 1532. Returns from mining investments allowed him to resume his theological studies in Witten-

---

[5] Josef Sudlovský and Vladimír Horák, *Kronika horního města Jáchymova a jeho hornictví v kontextu dějin zemí koruny česky* (Ústí nad Labem, 2009), 27–8; Pavel Kašpar and Vladimír Horák, *Schlikové a dobývání stříbra* (Prague, 2009), 19–24, 43.

[6] Kašpar and Horák, *Schlikové* (cit. n. 5), 43.

berg in 1538, where he became one of Luther's close associates. After submitting a dissertation entitled *Quaestio de rebus metallicis*, he returned to Joachimstal as a Lutheran preacher in 1540, and he was made pastor in 1544.[7]

Mathesius's *Sarepta, oder Bergpostill*, first printed in 1562, contains augmented versions of special sermons on mining topics that he delivered each year at carnival time, between 1553 and 1562.[8] These sermons, which dealt with subjects such as mining history, individual metals and ore minerals, minting, glassmaking, and the generation of metals and their ores, should have been of direct interest to the miners of Joachimstal, who naturally made up the majority of Mathesius's congregation. It is noteworthy that the sermons about the metals involved not only the most common ones, like gold, silver, copper, and iron, but also less familiar ones that were more difficult to characterize, and which occurred in the Joachimstal ores, including bismuth, antimony, and cobalt. It will be shown that these "impure" metals factored in the interpretation of *sulfur-mercurius* metallogenesis.

Recent studies have noted the prevailing religious character of the *Sarepta*, which is rich in biblical citations and analogies and never forgets that God's will is the first cause in the generation of metals.[9] Mathesius even composed a hymn, the second line of which appears in the title of this article, that mentions metallogenesis within the veins of the Earth from *sulfur-mercurius*, names several metals and minerals as God's blessings, and concludes that spiritual love of God is the greatest treasure of all.[10]

While it is certainly true that Mathesius addressed these subjects and designed his rhetoric so as to positively influence his congregation of mine workers, his own interest in minerals and metals, and in the secondary causes by which they are formed, should not be underestimated. Mathesius admitted that Georgius Agricola's (1494–1555) first book on mining and mineral subjects, *Bermannus, sive de re metallica* (1530), was very inspirational to him, marking the time when, as he wrote, "I began to read about, ask about, and study mining myself."[11] In the *Bermannus*, two physicians who were interested in minerals and their nomenclature explored the mines of Joachimstal with the mining administrator, Bermannus. The three interlocutors thus had the opportunity to view the natural occurrence of various ore minerals, and to discuss their colors, associations, compositions, and terminology. Indeed, the text seems to be a fictionalized account of how Agricola himself began to learn through his friendship with the mining foreman Lorentz Bermann (or Wermann; ca. 1490–1532), for whom the book is named, when Agricola was town physician in Joachims-

---

[7] Jaroslav Jiskra, *Těžba stříbrných rud v Jáchymově v 16. století s jáchymovskými osobnostimi a první báňskou školu* (Pilsen, 2009), 93–5. On Mathesius's *Quaestio*, see J. R. Partington, *A History of Chemistry* (London, 1969), 2:64. For further information on Joachimstal, Mathesius, and the *Sarepta*, see Warren Alexander Dym, "Mineral Fumes and Mining Spirits: Popular Beliefs in the *Sarepta* of Johann Mathesius (1504–1565)," *Reform. & Renaiss. Rev.* 8 (2006): 160–85, on 160–9.

[8] Dym, "Mineral Fumes" (cit. n. 7), 167.

[9] Dym, "Alchemy and Mining" (cit. n. 2), 242; Alfonso-Goldfarb and Ferraz, "Gur" (cit. n. 4), 3. See also Dym, "Mineral Fumes" (cit. n. 7); and John A. Norris, "The Providence of Mineral Generation in the Sermons of Johann Mathesius (1504–1565)," in *Geology and Religion: A History of Harmony and Hostility*, ed. M. Kölbl-Ebert (London, 2009), 37–40.

[10] The *Sarepta* went through numerous editions, the foliations of which differ. For convenience, I cite the second and third editions. Johann Mathesius, *Sarepta, oder Bergpostill* (Nuremberg, 1564), CCCXVI[a]; Mathesius, *Sarepta* (Nuremberg, 1571), CCXXVI[a].

[11] Mathesius, *Sarepta* (1564), $a_6{}^b$; (1571), $)(_3{}^b$ (both cit. n. 10). See also Dym, "Mineral Fumes" (cit. n. 7), 167. The symbol ")(" refers to the leaf signature given to the unnumbered preliminary leaves in the 1571 edition of the *Sarepta*.

tal from 1527 to 1531. Like Agricola before him, Mathesius likewise learned and obtained specimens directly from Joachimstal miners.[12]

MATHESIUS'S SOURCES AND THE *SULFUR-MERCURIUS* BACKGROUND

Mathesius's observations and comments generally interpret issues of metal composition and generation in terms of the *sulfur-mercurius* theory. However, this view was not universally shared among authors who had experience with mining and wrote on mineral subjects. Most notably, Georgius Agricola strongly objected to it, detailing his criticisms in *De ortu et causis subterraneorum* (1546). Nobis wrote that Agricola's contention depended largely on the ineffectiveness of efforts to accomplish metallic transmutations based on this theory.[13] This was certainly among Agricola's grievances,[14] but more germane to the present discussion is his opposition on terminological grounds. It has been noted elsewhere that Agricola was intolerant of the use of the words *sulfur* and *mercurius*, or their German equivalents, for anything other than common sulfur and mercury.[15] He pointed out that neither native sulfur nor metallic mercury occurred in the mines of the other metals, and that no involvement could therefore be claimed for these substances in metallogenesis.[16] This was not only a terminological fallacy of the chymists, for Agricola also bemoaned its currency among miners.[17] An example of what Agricola may have meant is provided by Leonhard Thurneisser (1530–96), a Paracelsian chymist and physician. Thurneisser had trained as a goldsmith and apothecary before being forced into mining and smelting work as a prisoner of war in the early 1550s, and he remained professionally involved in mining for several years following his release.[18] Based on his experience, he expressed his belief as follows: "That all metals have their origins from sulfur and mercury is so obvious and visible, that I don't believe anyone could deny it."[19] Thurneisser qualified his view with the common caveat, which Agricola would not permit, that the sulfur and mercury of metallogenesis were not the common substances of those names, but analogous compositional principles that combine in various proportions and degrees of purity to form metals and their ores. It should be noted that, aside from intolerance of *sulfur-mercurius* terminology, and disagreement about the efficacy of mineral vapors, Agricola's metallogenesis was very similar to the gur theories: a thick liquid forms when diverse earthy matters mix with water within the Earth's subterranean heat, gradually congealing into a metal or mineral of a corre-

---

[12] Mathesius. *Sarepta* (1564), a₆ᵇ; (1571),)(₃ᵇ (both cit. n. 10).

[13] Heribert M. Nobis, "Der Ursprung der Steine: zur Beziehung zwischen Alchemie und Mineralogie im Mittelalter," in *Toward a History of Mineralogy, Petrology, and Geochemistry,* ed. Bernhard Fritscher and Fergus Henderson (Munich, 1998), 29–52, on 47–50.

[14] Georgius Agricola, *De ortu et causis subterraneorum,* 2nd ed. (Basel, 1558), 64–6. The 1558 edition is used here because it contains additional text prepared by Agricola before he died in 1555.

[15] John A. Norris, "Dolování a představy o metalogenezi v Čechách 16. století," in Purš and Karpenko, *Alchemie* (cit. n. 1), 657–70, on 662–4.

[16] Agricola, *De ortu* (cit. n. 14), 66–7. Agricola noted the existence of mercury mines in Europe, which were almost always devoid of other metals.

[17] Ibid., 64, 67. Also noted in Dym, "Alchemy and Mining" (cit. n. 2), 237.

[18] J. C. W. Moeshen, "Leben Leonhard Thurneissers zum Thurn, Churfürst. Branderburgischen Leibarztes," in *Beiträge zur Geschichte der Wissenschaften in der Mark Brandenburg von den ältesten Zeiten an bis zu Ende des sechszehnten Jahrhunderts* (Berlin, 1783), 57.

[19] Leonhard Thurneisser, *Zehen Bücher von kalten, Warmen, Minerischen und Mettalischen Wassern* (Strasbourg, 1612), 11. This is the second edition of Thurneisser's *Pison* of 1572. Thurneisser believed not only in *sulfur-mercurius* but in the Paracelsian *tria prima,* including a saline principle.

sponding composition.[20] But despite Agricola's criticisms, we see in Thurneisser that the idea of *sulfur-mercurius* principles could seem evident to those who had actually done some digging.

This disparity of ideas was addressed in the earliest printed work on mining, the "Bergbüchlein" (ca. 1500),[21] by Ulrich Rülein von Kalb (or Calw; 1465–1523). In Rülein's text, the *sulfur-mercurius* theory, involving astral influences, mineral vapors, and a qualitative confluence of active (*sulfur*) and passive (*mercurius*) agents, is called the "opinion of the Wise." It is immediately juxtaposed with the opinion of those who, like Agricola, denied that metals were comprised of *sulfur-mercurius* on the grounds that no metallic mercury occurs where the other metals are mined but held a gur theory instead (though the term is not used here). The protometallic substance is described as "a moist and cold material, slimy and completely without sulfur, that is drawn out from the Earth as if it were its sweat, which mixes with sulfur, and forms all the metals." Rülein then attempted to integrate these two apparently opposing views, stating that the slime mentioned above is to be understood as the *mercurius* principle.[22] Such a forced reconciliation shows that some degree of contention existed even before the publication of Agricola's criticisms in 1546.

This identification of *mercurius* with gur is again encountered in the *De re metallica* (ca. 1551) of the Thuringian pastor Christoph Entzelt (or Encelius, 1520–86). In the chapter on mercury, Entzelt distinguished between the natural and the artificial. The former, he wrote, is exuded from ore veins within the Earth and constitutes the primal matter of metals (i.e., gur). Artificial mercury, in his opinion, is that which was extracted from the mineral cinnabar (i.e., true metallic mercury).[23] He explained that the metals form when this moist, slimy natural mercury meets with earthy, hot, naturally occurring sulfur in the depths of the Earth. Similar to Rülein's treatment, Entzelt discussed the passive/active, female/male analogies of the mixing of *sulfur* and *mercurius*, though in greater detail and adding the importance of compositional purity in these principles. Again echoing Rülein, and perhaps in direct response to Agricola's arguments in *De ortu*, Entzelt also noted that some deny this view of metallogenesis because of the absence of sulfur and mercury in metallic ore mines. He merely stated that these objections are wrong.[24]

Though the *sulfur-mercurius* theory was accepted into European mining, it certainly did not have its origin there. The presentation in Rülein's book appears to be an early integration of the ideas of the philosophers (*sulfur-mercurius*) with those probably originating within the mining industry (gur). It was perhaps in such a way that

[20] Agricola, *De ortu* (cit. n. 14), 80–1. See also John A. Norris, "Early Theories of Aqueous Mineral Genesis in the Sixteenth Century," *Ambix* 54 (2007): 69–84, on 73–6.

[21] The title of this work varied among editions. I adopt the convention of citing it as the "Bergbüchlein" from David. E. Connolly, "Problems of Textual Transmission in early German Books on Mining: 'Der Ursprung gemeynner Berckrecht' and the Norwegian 'Bergordnung,'" (PhD diss., Ohio State Univ., 2005).

[22] I have used the later version of this text appearing in *Uhrsprung gemeiner Berg-Rechte*, in *Corpus Juris & Systema rerum Metallicarum, Oder: Neu-verfaßtes Berg-Buch,* ed. Johann David Zunner (Frankfurt am Main, 1698), 14–5. See also Connolly, "Problems" (cit. n. 21), 524; and Anneliese Grünhaldt Sisco and Cyril Stanley Smith, eds., *Bergwerk- und Probierbüchlein* (New York, 1949), 19–21.

[23] I have used the following German edition: Christoph Entzelt, "Tractat von Metallischen Dingen," in Zunner, *Corpus Juris* (cit. n. 22), 5. The first edition of Entzelt's *De re metallica* is undated but contains a letter from Philipp Melanchthon to the Frankfurt printer Christian Egenolff dated 1551. The second edition is dated 1557.

[24] Entzelt, "Tractat" (cit. n. 23), 2–3.

the *sulfur-mercurius* theory entered the reasoning of early modern European miners from its distant origins in medieval Arabic science.

In the *Sarepta*, Mathesius did not cite any literary sources on the *sulfur-mercurius* theory, though he did mention three authors who we know wrote about it: Albertus Magnus (ca. 1200–80), Georgius Agricola, and Christoph Entzelt.[25] In the *Liber mineralium*, Albertus discussed the generation of metals and stones from vapors similar to those first described in Aristotle's *Meteorologica* and remarked that these vapors were equated with *sulfur* and *mercurius* in the works of Avicenna, while a more direct use of this compositional dyad is made in his discussions of specific metals and minerals. He also described a laboratory procedure in which sulfur and mercury were combined and sublimated in glass vessels, with the resulting vapors rising to the top of the apparatus. He concluded that this was analogous to the natural process in which *sulfur* and *mercurius* combine within the Earth to form metals,[26] and Mathesius made similar analogies to the subterranean distillation of moist (*mercurius*) and fatty (*sulfur*) vapors.[27] We have seen that Agricola wrote disapprovingly about *sulfur-mercurius* and mineral exhalations in his *De ortu*. Though direct mention is made only of the *Bermannus*, in which the *sulfur-mercurius* theory is not discussed, a surviving letter written by Mathesius in 1547 reports that he had read *De ortu* and *De natura eorum* from the group of Agricola's writings first published together in 1546.[28] Moreover, as Mathesius and Agricola became personally acquainted in 1550,[29] it is likely that they discussed these ideas among themselves. It was noted above that Entzelt considered the generation of metals in terms of *sulfur-mercurius*. Of the three sources mentioned here, Mathesius's descriptions seem to have the most in common with Entzelt's, even down to the general disdain for the ideas and aims of the chymists. And though it is not mentioned, it is not unlikely that Mathesius would also have read the "Berg-büchlein," of which several editions were printed during his lifetime.[30]

Indeed, Mathesius could have encountered the *sulfur-mercurius* theory in numerous works on alchemy or natural history. Conversely, consideration of the few sources he mentioned is enough to show that he could have learned about it without having read a single alchemical text. In addition to the literary sources mentioned above, he claimed to have become familiar with all aspects of mining from "deeply learned, experienced, and skillful miners,"[31] and not only in Joachimstal. The strong impulse he received from reading Agricola's *Bermannus* also led him to correspond with people at mining centers throughout central Europe, from whom he even obtained spectacular specimens "the likes of which Doctor Agricola, who was freely known in my house, had never seen before."[32]

What is certain is that his interpretation of ore and metal occurrences in terms of the *sulfur-mercurius* theory is based on direct observations, not on Aristotelian

---

[25] Mathesius, *Sarepta* (1564), a₅ᵇ; (1571),)(₃ᵇ (both cit. n. 10).

[26] Dorothy Wyckoff, "Albertus Magnus on Ore Deposits," *Isis* 49 (1958): 109–22, on 116–21; John A. Norris, "The Mineral Exhalation Theory of Metallogenesis in Pre-Modern Mineral Science," *Ambix* 53 (2006): 43–65, on 49–52.

[27] Mathesius, *Sarepta* (1564), Lᵃ; (1571), XXXVᵇ (both cit. n. 10).

[28] Frank Dawson Adams, *The Birth and Development of the Geological Sciences* (Baltimore, 1938), 198.

[29] Jiskra, *Těžba* (cit. n. 7) , 99.

[30] Sisco and Smith, *Bergwerk-* (cit. n. 22) , 54–62.

[31] Mathesius, *Sarepta* (1564), a₅ᵇ–a₆ᵃ; (1571),)(₂ᵇ (both cit. n. 10).

[32] Mathesius, *Sarepta* (1564), a₆ᵇ; (1571),)(₃ᵇ (both cit. n. 10).

physics. And unlike Rülein's and Entzelt's discussions, the *Sarepta* explains how the theory seemed consistent with what was seen in the mines. In what follows, I will examine the ways in which Mathesius's examples of *sulfur-mercurius* in the mines of Joachimstal made sense to him and to the miners about whom Agricola complained.

MANIFESTATIONS OF *SULFUR-MERCURIUS* IN THE SILVER MINES OF JOACHIMSTAL

Gur and the Mixing of Sulfur-Mercurius

The sixth sermon of the *Sarepta*, concerning silver, begins with what Mathesius called "a beautiful wonder of mining history." A foreman of the St. Lorenz mine in Abertam, about 10 kilometers from Joachimstal, found that a wooden support column had become thickly encrusted with rock. Upon closer examination, he noticed that the newly formed rock contained flakes of silver, which assaying proved to be quite pure. Further inspection confirmed that the rock originated from a gur that had trickled down from an overlying abandoned tunnel and had solidified along the column and onto the floor. Picking through the sintered column, the foreman found more pieces of native silver. This inspired him to mine out the overlying material through which the gur was seeping. Here he found relatively rich silver accumulations that were duly extracted, leaving a passage through which "the sulfurous and quicksilvery moisture could come though" and solidify for further mining. This presumed metallogenic event had occurred within a period of twenty years, which is how long the now mineralized support had been emplaced, and it continued to develop. Mathesius also claimed to have found such silver occurrences himself, likewise embedded in the wooden structures of mine tunnels.[33]

Such observations were not uncommon in mines. Writing in 1770, Johann Thaddäus Peithner (1727–92), professor of mining in Prague, reported seeing such a wooden support from the old, largely abandoned silver mines of Kutná Hora that was entirely permeated with native silver.[34] Unlike the numerous examples from mines and waste heaps of greater or unknown age, and overused reports of ore regeneration on the remote island of Elba, Mathesius reported a case in which the new mineral growth could be observed and dated. According to Mathesius, experiences like these convinced miners that the generation of rocks, minerals, and metals was indeed a continuous process, and that *sulfur-mercurius* in the form of gur was the substance from which they formed.

Gur has recently begun to receive more attention from science historians.[35] This substance, characterized as a viscous liquid with a noticeable metallic and mineral content, sometimes described as having a sulfurous odor or vapor, was commonly believed to represent a relatively early stage in the generation of metals and minerals. It would now be recognized as the product of mineral decay. The ores of Joachimstal are mostly sulfides and arsenides of silver, lead, bismuth, antimony, and cobalt,

---

[33] Mathesius, *Sarepta* (1564), LXXXVI[b]; (1571), LXII[a–b] (both cit. n. 10).

[34] Johann Thaddäus Anton Peithner, *Erste Gründe der Bergswerkwissenschaften* (Prague, 1770), 28–9.

[35] For example, Alfonso-Goldfarb and Ferraz, "Gur" (cit. n. 4); William R. Newman, "Geochemical Concepts in Isaac Newton's Early Alchemy," in *The Revolution in Geology from the Renaissance to the Enlightenment,* ed. Gary D. Rosenberg (Boulder, Colo., 2009), 41–9; Dym, "Alchemy and Mining" (cit. n. 2); Norris, "Early Theories" (cit. n. 20).

with abundant pyrite.[36] When such minerals oxidize in the presence of air and water, they break down into forms that are easily transported in groundwaters. This reaction is exothermic and releases sulfuric acid, which contributes an acrid stench and promotes degradation of the surrounding rocks. The result is a mildly acidic mud, containing dissolved metallic salts, with fragments of ore minerals and metal. This combination of viscous liquidity and metallic content seemed to encompass the properties of *mercurius*. A sulfurousness is also sometimes mentioned, leading some, like Mathesius, to consider gur as an early stage in the combination of the two principles. Common among the soluble metallic salts associated with gurs from metallic sulfide deposits is vitriol,[37] which forms when pyrite, usually abundant in such deposits, undergoes the oxidation reaction mentioned above. This explains the frequent association of vitriol with gur.[38] At Joachimstal, the ore veins have been eroded to the surface, making them susceptible to weathering and consequent secondary, or supergene, enrichment. The vitriol solutions generated from the oxidation of pyrite in turn oxidize silver sulfides, causing the downward-directed enrichment of secondary silver minerals and the eventual deposition of native silver.[39] Dramatic occurrences of this native silver, reported by Mathesius[40] and the secondary literature, appeared in dendritic or wiry structures and large masses, some weighing 100–200 kilograms. The rapid development and relatively fast decline of the mining boom in Joachimstal was due to the proximity of this enriched zone to the surface.[41]

This geological situation was favorable to the occurrence of gur liquids within the Joachimstal ore complex, several examples of which are given in the *Sarepta*. When the processes of metallogenesis were generally believed to be amenable to observation, one can see how the sulfur-smelling, metal-bearing, mildly acidic gurs that oozed from the ore veins could seem related to this process. In the sermon on the generation of metals, Mathesius further explained the role of gur in metallogenesis and its relation to *sulfur-mercurius*. Metals are earthy bodies. They form in veins and fissures from a subtle or distilled earth and a thick, fatty vapor generated through the Earth's natural heat. In this way, the earthiness and moisture are combined, forming a gur. This is the sulfurous and quicksilverish "seed" from which all rocks and metals arise. This gur continues to accumulate and begins to coagulate, becoming a thicker mass and gradually transforming into a better, or more pure, composition, until it reaches perfection naturally in the subterranean heat or is purified metallurgically.[42]

Agricola was correct that mercury does not occur among these deposits, where native sulfur is also scarcely present. The liquidity of gur, and perhaps its colors and the inclusion of metallic particles, made it comprehensible as the metallogenic *mercurius*, just as Rülein and Entzelt believed. This is probably what Mathesius meant when he noted that an alchemist asked him for sulfur and self-growing mercury (i.e.,

---

[36] FeS_2.

[37] Iron and copper sulfate, in solid forms called melanterite ($FeSO_4 \cdot 7H_2O$) and chalcanthite ($CuSO_4 \cdot 5H_2O$).

[38] Norma E. Emerton, *The Scientific Reinterpretation of Form* (Ithaca, N.Y., 1984), 214–8.

[39] For a general description of these processes, see J. M. Guilbert and C. F. Park Jr., *The Geology of Ore Deposits* (New York, 1986), 802–12.

[40] Mathesius, *Sarepta* (1564), XL[a]; (1571), XXVII[a] (both cit. n. 10).

[41] Petr Pauliš, *Nejzajímavější mineralogická naleziště Čech II* (Kutná Hora, 2003), 19–21, 24.

[42] Mathesius, *Sarepta* (1564), XXXIX[a–b]; (1571), XXVII[b] (both cit. n. 10).

mercurius) from the mines of Joachimstal.[43] The *sulfur* principle, however, seemed evident in other ways.

> That no rich mine is without sulfur or *bleischweiff*[44] is a matter of experience. Many *berg-arten*[45] burn, smoke and stink like coal, [so] there must be a sulfurous fattiness [therein]. The stench in smelting also comes from the impure and mixed sulfur. When, in a tunnel that comes from a rich vein, the sulfur is warmed by the sun after the summer rains, one smells it from numerous masses.[46]

Mathesius recorded instances of sulfur seeming to be present in a vaporous state, even mentioning the case of someone who, when considering whether or not to purchase shares in a mine, would climb inside to test the strength of the sulfurous smell with his own nose.[47]

Because of its conspicuous presence at places of volcanic activity, sulfur was generally associated with subterranean heat. Mathesius cited thermal springs as evidence of ubiquitous underground warmth, along with the occurrence of coal and petroleum, the combustibility of which was also assumed to be related to sulfur.[48] All of this was consistent with the general idea that mineral- and metallogenesis should be a type of fermentation process. Some observations from the Joachimstal mines likewise seemed to indicate the action of sulfurous heat and mineral fermentation. In the fifth sermon, Mathesius described how miners often noticed a white or vaporous moisture that seemed to ferment from silver and congeal.[49] The Earth's interior, Mathesius wrote, contains God's "wondrous *laboratoria* and distillation oven," as evidence of which he described certain fissures among the ore veins that contained slimy, soapy matter that seemed to ferment and bubble "like beer in a vat. Thence come many types of stick-like, hair-like, and root-like silver."[50] In the same passage, Mathesius wrote,

> But it is natural and normal that the Earth has its natural heat and action, and not only through its wondrous and inborn fire that often erupts on the surface. In moist veins, fat and sulfurous vapors operate, from which all *bergkart*, metal and juices come to be, when they congeal or sinter together, as the miners report.[51]

Such observations of gur liquids and fermentation appeared to involve the *sulfur-mercurius* principles and the generation of metals and minerals. About this, Mathesius was sure that he and the miners of Joachimstal were merely reporting their experience.[52] Even with knowledge of Agricola's criticisms, such evidence led him to declare:

---

[43] Mathesius, *Sarepta* (1564), XLIII[a]; (1571) XXX[b] (both cit. n. 10).

[44] This was a clayey, sulfurous earth containing lead and silver, probably from weathered, silver-bearing galena (PbS). See E. Göpfert, "Die Bergmannsprache in der Sarepta des Johann Mathesius," *Z. Deut. Wortforsch.* 3, supplement (1902): 1–107, 15.

[45] This term denoted minerals devoid of notable metallic content, though associated with ore veins. See ibid., 10.

[46] Mathesius, *Sarepta* (1564), XLIII[b]; (1571), XXXI[a] (both cit. n. 10).

[47] Mathesius, *Sarepta* (1564), XLIII[b]; (1571), XXXI[a]; see also (1564), LXXIX[a]; (1571), LVI[b] (both cit. n. 10).

[48] Mathesius, *Sarepta* (1564), LXXIX[b]; (1571), LVI[b] (both cit. n. 10).

[49] Mathesius, *Sarepta* (1564), LXXIX[b]; (1571), LVI[b] (both cit. n. 10).

[50] Mathesius, *Sarepta* (1564), LXXIX[a–b]; (1571), LVII[a] (both cit. n. 10).

[51] Mathesius, *Sarepta* (1564), LXXIX[a]; (1571), LVI[b] (both cit. n. 10).

[52] Mathesius, *Sarepta* (1564), LXXXVII[a]; (1571), LXII[a] (both cit. n. 10).

I cannot then without special reason dispense with the opinion of the old wise people, alchemists, and experienced miners, who unanimously testify that all metals come from quicksilver and sulfur [i.e., *mercurius* and *sulfur*]. . . . In both one finds a cold moisture mixed with a certain earthy and heating fattiness, such that earth and water come together and become purified into a metallic body.[53]

As if to summarize, he recounted an epigram penned by Philipp Melanchthon (1497–1560), apparently in direct reference to the Joachimstal ores:

Where milky mercuries mixed with sulfurous fumes are cooked,
there are the first seeds of a new vein.[54]

Smelting Smoke as a Manifestation of **Mercurius**

In addition to gur, another perceived manifestation of *mercurius* was the toxic smoke released from ores when they were roasted or smelted. As the ores of Joachimstal are abundant in silver, lead, arsenic, antimony, bismuth, and cobalt, all of which are harmful when inhaled, the threat posed to anyone in the vicinity of the ore roasting ovens was potentially great. Mathesius had much to say on this issue.

Seldom is there a *bergart* or ore without poison, and especially pyrite, cobalt, bismuth and *rotgüldig* ore.[55] And not only from lead but from all kinds of metals a poisonous smelting smoke wheels at the oven and is deposited in the refining furnace. Now Mercurius, be it sublimated or not, is a strong poison, as besides experience an old verse shows:
They report [that] earthy metals are produced by Mercurial venom,
They bear, it is true, nothing but poison.[56]

The examples Mathesius gave of the minerals most infamous for bearing the poisonous *mercurius* are of considerable interest. None of them contain metallic mercury, nor is that metal present in the Joachimstal ore complex. As such, it is certain that this is a reference to the compositional *mercurius* principle. The toxicity and volatility of common mercury had been well known for centuries from laboratory and metallurgical procedures. Like its common metal namesake, *mercurius* was also considered to be highly toxic, or rather, the poisons encountered in metallic ore minerals were interpreted as corresponding to the toxicity of metallic mercury, thus contributing a further level of correlation between the two.

Mercurius is strongly associated with mineral toxins in the work of Mathesius's older contemporary, Theophrastus von Hohenheim, called Paracelsus (1492–1541). In *Von der Bergsucht und anderen Bergkrankheiten* (first printed in 1567 but probably written during the 1530s),[57] Paracelsus considered the diseases contracted by smelters, including a well-developed correlation between toxic fumes and *mercurius*. He explained how the *mercurius* enters the sinuses and lungs of the workers through the inhalation of smoke and described its corroding effects on the interior

---

[53] Mathesius, *Sarepta* (1564), XLIII[a]; (1571), XXX[b] (both cit. n. 10).
[54] Mathesius, *Sarepta* (1564), XLIII[b]; (1571), XXXI[a] (both cit. n. 10).
[55] *Rotguldig ertz*, probably pyrargyrit (Ag_3SbS_3).
[56] Mathesius, *Sarepta* (1564), XLIII[b]; (1571), XXXI[a] (both cit. n. 10).
[57] Paracelsus, "On the Miners' Sickness and Other Miners' Diseases," trans. George Rosen, in *Four Treatises of Theophrastus von Hohenheim called Paracelsus,* ed. Henry E. Sigerist (Baltimore, 1941), 46–7.

tissues of the body and brain. Paracelsus carefully distinguished this *mercurius* poisoning from that contracted by working with real mercury[58] (which he discussed elsewhere in the same book).

This association of toxic smelting fumes with *mercurius* lasted into the eighteenth century, as did the distinction of diseases caused by *mercurius* from those of metallic mercury. In his book on miners' diseases, the German physician and mineral chymist Johann Henckel (1678–1744) considered the illnesses associated with the smoke from metallic ores and listed symptoms consistent with those described by Paracelsus.[59] He noted Paracelsus's differentiation of diseases related to *mercurius* from those of metallic mercury and commented that this division continued even into his day. *Mercurius*, he wrote, denoted a volatile substance released from the smelting of metallic ores, though it was not itself mineral or real quicksilver (metallic mercury). And though there was no metallic mercury in the mines of his region, the *mercurius*-caused diseases that originated at the ore furnaces were unfortunately all too common. Interestingly, Henckel complained that this reference to *mercurius* was quite ambiguous by his time, when the *sulfur-mercurius* theory was no longer current. In his opinion, this confusion resulted from the erroneous identification of metallic mercury with the Latin *Mercurius* instead of with its proper name, *Argentum Vivum* (quicksilver).[60] It nonetheless shows just how deeply engrained this association was in the concepts and language of mining.

Returning to Mathesius, we see that the noxious fumes given off when roasting these ores were also directly associated with *mercurius* in the *Sarepta*. This toxic *mercurius* was nearly unavoidable in the Joachimstal ores:

> Now it is well known that a metal seldom occurs pure or massive, so if one finds native gold, silver, copper, tin and lead, there occurs there much more ore that carries much impurity, wildness, and the wrong things with it. Our silver occurs generally in cobalt, pyrites, bismuth, galena, *glockenspeyß*,[61] [and] black, yellow, and other [colored] mineral types that are full of sulfur and visible and imperceptible quicksilver [i.e., *sulfur* and *mercurius*].[62]

Minerals and metals that seemed to represent relatively impure compositions also seemed to be the most poisonous. Though compositional refinement could be accomplished in the smelting furnace, Mathesius warned that it was better to leave these minerals alone until the natural heat of the Earth consumed the poison, leaving a metal of a more pure and fixed composition.[63]

Antimony and bismuth were among the most toxic metals occurring in the Joachimstal ores mentioned by Mathesius.[64] When these *mercurius* venoms were unleashed, they constituted a threat to all of nature. The antimony ore,[65] for example, is called

---

[58] Paracelsus, "Von der Bergsucht und anderen Bergkrankheiten," in *Opera, Bücher und Schriften* (Strasbourg, 1603), 652–6; see also Paracelsus, "Miners' Sickness" (cit. n. 57), 80–91.

[59] Johann Friedrich Henckel, *Medicinischer Aufstand und Schmelz-Bogen Von der Bergsucht und Hütten-Katze* (Dresden, 1745), 139.

[60] Ibid., 162–3.

[61] A recalcitrant, apparently cobaltish mineral that might hold a little silver. See Göpfert, "Bergmannsprache" (cit. n. 44), 38.

[62] Mathesius, *Sarepta* (1564), CLII^a; (1571), CVIII^b (both cit. n. 10).

[63] Mathesius, *Sarepta* (1564), XLIII^b; (1571), XXXI^a (both cit. n. 10).

[64] Mathesius, *Sarepta* (1564), CLIII^b; (1571), CIX^b (both cit. n. 10).

[65] *Spießglaß*, stibnite: Sb_2S_3.

a "poisonous pyrite." Antimony was, in fact, widely recognized as a poisonous sub-stance in medical practice, where small doses were administered as a strong purga-tive. For Mathesius and the miners and smelters of Joachimstal, however, the danger lay in processing the ores of this metal, "the smoke and stench from which corrupts foliage, grass, hops, and grain, and the water that comes off from purifying it is very poisonous, as many people and livestock died from it."[66]

Bismuth was a metallic oddity that occurred in relative abundance in the Joachim-stal ores. This metal, which experience had shown to be spatially associated with sil-ver [sulfide] ores, was cited in the *Sarepta* as providing evidence for the continuous generation and compositional refinement of the metals, as it seemed to be a compo-sitionally immature form of silver. Mathesius wrote that as long as the bismuth ore remains underground, in the natural heat of the Earth, the *sulfur* and *mercurius* of its composition would continue to be refined toward those of silver:[67] "The miners say they threw bismuth that did not hold any silver into a tunnel, though silver was found in it some years later, as . . . bismuth transforms into silver, most of all in unmined fields where it has its nourishment from sulfur and quicksilver or fat vapors."[68] As Dym has pointed out,[69] the uncovering of bismuth ore caused miners to conclude that they had begun digging the ore before it had completed its compositional processing. They also knew that upon striking bismuth ore, they were likely to soon reach silver ore as well.[70] The especially noxious smoke from roasting ores containing this metal, along with its low melting temperature, seemed to indicate that it was particularly rich in *mercurius*:

> Because this poisonous metal has much quicksilver [i.e., *mercurius*] in it, it is very fus-ible and flowing in the fire. As it occurs amongst the ore, it all goes away in the fire, and makes a very poisonous smelting smoke. . . . I have also seen a pure bismuth that was already flowing from the natural heat and operation in the mines.[71]

These citations from Mathesius's *Sarepta* show how very real the *mercurius* prin-ciple could seem in early modern mining and smelting. The noxious *mercurius* of the metals, of which there was daily experience, endangered the lives of individuals and entire communities.

CONCLUSION

Though the opinions of early modern miners about the metals and minerals with which they worked were potentially quite variable, both Georgius Agricola and Mathesius reveal the utilization of the *sulfur-mercurius* theory within mining culture. Whether their testimony may have pertained mostly to educated persons involved in mining, or also to the largely illiterate, common mine workers, might never be known. We can only take Agricola's and Mathesius's claims as valid in stating that this theory was to some unknown degree believed among miners. Both authors learned about

---

[66] Mathesius, *Sarepta* (1564), CXXXIX[b]; (1571), XCIX[b] (both cit. n. 10).

[67] Mathesius, *Sarepta* (1564), LXXXVII[a]; (1571), LXII[b] (both cit. n. 10).

[68] Mathesius, *Sarepta* (1564), CXLI[a]; (1571), CI[a] (both cit. n. 10).

[69] Dym, "Mineral Fumes" (cit. n. 7), 180–1; Mathesius, *Sarepta* (1564) LXXXVII[b], (1571) LXII[b] (both cit. n. 10).

[70] Mathesius, *Sarepta* (1564), CXLI[a]; (1571), C[b] (both cit. n. 10).

[71] Mathesius, *Sarepta* (1564,) CXLI[a]; (1571), C[b] (both cit. n. 10).

minerals and metals from miners in Joachimstal, though they themselves differed on the validity of interpreting observations in terms of *sulfur-mercurius*. It is of interest to note that Agricola also briefly scoffed at the opinion that all metals and ore veins were contemporaneous with the Creation but did not specifically ascribe this to miners. Mathesius also mentioned this belief but countered it with numerous instances of continual ore generation that (according to him) the miners reported and believed.

Despite Agricola's scholarly refutation, Mathesius's text demonstrates that the *sulfur-mercurius* theory had considerable explanatory power when applied toward understanding the metallic ores and other related phenomena observed within the mines. To whatever extent the theory was believed among miners, the activity of *sulfur* and *mercurius* was evident in a number of apparently self-consistent ways. True sulfur is found only in minor amounts as a secondary product in the Joachimstal mines, but sulfurous matter was clearly present in the general odor within the tunnels and from the release of sulfur when roasting the ores. Sulfur's traditional association with the Earth's internal heat was consistent with the perceived vaporous presence of *sulfur* in the mines; and this relation helped make sense of mineral reactions and textures that seemed to indicate the fermentation processes expected of metallogenesis. As we saw in the "Bergbüchlein," the *mercurius* principle may have seemed less evident than that of *sulfur*, though the viscous gur liquids were said to answer for this component, an opinion shared in the *Sarepta* and by other authors. Indeed, the attempt in the "Bergbüchlein" to equate the *sulfur-mercurius* theory with a gur theory may be a sign of the gradual introduction of the former idea into mining culture.

However, the *mercurius* principle seemed all too real in the roasting and smelting of (sulfide) ores, where it was manifested in the toxic fumes emitted from these processes. This association is present in Paracelsus's works, and it is significant that he also claimed generally to have learned much about nature from common people, as opposed to scholars or the books of classical authorities. He had been professionally involved at mining and smelting works at least twice, and his father had been a physician in the Alpine mining town of Villach when Paracelsus was a youth.[72] As with his thoughts on the generation and composition of minerals and metals, his ideas about miners' diseases are given in terms of his own physical worldview. Whether the correlation of toxic smelting smoke with the *mercurius* principle was among Paracelsus's conceptual innovations or a relatively common belief that he adopted, it was certainly a common idea for two centuries after his lifetime. Mathesius, who was particularly concerned about this issue, does not present it as a new or controversial idea, and it has been shown that this association persisted into the eighteenth century, even when the terminology of *sulfur-mercurius* had become outmoded.

The *Sarepta* encapsulates a valuable glimpse into the early modern interpretation of Earth processes that were uncovered in the new, extensive, and deep mines of Joachimstal. Within its pages, Mathesius showed that metallogenic *sulfur* and *mercurius* could seem very real to those who were actively involved in the mining and processing of ores and metals. It therefore cannot be said that the idea was mainly the property of laboratory chymists, nor was it mere mining lore, but an explanation of generation and composition that seemed to be confirmed in the tunnels and refining sites of an active and freshly hewn mining town.

---

[72] Paracelsus, "Miners' Sickness" (cit. n. 57), 46.

Eloquence in the Marketplace:

Erudition and Pragmatic Humanism in the Restoration of Chymia

by Bruce T. Moran*

ABSTRACT

This chapter focuses upon the relation between textual and social practices that influenced the formation of a communal approach to acquiring chemical knowledge in the early seventeenth century. It also describes the utilitarian purpose of a humanist-inspired program of chemical learning that blended practices of textual/linguistic expertise and artisanal know-how. Humanism, made pragmatic, sought to define the principles for "making things well." In the design of Andreas Libavius (ca. 1555–1616), interpretive intuitions resulting from practiced reading of ancient and medieval texts combined with a knowledge of workshop language to build consensus about chymia's tools, procedures, and materials and to define its *principia artificialia*.

INTRODUCTION

A large and expanding body of scholarship has established disciplinary crosscurrents in the early modern world that merged text-based studies, empirical inquiries, and technical traditions within humanist cultures.[1] At the same time, the social effervescence resulting from a mixture of elite consumption, collecting, and commerce has muddled categories long considered distinct.[2] Studies of secrets, vernacular cognition, and curiosity in particular have offered insights into the processes of producing and sharing knowledge, blurring the realms of public and private while reshaping the domains of artisans, merchants, and scholars and blending the rhetoric of wonder with the language of utility.[3] The same blending of practices in the production of

* Department of History, University of Nevada, Reno, NV 89557, USA; moran@unr.edu.

My thanks to Dr. J. Mark Sugars for assistance with difficult passages in Latin and Greek.

[1] Fundamental to other discussions is Paula Findlen, *Possessing Nature: Museums, Collecting, and Scientific Culture in Early Modern Italy* (Berkeley, Calif., 1994).

[2] Lorraine Daston, "Curiosity in Early Modern Science," *Word & Image* 11 (1995): 391–404; Deborah E. Harkness, *The Jewel House: Elizabethan London and the Scientific Revolution* (New Haven, Conn., 2007); Harold Cook, *Matters of Exchange: Commerce, Medicine, and Science in the Dutch Golden Age* (New Haven, Conn., 2007).

[3] E.g., William Eamon, *Science and the Secrets of Nature: Books of Secrets in Medieval and Early Modern Culture* (Princeton, N.J., 1994); Alexander Marr, "*Gentille Curiosité*: Wonder-Working in the Culture of Automata in the Late Renaissance," in *Curiosity and Wonder from the Renaissance to the Enlightenment*, ed. R. J. W. Evans and Alexander Marr (Aldershot, 2006), 149–70; Elaine Leong and Alisha Rankin, eds., *Secrets and Knowledge in Medicine and Science, 1500–1800* (Farnham, 2011).

knowledge in the mechanical and liberal arts fashioned the contours of early modern chymia and in one important instance combined humanist erudition with artisanal know-how in a program aimed at shaping communal knowledge with a utilitarian purpose. The plan was that of the German alchemist, physician, and pedagogue Andreas Libavius (ca. 1555–1616), whose model of humanist eloquence embraced linguistic proficiency in regard to both ancient and medieval texts as well as vernacular expertise in the language and operations of the workshop.

Libavius's significance in early modern science has been largely discussed in terms of a clash of world views and their respective languages. Owen Hannaway aligned the Libavian advocacy of openness and didactic efficacy in chemical language with the ideals of civic humanism, a campaign aimed in particular, he argued, at opposing the secrecy and magical temperament of Paracelsian medical cosmology.[4] The condemnation of Paracelsian claims to natural knowledge is a noticeable thread throughout many Libavian texts. However, just as important are those writings in which the Paracelsian magical universe is not the issue. There, clarity about what others had written concerning the structure and utility of nature mattered most. In fact, one of the main criticisms of Paracelsian practice was that it ignored the collective experience of ancient and medieval practitioners set down in texts by, among others, Hermes, Avicenna, Rhazes, Roger Bacon, and Rupescissa. Libavius felt well equipped to pronounce authoritatively about the sense and meaning of these texts. "Whoever will extract the kernel from this nut," he observed, "will explain and teach the art [of chymia] methodically [*diserte*]."[5] Impenetrable enigmas aside, what seemed obscure to others appeared clear to him by means of an implicit comprehension of meaning. Textual consensus required clear explications of procedure and transparency in regard to terminology. Through language that defined the theoretical limits and practical procedures of the *ars chymia*, artisanal experience could then be reflected upon, defined, expressed in mutually agreed upon terms, and communicated. What we might call artisanal *Fingerspitzengefühl* (a talent for doing or making) and humanist *Bedeutungengefühl* (insight into meaning and interpretation) combined to define chymia's *principia artificialia* and brought the techniques of chemical praxis within the custody of an interacting chemical community. Scholia, paraphrase, and explication were the tools with which he worked. These were humanist techniques, and they counted among the practices of early modern chemistry. With such tools, the theoretical discussions of well-known Paracelsians like Oswald Croll, Petrus Severinus, and Johannes Hartmann were dissolved and recast into comprehensible parcels sent out thereafter through print for others to judge. Similarly, the texts of authors whose focus centered more precisely upon laboratory procedure, analysis, and transmutation passed also through the mangle of interpretation and clarification. Some authors have names too obscure to recognize easily; others are better known—George Ripley, Michael Sendivogius, and Cornelius Drebbel among them. Humanist practices of reading and interpreting helped especially unwrap the texts to two medieval authors that Libavius valued most highly, Arnold of Villanova and Raymond Lull. In this chapter, paying attention to the circumstances in which

[4] Owen Hannaway, *The Chemists and the Word: The Didactic Origins of Chemistry* (Baltimore, 1975).

[5] *Rerum chymicarum epistolica forma ad philosophos et medicos quosdam in Germania excellentes descriptarum liber primus . . .* (Frankfurt am Main, 1595), 144.

Libavius brought his textual and procedural tools to bear upon writings related to pseudo-Villanova and pseudo-Lull will suggest ways in which pragmatic humanism functioned within the public sphere, including the marketplace, to help fashion the principles of artifice and to shape the social conditions necessary for the communal practice of chemistry.

ALCHEMY AND WAYS OF DOING THINGS

In 1599, a senator of the German city of Rothenburg ob der Tauber named Adam Gevsius received a book that a family friend, Andreas Libavius, had, in part, dedicated to him. Libavius had known Adam's uncle, "a good man," he wrote, "born to come to the aid of his country," and who had, for that reason, studied the alchemical art for the benefit of humanity. The alchemical muses that had inspired Adam's uncle now seemed to excite Adam as well, and for that reason Libavius thought it appropriate to offer alchemical encouragement and advice. He did so in a book called *A New Treatise on the Medicine of the Ancients, both Hippocratic and Hermetic* (1599), where, in a separately titled second part, he offered explications of alchemical procedures found in works attributed to Raymond Lull and Arnold of Villanova on the basis, he proclaimed, of reason, experience, and textual insight.[6]

Adam, Libavius observed, had been well educated, was protected from vain desires by pious virtues, and could easily admire the divine powers that existed in nature. He was also well off and thus might be less interested in seeking riches through the transmutation of metals and more inclined to study and practice alchemy for the purpose of aiding health and curing illness. This was a far more noble goal, since, Libavius explained quoting Horace as he often did, "no house or land, no pile of bronze or gold, has ever freed the owner's body of fevers, or his sick mind of cares."[7]

Besides combining Hippocratic and Hermetic medicine through the alchemy of pseudo-Lull and pseudo-Villanova, Libavius pursued another agenda in the part of the book dedicated to Adam. He wanted to make clear what kind of art alchemy was, and what sort of practices it required. "No art," he explained, "is so common that it does not keep outsiders at a distance by means of its own ways of doing things and its own professional habits. Then what must be said about alchemy, in relation to which I know of no other [art] that requires a more exact mind, greater diligence, and careful attention [to detail]?"[8] Alchemy's ways of doing things required skills both academic and artisanal. It was a feast, he declared, prepared not mysteriously in some far off land, but cooked up for all to see by means of questioning and disputation, and with the help of physical labors known to craftsmen.[9] Some discussed the art in terms

---

[6] *Medicinae hermeticae artificibus catholicae ad hominis sanitatem tuendam adversamque valetudinem profligandam; et hydrargyrum, imperfectaque metalla in aurum vel argentum transmutanda, expositio fidelis, sincera et dilucida, Raymundi Lullii, et Arnoldi Villonovani; cum scholiis et interpretationibus apertissimis, Andreae Libavii Med. Rotemburgotuberani* (Frankfurt am Main, 1599), 244. This is the second part, paginated consecutively (241–567), of *Nouus de medicina veterum tam Hippocratica, quam Hermetica tractatus. In cuius priore parte dogmata plaeraque inter utriusque professores recentes controversa, adversus ultimum per Iosephum Michelium Paracelsitarum conatum discutiuntur; in posteriore universale alchymistarum, autoribus Lullio et Arnoldo, quam liquidissime exponitur . . .* (Frankfurt am Main, 1599).

[7] *Medicinae hermeticae artificibus catholicae* (cit. n. 6), 246; Horace, *Epistles*, bk. 1, epistle 2. *Horace: Satires, Epistles, Ars Poetica*, trans. H. R. Fairclough (Cambridge, Mass., 1926), 267.

[8] *Medicinae hermeticae artificibus catholicae* (cit. n. 6), 257.

[9] Ibid., 245.

of transmutation, but its province reached further than that and entailed an assortment of procedures by means of which one made a variety of useful things. In that regard, the procedures of Lull and Arnold, he told Adam, were among the clearest descriptions of alchemical processes, and these served the goals of physicians as well as the aims of metallurgists. In fact, Lull and Arnold had written so clearly that, Libavius warned, anyone who did not understand what they said should give up hope of learning the art altogether.[10]

While apparently a minor, even local, matter, Libavius's remarks to Gevsius, and the controversy within which they were embedded, indicate the extent to which practices of erudition, and the social and ethical standards that accompanied late scholastic and humanist scholarship, possessed practical momentum within a public sphere. The means by which "shared practical understanding" enters the communal domain has, over the last several decades, become the focus of an entire genre of "practice" literature. As part of that discussion, nonlinguistic practices relating to personal know-how and an embodied cognition of materials have gained a more central place in debates about how skills get shared and procedure leads to knowledge.[11] And yet, the construction of chymia, no matter how much tied to the transmission of practical understanding, was never entirely a posthumanist project. Communication of what may have been originally tacit artisanal knowledge was still a matter inevitably tied to language, either visual or verbal. In this regard, the book that Libavius dedicated to Adam Gevsius stands, in its entirety, as a striking example of humanist strategies working within a public network of artisanal procedures and social practices.

SOCIAL PRACTICES AND PRACTICAL ERUDITION

Libavius's explication of the *practica* of Lull and Arnold comes in a book whose first part is concerned with neither author. It is addressed instead to refuting the views of an elderly Italian physician, Joseph Michael, living in the little town of Middelburg. Neither the person nor the place mattered. What Michael had written, and sent out into the world with the consent of the city senate, did. Michael had composed what he called an *Apologia Chymica*[12] aimed at Libavius's debut volume, a discussion of a wide variety of chymical matters written in the form of letters (*Rerum chymicarum epistolica forma . . . liber primus*, 1595). The *Apologia* not only dismissed Libavius's construction of chymia as an art but also scorned his elitist understanding of chemical practice.

To the senate and people of Middelburg, Libavius explained in his reply that he had indeed, several years earlier, published a volume about chymistry that was deliberately written in the form of letters to friends and other learned men in Germany. It was, he said, a first attempt [*rudimentum*] at a chymical primer and had also included a little commentary on the art and methods of alchemy. In responding to what Paracelsian physicians had claimed about the art, he had adopted a literary strategy of simultaneous prosecution and defense, attacking chymia of the Paracelsian sort, while

---

[10] Ibid., 246–7.

[11] See, among others, Stephen Turner, *The Social Theory of Practices: Tradition, Tacit Knowledge, and Presuppositions* (Chicago, 1994); Theodore R. Schatzki, Karin Knorr Cetina, and Eike von Savigny, eds., *The Practice Turn in Contemporary Theory* (New York, 2001). See also Pamela Smith, *The Body of the Artisan* (Chicago, 2004).

[12] *Apoligia chymica, adversus invectivas Andreae Libavi calumnias . . .* (Middelburg, 1597).

defending genuine chymia as an art based in the correct understanding of artisanal precepts and practices. He had drawn assistance in the project from "humane studies and languages" and had thus embraced an epistolary genre, preferring not to write in bare prose but, as the genre required, to use eloquence when opposing one thing and defending another.[13]

Letters were often used to air controversial opinions in the early modern era, and they were usually written for an audience that held similar points of view, shared a similar vocabulary, and even recognized similar enemies as the letter writer.[14] Michael probably understood the strategy and responded that an expertise in Greek and Latin had been used solely to flaunt talents for subtle reasoning and sophistic quibbling.[15] It was a condemnation of a certain kind of erudition, and Libavius made sure to point out that he had made use of his knowledge of classical languages not to show off [*neutiquam tamen istiusmodi rerum ostentandarum caussa*], but as a way to combine eloquence with a knowledge of practical procedure. Michael might have been confused by this, since this was, admittedly, a linguistic practice that differed from other erudite methods of textual interpretation. To make the point, he referred to what Jerome Cardano (1501–76) had written in his *De Rerum Varietate* (1557)[16] about a Parisian scholar named Emarus Ranconetus (Aimar de Ranconnet, 1498–1559) who had interpreted a Greek sibylline verse that disclosed the identity of the philosophers' stone by means of a grammatical, numerical, and alphabetical riddle. Ranconetus claimed to have solved the riddle and concluded that the stone was, in fact, arsenic.[17] As clever as it seemed, this was erudition uninformed by practical experience. While arsenic, in certain instances, was useful as a tinging agent, Libavius observed, no one should dream of sublimations and fixations of vulgar arsenic as a means of producing a universal elixir.[18]

Chymical language had indeed become infiltrated with obscure novelties and occult jargon,[19] but cutting through the forest of enigma and ambiguity required, in addition to talents related to ancient languages, an understanding of the language of artisanal praxis. "If you do not understand the discourse of artificers," Libavius instructed in one of his letters, "you have not yet completed your apprenticeship." Understanding the habits of the workshop helped one get clear about what some ancient and medieval authors were saying. "For custom has made law such that one skilled in chymical works, and their organization, may easily understand to what end an author is proceeding."[20] In his letters and in his explications of the procedures

[13] *Nouus [tractatus] de medicina* (cit. n. 6), preface.

[14] Chaim Perelman and L. Olbrechts-Tyteca, *The New Rhetoric* (Notre Dame, Ind., 1971), 19–35.

[15] *Nouus [tractatus] de medicina* (cit. n. 6), preface.

[16] For various editions of this text, see David F. Larder, "The Editions of Cardanus' *De Rerum Varietate*," *Isis* 59 (1968): 74–7. On Ranconnet, see F. Secret, "Jérôme Cardan en France," *Studi francesi* 30 (1966): 480–2; Germana Ernst, "'Veritatis amor dulcissimus': Aspects of Cardano's Astrology," in *Secrets of Nature: Astrology and Alchemy in Early Modern Europe*, ed. William Newman and Anthony Grafton (Cambridge, Mass., 2001), 39–68, on 41. Also, Anthony Grafton, *Cardano's Cosmos: The Worlds and Works of a Renaissance Astrologer* (Cambridge, Mass., 1999).

[17] *Hieronymi Cardani Mediolanensis medici, de rerum varietate libri XVII* . . . (Avignon, 1558), 530–1.

[18] *Medicinae hermeticae artificibus catholicae* (cit. n. 6), 398.

[19] *Rerum chymicarum . . . liber primus* (cit. n. 5), epistle XVIII: "De obscura chymicorum locutione," 162–70.

[20] "Consuetudo enim legem fecit, adeo ut peritus operum chymicorum et ordinis facile intelligat quorsum tendat autor"; ibid., epistle III: "De difficultatibus quibusdam in usu etiam verae chymiae," 48–56, 48, 49.

of Lull, Arnold, and others, Libavius sought to create a common agreement about what chymia's customs, its ways of doing things, were. Erudition, in this sense, meant investigating and explicating texts by means of identifying those things that resembled familiar laboratory practice. From this, general precepts might be derived. The approach was unambiguous. The aim of his book of chymical letters was nothing less than to discover the genuine foundations of chymia [*institutiones chymias*]. To do this required turning to those who had learned precepts through practice and by imitation of artisans. The most notable were Geber, Raymond Lull, and Arnold of Villanova.[21] To describe their praxes, however, it was sometimes necessary "to explain a secretive way of speaking" [*Studui sane arcanam locutionem . . . explanare*] and in this manner "to open a path to judgment" [*et sic viam ad sententiam aperire*].[22]

SOCIAL PRACTICES AND PRECEPTS FOR "MAKING THINGS WELL"

On the one hand, then, fashioning the precepts of practice in the service of chymia depended upon expertise in textual explication by means of linguistic skill and a knowledge of craft procedures. On the other, it meant establishing agreed upon definitions of materials and procedures within a community of artisans and philosophers. Among an array of practices required of the *chymicus*, some were textual and some were social, and it is by making textual traditions relevant to social and procedural practices in the public domain that pragmatic humanism helped also to give an outline to the fuzzy discipline of chymia.

The genuine chymist, Libavius announced in his first book of chymical letters, operated in his own workshop [*officina*], one well stocked with supplies of natural substances. When a work was completed, he did not hide it but brought it out for public use. He expounded the theory behind his wonders so that learned custom [*humana consuetudo*] would be able to assess whether the explanation was satisfactory. He required faithful witnesses of his works, sharing his products and submitting to expert judges. He entered the laboratory not in a way that would prove useless to other aspects of life but cultivated the chymical art in such a way that he did not neglect what pertained to the divine, did not reject the good of the republic, and did not desert his home. The true *chymicus* did not turn away those who were wise from his workshop and did not make use of obscure words. He cherished humanity, composing, by means of the manual arts, goods for the convenience and health of all. Chymical artificers knew that solitary speculation did not profit human society if replicable practices were not combined with it and that no art was divine and excellent in which theory was not followed by action. Thus the conscientious chymist did not withdraw from people, except to separate himself from the unskilled, and preferred the crowns of eminent philosophers to the magnificence of palaces.[23] These were social practices that encouraged criticism and refinement by means of keeping procedures open to scrutiny. "I dutifully ask you," Libavius wrote at the beginning of his explication of Lull and Arnold, "that you do not become my enemy before you become my diligent advisor *(monitor sedulus)*. Show me, by the evidence of reason and labor (*rationum*

---

[21] *Rerum chymicarum . . . liber secundus* (Frankfurk am Main, 1595), 1–5: "We [I] searched much in Geber; even more in Lull, Arnold and others. I hunted down the elusive art across almost the entire world," 2.

[22] *Medicinae hermeticae artificibus catholicae* (cit. n. 6), 250–1.

[23] *Rerum chymicarum . . . liber primus* (cit. n. 5), 23–8, 41–2.

et operum), wherever I have gone astray. You will be complying with the [pursuit of] truth."[24]

Given his attention to correct social practice and proper moral behavior in establishing a cooperative network of chymical artificers, the apparently trivial matter related to the attack of Joseph Michael became a matter of crucial significance. Beyond the personal assault, the attack publicly echoed symptoms of a serious professional dis-ease that, if not aggressively treated, had the momentum to harm both the civic domain and the common good. In his response, Libavius linked humanist scholarship with practical experience, using pragmatic erudition to expose what was false and to act upon the cultural and social sources of communal disarray. Especially claims following from erroneous and naïve textual understanding required expressiveness and vigor in shattering false opinion. Humanist expression, at this point, turned caustic. It served, however, as a channel for social action. Once the demolition of pretense was complete, true proofs would follow from correct precepts "made known for the common good and fortified by the evidence of things."[25]

Michael had misunderstood and misrepresented what was clear to skillful textual scholarship. He had defined chymia solely as an art that taught how to make a single, universal medicine,[26] a view linked to misunderstanding Roger Bacon's earlier description of alchemy as the knowledge of making a medicine called elixir. However, Libavius explained, the term "medicine" in this sense did not derive its meaning from the medical arts. Rather, it was intended to be understood in Aristotelian terms as that which possessed a sensible and material substance.[27] Thus the compass of chymia was far greater than any single medicine. It was the experience of the world and the consensus of artisans, he maintained, that chymia taught how to make many things, accommodated to various circumstances, not only medicines but works of craftsmanship (*artificia*) designed to serve a variety of human needs.[28]

Without literary expertise, Michael had become fixed upon the sixteenth-century alchemical text, the *Rosarium Philosophorum*, and especially upon a passage instructing that, "if nature is understood," art required only one thing, one stone, one medicine, one vessel, and one regimen. But this was another misunderstanding following from inexperienced textual practice. Michael had used the *Rosarium*, Libavius scolded, to dream up puzzles and fables (*aenigmata et fabulas sibi somniet*). When interpretation was assisted by commentaries and practice (*commentariis usuque*), it became clear that the *Rosarium* did not reduce chymia to a "single, continuous, uninterrupted universal operation," and it was also clear that the text referred to numerous processes, vessels, and materials as appropriate to a variety of procedures, including the alteration of metals.[29] Knowing how to use instruments was necessary for anyone who wished to preserve the precepts of, and one's competence in, the art. "And I write this," Libavius added, "knowingly" (*haec sciens scribo*).[30]

---

[24] *Medicinae hermeticae artificibus catholicae* (cit. n. 6), 250.

[25] Ibid., 249.

[26] *Apologia chymica* (cit. n. 12), 5.

[27] Libavius, *Syntagmatis selectorum undiquaque et perspicue traditorum alchymiae arcanorum, tomus primus [-secundus]* . . . (Frankfurt am Main, 1613–15), *tomus primus, Liber Primus, De Magisteriis formalibus*, 5–6, and n. a, 7–8.

[28] *Nouus [tractatus] de medicina* (cit. n. 6), 7–8.

[29] Ibid., 128–9.

[30] Ibid., 127.

ART, NATURE, AND THE MARKETPLACE

The art of chymia followed from principles that allowed for the extension of general knowledge to a variety of specific endeavors. Principles, Libavius explained, were "like a mother to other arts, since from her breast come forth things almost innumerable, which then withdraw into unique companies, so that while the bronze worker [for instance] pursues his own study, he has originally received his tenets from *chymia*."[31] Treating art solely as a means of perfecting nature, as many Paracelsians had insisted, was in this regard a crucial error. Nature might seek perfection, but art did not have to.[32] In fact, decoupling nature from art in forms of practice, as Aristotle had insisted,[33] gave great advantage to the art of chymia for making things better and for serving the civic good. Refashioning, rather than correcting, nature was the aim of chemical praxis, although sometimes art assisted nature and both had the same end.[34] At other times, art had one end and nature another. In such cases, Libavius proclaimed, art could change even cadavers into useful things.[35]

The art of chymia, from a humanist perspective, looked a lot like another art, one that almost every schoolboy studied and that also admitted of both a sound and a shady past, namely, the art of rhetoric. Here, what was there by nature and what was there by art were of special concern as each provided a structure, in its positive application, for genuine debate within a broad community. The Roman rhetorician Quintilian noted that while rhetoric had its origin in nature, as did other arts like medicine and architecture, it became an art when reason, purpose, and efficient modes of learning were applied.[36] In practical terms, chymia, like rhetoric, provided the principles for how things got made, whether arguments or objects. The *chymicus*, Libavius declared, supplied from his own provisions that which other arts required. He discharged from his own learning those who worked as metallurgists, apothecaries, gold-beaters, glass and gem makers, among a host of others, since in his possession were the precepts for making things well [*de bene formandis rebus*].[37] Even the chymist who pursued the philosophers' stone might, by removing fixity, introducing volatility, and arousing motion, make something otherwise useful from nature if, Libavius observed, the artisan followed processes that were built up by definite, unchanging experience and that were confirmed by examples open to everybody.[38]

Precepts open to everybody did not, however, matter much to the practice of chymia in the marketplace, and here too there loitered another source of profes-

---

[31] *Syntagmatis . . . alchymiae arcanorum . . . tomus primus, liber primus, De magisteriis formalibus* (cit. n. 27), 4.

[32] *Nouus [tractatus] de medicina* (cit. n. 6), 18–9.

[33] On the broader importance of the art-nature distinction in the history of science, see William Newman, "The Place of Alchemy in the Current Literature on Experiment," in *Experimental Essays—Versuche zum Experiment*, ed. Michael Heidelberger and Friedrich Steinle (Baden-Baden, 1998), 9–33. Also Anthony Grafton, "Renaissance Histories of Art and Nature," in *The Artificial and the Natural: An Evolving Polarity*, ed. Bernadette Bensaude-Vincent and William Newman (Cambridge, Mass., 2007), 185–210.

[34] *Nouus [tractatus] de medicina* (cit. n. 6), 29.

[35] Ibid., 32.

[36] *The Institutio Oratoria of Quintilian*, trans. H. E. Butler (Cambridge, Mass., 1953), bk. 2, chap. 17, 1:325–45.

[37] *Syntagmatis . . . alchymiae arcanorum . . . tomus primus, liber primus* (cit. n. 27), 43.

[38] Ibid., 158–60.

sional disorder. Some thought of chymia solely as a manual craft. Cardano, for instance, had very few words to describe the subject and simply referred to the chymical art [*chymistica ars*] as the knowledge and skill involved in making things. Some of those things, he noted, were admirable, some worthless, some dubious, some beautiful, some healthful, some efficacious, and some almost divine. His, however, was a purely empirical definition. Chymia, in the sense that Cardano had used the term, was an umbrella reference to a variety of artisanal creations and performances. Examples included stretching glass into long strands, interweaving glass with white threads,[39] hardening glass, softening glass without fire, making false gems (from glass), engraving or etching images into glass, making artificial amber, mixing and altering metals, making white gold, or electrum, purifying camphor, producing waters and oils by means of alcohol, bleaching silk and whitening flowers by means of sulfuric vapors, making true purple, making waters for dissolving gold and silver, and composing the finest waters for penetrating spaces [*dimensiones*] judged impossible by nature. By means of their art, Cardano's chymists produced saleable goods, softening horns and bone to make handles and sword hilts and making inks, cosmetics, combs, cases, and other containers.[40] Libavius did not doubt that these things were valuable, often beautiful, and required skill to make, but Cardano had not considered what the nature of the chymical art was, per se, nor had he instructed how the processes of making things were to be communicated.[41] In Cardano's view, the "art" of chymia was defined by way of artifact, dexterity, and popular opinion. These were views of the marketplace where the subject of chymia took its meaning from what people bought and sold, and popular judgment attributed more of the art to little old women, market girls, and unguent sellers than to any philosopher or doctor.[42] Chymia in this context was a kind of local, even domestic, talent. However, making things did not make an art. A real art had nothing to do with luck, or accident [*fortuna*], or even an aptitude for technique. Someone who stitched together a shoe, he argued, could not be considered an *artifex* unless he understood the principles and causes for doing what he did.[43] It was through a knowledge of precepts and axioms that chymia became worthy of a liberal person and a sign of true expertise. This was the mark of expert artisans [*solertes*], many of whom, Libavius observed, had gained the support of flourishing republics like Denmark and England.[44]

[39] Cardano is referring to the production of filigrano or filigree glass. See Patrick McCray, *Glassmaking in Renaissance Venice: The Fragile Craft* (Aldershot, 1999), esp. 124–5. Also, Marco Beretta, *The Alchemy of Glass: Counterfeit, Imitation, and Transmutation in Ancient Glassmaking* (Sagamore Beach, Mass., 2009).

[40] *Hieronymi Cardani medici mediolanensis De subtilitate libri XXI* . . . (Paris, 1551), 269v–271r. For alchemy in commercial and industrial contexts, see Tara Nummedal, "Practical Alchemy and Commercial Exchange in the Holy Roman Empire," in *Merchants and Marvels: Commerce, Science, and Art in Early Modern Europe*, ed. Pamela H. Smith and Paula Findlen (New York, 2002), 201–22. Also, Pamela H. Smith, *The Business of Alchemy: Science and Culture in the Holy Roman Empire* (Princeton, N.J., 1994).

[41] *Syntagmatis* . . . *alchymiae arcanorum* . . . *tomus primus, liber primus* (cit. n. 27), 4.

[42] Ibid., 51.

[43] *Rerum chymicarum* . . . *liber secundus* (cit. n. 21), 524.

[44] *Rerum chymicarum* . . . *liber primus* (cit. n. 5), 51–2. For a recent discussion concerning the idea of expertise, see Eric Ash, "Introduction: Expertise and the Early Modern State," in *Expertise, Practical Knowledge, and the Early Modern State*, ed. Eric Ash, vol. 25 of *Osiris* (2010). Also, Ash, *Power, Knowledge and Expertise in Elizabethan England* (Baltimore, 2004).

AN ART THAT SHAPES ITSELF THROUGH INTELLECT AND PRACTICE

What Libavius called alchymia, and sometimes chymia, was a form of learning first acquired by craftsmen and thereafter advanced by sharp intellect and long practice, "so that it is as if it shaped itself" [*ut sit quasi suiipsius informatur*].[45] As an art that shaped itself through practice, it required clear exposition of its procedures. Thus, the second part of Libavius's treatise *Concerning the medicine of the ancients* becomes more specific in its purpose and less polemical in its style. Its separate title is "Concerning the preparations of the universal, hermetic medicine for preserving human health and for overcoming illness, and also for transmuting mercury and imperfect metals into gold and silver; a faithful, sound, and plain explanation of Raymond Lull and Arnold of Villanova with most clear commentaries and interpretations."[46]

By the time Libavius composed his text, Lull's *Clavicula* and Arnold's *Rosarium Philosophorum* already existed in several variations and printed editions. The *Rosarium* that he refers to, however, appears to be a variant of what came to be called the *Thesaurus Thesaurorum et Rosarium Philosophorum*.[47] Libavius admitted that he did not know who the author was, and that some held that the second part, concerning *practica*, had been written by Lull himself. Somewhat out of character, he added that he did not wish to fight about it. Whatever position he took privately concerning authorship, it is clear that Lull and the *Clavicula* (as well as references to pseudo-Geber) interlaced the discussion of the *Rosarium*. Direct links between the two occur routinely: "The last chapter taught almost the same thing as Lull's chapter nine of the *Clavicula*"; "chapter twenty-six here agrees with the precepts of the tenth chapter [of Lull], which is entitled 'the fixing of sulfur."[48] "The praxis of the fermentation for redness ... corresponds to Lull's chapter two of the *Clavicula*." "These multiplications are simply Lull's fixations in chapter ten."[49] As if to drive home the relationship, the commentary on the *Rosarium* is followed immediately by "A Summary of the art, ascribed to Lull" in which general references from Lull are associated with specific chapters from the *practica* of the *Rosarium*.[50] The associations give further evidence of what William Newman, Michela Pereira, and Michael McVaugh[51] have already pointed out, namely, that the texts of pseudo-Villanova bear close relationships to discussions in pseudo-Lull and pseudo-Geber (whom we now

[45] *Medicinae hermeticae artificibus catholicae* (cit. n. 6), 258–62.

[46] See n. 6.

[47] See Sebastià Giralt i Soler, *Arnau de Vilanova en la impremta renaixentista; Publicacions de l'Arxiu Històric de les Ciències de la Salut (PAHCS)* (Barcelona, 2002); Antoine Calvet, "Mutations de l'alchimie médicale au XVe siècle: A propos des texts authentiques et apocryphes d'Arnaud de Villeneuve," *Micrologus* 3 (1995): 185–209; also Calvet, "Les alchimica d'Arnaud de Villeneuve à travers la tradition imprimée (XVIe–XVIIe)," in *Alchimie: Art, Histoire et Mythes*, ed. Didier Kahn and Sylvain Matton (Paris, 1995), 157–90; Lynn Thorndike, *A History of Magic and Experimental Science* (New York, 1934), 3:55–9; Michela Pereira, *The Alchemical Corpus Attributed to Raymond Lull* (London, 1989). For comparison, I have used Jean Jacques Manget, *Bibliotheca Chemica Curiosa . . .* (Cologne, 1702; rept., Bologna, 1976–7), Arnoldus de Villanova, *Thesaurus thesaurorun et rosarium philosophorum*, 1:662–76; and Lull, *Clavicula*, 1:872–5.

[48] *Medicinae hermeticae artificibus catholicae* (cit. n. 6), 544.

[49] Ibid., 525.

[50] Ibid., 563–7.

[51] William R. Newman, *The "Summa Perfectionis" of Pseudo-Geber* (Leiden, 1991); Michela Pereira, "Arnaldo da Vilanova e l'alchimia: Un indagine preliminare," in *Arxiu de Textos Catalans Antics* 14 (1995): 165–74; Michael McVaugh, "Chemical Medicine in the Medical Writings of Arnau de Vilanova," *Arxiu de Textos Catalans Antics* 23–4 (2004–5): 239–64.

know as Paul of Taranto). But it is also clear that Libavius sought to sort out the specific components of the amalgam and attempted to bring about a consensus regarding recommendations for practical procedures. Such procedures were, of course, always open to adjustment. In his own writings, especially in what he had previously written about mercury, *aqua permanens*, and the philosophers' stone, he admitted to having taken things a little differently than described by Lull and Arnold. This was by no means a reckless thing to do, he declared. "For although the art has one and the same rule, the same progression, and one *materia*, yet one *artifex* is more talented than another in matters of procedure and in finding short-cuts in the work" [*in praeparatione et abbreviatione operis*].[52]

One reason why the *Rosarium* was important to get right was that Libavius viewed it as the link between metallic transmutation and medical alchemy. Among the many uses of alchemy, making medicines stood out as most important in serving the public good. While Lull was insightful about the medical uses of a fifth essence, his method for obtaining it remained obscure. Arnold, on the other hand, "by analogy between metals and our body, transferred the use of this stone to human medicine and obtained astonishing results."[53] Both medicine and metallurgy dealt with fluids and solution. Hydrargyrus, Libavius believed, could be considered almost the humor of humors.[54] Of his own role in bringing procedures of chymia to bear upon the production of medicines, he explained, "I also wanted to serve discovery . . . and to declare the formulas of the skilled through collections, disputations, and natural and chemical principles. Therefore I display the memoranda of various teachers in which, if I did not actually take the tiller [*clauum tetigi*] at least I tried to wander no longer. . . . Anything I could pursue by reading, meditating, collecting, or working with, whether through me or others, I made use of [*insumsi*] . . . not for my own sake, but for the sake of others."[55]

In his explications of Lull and Arnold, Libavius proceeded in the traditional manner by adding *scholia* to passages, sometimes to individual words. Along the way, however, he informed his reader, assumed to be a beginner in the art of chymia, how to navigate the text, giving instruction for what chapters to read first, in what order, and what chapters to refer back to when one needed help. The reader should first get clear about materials, the order of procedure, and what each operation entailed. Studying the theories of the first part of Arnold's text, students would recognize that mercury should be used to extract from perfect metals a pure mercurial substance. Mercury took up things that were familiar to it and rejected those that were not. Arnold, he explained, taught that mercury (argent-vive) did so by penetrating into the innermost parts of bodies, mixing and clinging to their smallest parts [*Miscetur enim et cohaeret per minima*]. The medium for extracting this mercurial substance from bodies was mercurial sulfur, an extract of mercury itself that was to be gotten from perfect bodies, namely, gold and silver.[56]

---

[52] *Medicinae hermeticae artificibus catholicae* (cit. n. 6), 251–2. In the seventh book, "De transmutationibus variis," of the first volume of his *Syntagmatis . . . alchymiae arcanorum . . . tomus primus [-secundus]* (cit. n. 27; 1613–15), Libavius again took up the question of metallic transmutations and brought to bear the same technique of procedural explication among a list of authors including Arnold in regard to the use of argent-vive in the transmutation of metals (esp. 273–4).

[53] *Syntagmatis . . . alchymiae arcanorum . . . tomus primus, Liber septimus* (cit. n. 27), "De transmutationibus variis," 273.

[54] *Nouus [tractatus] de medicina* (cit. n. 6), 136.

[55] *Syntagmatis . . . alchymiae arcanorum . . . tomus secundus* (cit. n. 27), 5–6.

[56] *Medicinae hermeticae artificibus catholicae* (cit. n. 6), 459–63.

Once students understood the mercury theory of metals, they could move on to the study of the *practicum opus* in the *Rosarium's* second part. There, *practica* was divided into specific procedures of solution, ablution, reduction, and fixation, and Libavius provided ample commentary regarding each, sometimes pointing out places of confusion and enigmatic impenetrability. Pseudo-Villanova's instructions concerning recombining elements by means of weight seemed especially obscure. To other seemingly impossible directives he gave alternative explanations.[57] Despite the problems, Libavius established a clear-cut route to reading the text—a roadmap of communal clarity. That was important if there was to be public discussion and debate of what heretofore had been the province of inspired adepts. The initial focus should be on chapters 2 and 3, which treated the material of the stone as well as its preparation and solution. Then, apprentices should pass on to chapter 15, which taught the practice of coagulation. From there one moved on to chapter 16 in order to learn how to use mercury in the "white" and "red" work (procedures involving transmutations of metals into silver and gold, respectively).[58] For more specific understanding of the "white" work, Libavius recommended chapter 25. Chapter 26 would help most in understanding what to do in regard to fixation, and chapter 27 would help to clear up questions about the "red" work, "the fixing and increasing of which is," he explained, "in chapter 28." Whenever one made a mistake, "unless it is irreparable," chapter 23 would help. The other chapters were "partly filled with allegorical information" and partly served "to turn away the unworthy and stupid."[59] Readers should know that the *Rosarium* contradicted itself concerning sublimation in chapters 12, 25, and 27.[60]

EXPLICATION, TRANSMUTATION, AND THE MINIMA

In his books of chymical letters, as well as in his explications of the procedures of pseudo-Villanova and pseudo-Lull, Libavius sought to find public consensus, through textual clarity and replicable procedures, about the precepts of praxis in chymia. At its core, this was a humanist project that required linguistic skills in combination with procedural and material know-how. There can be little doubt that Libavius participated in workshop *practica* (although the extent of that involvement remains to be more accurately determined). Commenting at one point concerning a certain ferment in Arnold's *Rosarium*, he noted, "I myself have employed this method successfully."[61] Referring to Lull's procedures, he proclaimed that "although I have not finished the work, I attest that I have seen with my own eyes many proofs of its truth. If God will grant me [continued] life, I will not hesitate to proceed further or, at least, to teach [the work] to my sons." Acknowledging frustrations left unexplained, he added, "there are some good reasons why I am not moving forward."[62]

Some projects also involved transmutation, generally conceived. In what was called the *Greater Rosary* of pseudo-Arnold, alchymia signified transmutation, and this, Libavius explained, was really about the passing away of a compound and its re-

---

[57] Ibid., 526–9.
[58] For further discussion of alchemical processes, see Lawrence M. Principe, *The Secrets of Alchemy* (Chicago, 2013); for pseudo-Arnold and pseudo-Lull, 67–73.
[59] *Medicinae hermeticae artificibus catholicae* (cit. n. 6), 560–1.
[60] Ibid., 516.
[61] Ibid., 542.
[62] Ibid., 426.

duction into component parts.[63] Reference to dissolution and constituent parts draws attention, of course, to one of the most interesting facets of medieval alchemy, a tradition of tiny corpuscles that could penetrate minute spaces within bodies. The idea can be traced to Aristotle, although Aristotle rejected the notion of atoms, and reoccurs especially in a text called the *Summa perfectionis* composed around the end of the thirteenth century by a Latin author also known as Geber. As William Newman explains, the *Summa* described a combination of *minimae partes* that formed strong compositions and made up the substances of sulfur and mercury. The corpuscularian tradition found in the *Summa* extended to the texts of pseudo-Villanova, especially the *Rosarium philosophorum*, and also to those of pseudo-Lull. Libavius was certainly aware not only of these discussions but also of the arguments concerning *minima naturalia* linked to Aristotle's *Physics* and repeated by a host of medieval authors. From his own discussions as well there is good evidence that, as Newman suggests, Libavius organized corpuscularian views within an Aristotelian natural philosophy while distancing himself from ancient atomist notions of void space and random combinations of particulate matter.[64]

Blending traditions by means of erudite interpretations opened possibilities of theoretical convergence, even between seemingly opposing views of nature, and is another way in which pragmatic humanism helped fashion debate within an emerging chemical community. Experience showed that mercury was the matter of metals and that mercury was the substance in which, Libavius observed, "nature has hidden her seed."[65] The reference to nature's hidden seeds alerts us to a mixture of theoretical influences, including notions of generative force, at work in Libavius's conception of matter and change. As he explained, God had constituted compounds from the elements and had sown in them seminary principles in conjunction with which each thing produced its own seed and engendered its like.[66] The shadowy fifteenth-century figure, Bernardus Trevisanus, had also referred to a generative force in nature constituted as seeds, and in the *Clavicula*, Lull had described an elixir, called by some the sulfur and coagulated philosophical argent-vive, "which the craftsman uses like a seed . . . as if he had gotten from nature the plastic force [*vim plasticum*] of the seed that acts to perfect metals."[67] Such views were opposed neither to those of Aristotle nor those of Galen. However, another view, that of the Danish Paracelsian Petrus Severinus (1540–1602), described the elements and seeds as originally united and in no way separated by different functions. What Severinus had referred to as seminary principles might, Libavius proposed, also be explained in Aristotelian and Galenic terms if conceived of as a humor (*khymos*) or an innate substantial heat, or as an essence, a term more familiar to chymists. Thus, he added, if any of the Paracelsians ought to be approved, his feelings would be along the lines of Severinus, provided, of

---

[63] *Nouus [tractatus] de medicina* (cit. n. 6), 112–3.

[64] Along with other articles, William Newman, "Experimental Corpuscular Theory in Aristotelian Alchemy: From Geber to Sennert," in *Late Medieval and Early Modern Corpuscular Matter Theories,* ed. Christoph Lüthy, John E. Murdoch, and William R. Newman (Leiden, 2001), 291–329. Also, Newman, *Atoms and Alchemy: Chymistry and the Experimental Origins of the Scientific Revolution* (Chicago, 2006).

[65] *Nouus [tractatus] de medicina* (cit. n. 6), 61, 57.

[66] In discussing mixing and generation, Libavius held always to guiding principles and excoriated views that he found "not very far away from the atoms of Democritus and the ravings of people like him who constitute everything rather by chance than by a definite presiding principle" (ibid., 97).

[67] *Medicinae hermeticae artificibus catholicae* (cit. n. 6), 358.

course, that he could supply this sort of interpretation [*Si itaque Paracelsitarum quis probandus est, senserim potius cum Severino addita scilicet interpretatione quam dixi . . .*].[68] There were limits, however, to making erudite philosophical compounds by means of the speculative dissolving agents of interpretation and inference, and Libavius made clear to his apprentice readers that his understanding in no way admitted of Severinus's notion that the elements themselves amounted only to incorporeal places or matrices within which seeds obtained vital powers. The Platonic metaphysics of Severinus's medical philosophy required material things to emerge from matterless *semina*, incorporeal seeds that contained the divine plan for the construction of natural bodies. To Libavius the notion was a bottomless metaphysical pit in which being emerged constantly from nonbeing and "in accordance with which nothing is in existence *per se* but exists only as a relation of a place to what is placed in it" [*relatio loci ad locatum*].[69] Chymia required more solid, Aristotelian footing for its principles, both in theory and in practice. When compounds were reduced to parts, they lost their Aristotelian forms, not their intermediate link to the divine.

Making knowledge in early modern chemistry involved more than observation, reasoning, and technical know-how. In understanding nature, there was also the matter of how knowledge was to be organized and expressed. In this regard, texts and practices enlightened one another. Thus chymia required a knowledge of both words and things. However, words, and especially names, Libavius insisted, needed always to follow from things themselves. Where there were no words for procedural knowledge, or where procedural descriptions seemed unclear, the chymical art relied upon "autopsy and the application of the hands" since material practice brought to light those things obscurely hidden away in the language of precepts.[70]

The epistemic foundation for the discipline of chymia required skills related to both language and manual procedure. Those practices helped to form communities whose members agreed about the meaning of terms and, through focused debate and replications of method, confronted differences while sharing the results of investigations, analyses, and discoveries.[71] As an art and discipline supported by a communal episteme, chymia, in the Libavian scheme, relied also upon practices that were social, embracing, along with expectations of openness and descriptive clarity, agreed-upon norms of ethical conduct. What has been referred to here as pragmatic humanism expanded notions of erudition and eloquence, mixing together scholarly practices, the language of the artificer, and familiarity with the tools, vessels, and materials of the workshop. The combination defined the principles for "making things well" and spanned communities of natural philosophers and artisans in efforts to effect various forms of well-being within the public sphere. While proficiency in both theory and practice defined post-Baconian natural philosophy, pragmatic humanism insisted early on that chymical artisans learn the philosophical principles of their labors and that philosophers learn the technical language of those who read the book of nature with their hands.

[68] *Nouus [tractatus] de medicina* (cit. n. 6), 107.

[69] Ibid., p. 121. On Severinus's philosophy of semina, see Jole Shackelford, *A Philosophical Path for Paracelsian Medicine: The Ideas, Intellectual Context, and Influence of Petrus Severinus: 1540–1602* (Copenhagen, 2004).

[70] *Rerum chymicarum . . . liber primus* (cit. n. 5), 50.

[71] See discussions in Pamela H. Smith and Benjamin Schmidt, eds., *Making Knowledge in Early Modern Europe: Practices, Objects, and Texts, 1400–1800* (Chicago, 2007).

Robert Boyle, Transmutation, and the History of Chemistry before Lavoisier:

A Response to Kuhn

by William R. Newman*

ABSTRACT

In an influential article of 1952, Thomas Kuhn argued that Robert Boyle had little or no influence on the subsequent development of chemistry. This essay challenges Kuhn's view on two fronts. First, it shows that Johann Joachim Becher developed his hierarchical matter theory under the influence of Boyle and then transmitted it to the founder of the phlogiston theory, G. E. Stahl. Second, this essay argues that transmutational matter theories were not necessarily opposed to the existence of stable chemical species, *pace* Kuhn. Boyle's corpuscular theory descended largely from the tradition of "chymical atomism," which often advocated both chrysopoeia and the reality of robust chemical substances.

INTRODUCTION: KUHN'S CLAIM THAT BOYLE HAD
A NEGLIGIBLE INFLUENCE ON CHEMISTRY

In the last twenty-five years, we have come a long way from the contemptuous view of alchemy as nothing but the province of charlatans and cheats. This viewpoint, broadcast to the public by the French academician Bernard Le Bovier de Fontenelle and others in the early Enlightenment, was adopted as a fundamental tenet in modern surveys of the Scientific Revolution, where it remained the dominant perspective until the final decades of the twentieth century.[1] Today alchemy is no longer considered an embarrassing outlier, and yet certain vestiges of the older historiography still live on. One example of this lies in the web of interpretation enveloping a central figure of the Scientific Revolution, namely, Robert Boyle. I do not refer here to the hoary old chestnut, still recapitulated by J. R. Partington, that Boyle represented a decisive break with alchemy or that he was "the founder of modern chemistry."[2] This overblown claim cannot be sustained in the light of the sophisticated chymical knowledge already available in the work of writers such as Daniel Sennert, Joan

* Department of History and Philosophy of Science, Goodbody Hall 130, Indiana University, 1011 East Third Street, Bloomington, IN 47405; wnewman@indiana.edu.

[1] William R. Newman, *Atoms and Alchemy: Chymistry and the Experimental Origins of the Scientific Revolution* (Chicago, 2006), 6–12. See also Lawrence M. Principe and William R. Newman, "Some Problems with the Historiography of Alchemy," in *Secrets of Nature: Astrology and Alchemy in Early Modern Europe*, ed. William R. Newman and Anthony Grafton (Cambridge, Mass., 2001), 385–431.

[2] J. R. Partington, *A History of Chemistry* (London, 1960), 2:496.

Baptista Van Helmont, Johann Rudolph Glauber, and George Starkey, whose expertise Boyle often tacitly appropriated.[3] Nor do I refer to the outdated idea that Boyle
was the inventor of the "modern concept of the element," a position that was already
discredited by Marie Boas Hall in the 1950s.[4] In reality, Boyle's definition of the element as that out of which bodies are made and into which they can be decomposed is
already found in Aristotle's *On the Heavens*, where it formed the basis of scholastic
commentaries that sometimes included a discussion of alchemy.[5] It has long been
known that Boyle himself was a firm believer in chrysopoetic transmutation, just as
his major sources in the chymical tradition had been, and that this belief in the alchemical ability to transform matter fit rather well with his mechanical philosophy.[6]
Yet it is Boyle's very belief in transmutation that has led to the historical error to
which I now refer: what I have in mind is the claim that Boyle exercised little if any
influence on the development of chemistry from the late seventeenth century up to
the period directly before Lavoisier.[7]

The challenges to Boyle's status as an influential innovator have a long history,
but certainly a major part of the credit must go to Thomas Kuhn's brilliant if unsat-
isfying article of 1952, "Robert Boyle and Structural Chemistry in the Seventeenth
Century."[8] Kuhn was already well aware of Boyle's transmutational aspirations when
he claimed there that Boyle's mechanical philosophy was not a significant contribu-
tor to the history of chemistry: indeed, Kuhn saw Boyle's belief in transmutation as
the main cause of his lack of influence.[9] Ironically, at the very moment that the history
of alchemy has finally shaken off its long oblivion and come to life, Kuhn's portrayal
of Boyle has also attained a renewed significance, having been adopted directly or
indirectly and expanded upon in the last twenty years by a number of prominent historians and philosophers, including among others Ursula Klein and Alan Chalmers.[10]

---

[3] William R. Newman, "The Alchemical Sources of Robert Boyle's Corpuscular Philosophy," *Ann.
Sci.* 53 (1996): 567–85. See also Newman and Lawrence M. Principe, *Alchemy Tried in the Fire:
Starkey, Boyle, and the Fate of Helmontian Chymistry* (Chicago, 2002).

[4] Marie Boas [Hall], *Robert Boyle and Seventeenth-Century Chemistry* (Cambridge, 1958), 95–6.

[5] Aristotle, *De caelo* 3 302a15–18 (in Guthrie's translation), "Let us then define the element in
bodies as that into which other bodies may be analysed . . . and which cannot itself be analysed into
constituents differing in kind." See Aristotle, *On the Heavens*, trans. W. K. C. Guthrie (Cambridge,
Mass., 1971), 283.

[6] Boyle's chrysopoetic interests were known to writers in the 1950s through the biography of Louis
Trenchard More, who had devoted a chapter to Boyle's alchemy in his *The Life and Works of the Hon-
ourable Robert Boyle* (London, 1944). Our knowledge of Boyle's involvement in transmutation has of
course been greatly amplified more recently by Lawrence M. Principe's *The Aspiring Adept: Robert
Boyle and his Alchemical Quest* (Princeton, N.J., 1998).

[7] I thank Seymour Mauskopf for pointing out that Herbert Butterfield went even further in denigrat-
ing Boyle's contribution to chemistry. Butterfield, who famously named one of the chapters in his
Origins of Modern Science "The Postponed Scientific Revolution in Chemistry," explicitly suggested
Boyle as one of the causes of the delay. See Butterfield, *The Origins of Modern Science, 1300–1800*
(New York, 1961), 136–7.

[8] Thomas Kuhn, "Robert Boyle and Structural Chemistry in the Seventeenth Century," *Isis* 43
(1952): 12–36.

[9] Kuhn explicitly refers to L. T. More's work on Boyle (More, *Life and Works* [cit. n. 6]) and alchemical transmutation as a support for his own argument about Boyle's uninspiring reception. See
Kuhn, "Robert Boyle and Structural Chemistry" (cit. n. 8), 21, n. 40: "More's evaluation of Boyle's
views on transmutation most nearly coincides with the opinion developed below."

[10] Ursula Klein has built on Kuhn's argument in her "Robert Boyle, Der Begründer der neuzeitlichen
Chemie?" *Philosophia naturalis* 31 (1994): 63–106. See her references to Kuhn on 63–4; in essence,
she extends Kuhn's claims about Boyle's lack of influence in the realm of material composition to
include seventeenth- and eighteenth-century ideas about chemical bonding as well. The Kuhnian
influences are particularly marked on 85–93. See also Klein's monograph *Verbindung und Affinität:*

A sort of cognitive dissonance results from this situation: on the one hand, belief in transmutation has been rescued from its general opprobrium as "the greatest obstacle to the development of rational chemistry," yet in the specific case of Boyle it is blamed for his putative absence of influence on the later development of chemistry.[11]

It is therefore worth our while to reexamine the role of transmutation and matter theory in Boyle's thought and to determine whether his belief in the ability of substances to undergo the fundamental metamorphosis required by chrysopoeia produced a decisive, negative impact among his colleagues during his own lifetime and in the following century.[12] In the following, I will first present Kuhn's position and then provide empirical evidence to show that important elements of Boyle's matter theory, despite condoning transmutation, were adopted by some of the most influential chymists in Europe. This will lead me to a discussion of Johann Joachim Becher and his debt to Boyle, for it was Becher who transmitted important elements of Boyle's matter theory to the father of the phlogiston theory himself, Georg Ernst Stahl. At the conclusion of this essay, I will also provide some brief thoughts concerning the part that transmutation played more generally in the framing of "chymical atomism."[13]

Kuhn's principal argument derives from Boyle's concept of *prima naturalia* or *minima naturalia* composed of "uniform catholic matter." In Boyle's philosophy these materially homogeneous and practically indivisible particles underlie the more complex corpuscles that make up chymical species as we know them. Obviously Boyle's belief in an underlying material simplicity opens up the possibility of transmuting more complex corpuscles, such as those making up silver, into other complex corpuscles, such as gold, by merely breaking the silver corpuscles down into their simpler components and rebuilding them as gold corpuscles. Basing his arguments primarily on *The Sceptical Chymist*, Kuhn maintained that Boyle's emphasis on transmutability undercut any possibility of having permanent elements underlying the phenomena and stressed that this de-emphasizing of elementarity led Boyle away from the most fruitful trends of eighteenth-century chemistry. As Kuhn himself put it, with his usual clarity, Boyle's mechanism

---

Die Grundlegung der neuzeitlichen Chemie an der Wende vom 17. zum 18. Jahrhundert (Basel, 1994), which develops the same arguments at greater length but with less focus on Boyle. For Alan Chalmers's rather strident claim that Boyle did not "set chemistry on its modern course," see his *The Scientist's Atom and the Philosopher's Stone* (Dordrecht, 2009), 139. Although Chalmers does not refer to Kuhn's 1952 article, his entire eighth chapter, "The Emergence of Modern Chemistry with No Debt to Atomism," is explicitly based on the approach of Klein, which finds its inspiration in Kuhn; see also 97–122, which he devotes to the inadequacies of Boyle's mechanical philosophy. For an extended criticism of Chalmers's views on Boyle, see William R. Newman, "How Not to Integrate the History and Philosophy of Science: A Response to Chalmers," *Stud. Hist. Phil. Sci.* 41 (2010): 203–13.

[11] The quotation comes from A. Rupert Hall, *The Scientific Revolution, 1500–1800: The Formation of a Modern Scientific Attitude* (Boston, 1962), 310.

[12] One should note that Kuhn's claim was already disputed by Antonio Clericuzio in an influential article. See Clericuzio, "A Redefinition of Boyle's Chemistry and Corpuscular Philosophy," *Ann. Sci.* 47 (1990): 561–89; see 563, 587–8. In a subsequent book, Clericuzio points out the similarity of Boyle's matter theory to Becher's but leaves the possibility open that Becher derived his ideas from Daniel Sennert rather than from Boyle. As this essay argues, terminological considerations make this possibility highly unlikely. For Clericuzio's position, see his *Elements, Principles and Corpuscles* (Dordrecht, 2000), 196.

[13] On the meaning and justification of this term, see William R. Newman, "The Significance of 'Chymical Atomism,'" in "Evidence and Interpretation: Studies on Early Science and Medicine in Honor of John E. Murdoch," ed. Edith Dudley Sylla and William R. Newman, special issue, *Early Sci. Med.* 14 (2009): 248–64.

promoted the opinions that there were no chemical elements, that any substance could be transmuted to any other, and that the object of the chemist was to fabricate novel substances by micro-mechanical operations upon the neutral corpuscles of base matter. This view of chemistry and the chemist was rejected by many of Boyle's contemporaries and successors, because it conflicted with the still prevalent conception of chemistry as an art of separation and combination whose ultimate objectives were the isolation of elements and the determination of combination. A retrospective glance at the history of seventeenth and early eighteenth century chemistry suggests that the true progenitors of Lavoisier's chemical revolution were necessarily among Boyle's opponents.[14]

Kuhn's position here is not entirely without foundation. No one can reasonably deny that Boyle believed in a uniform catholic matter, or that he integrated this belief into his chrysopoetic endeavors. But there are other features of Kuhn's pronouncement that are troubling. First, does a mere "retrospective glance" at eighteenth-century chemistry really do justice to the question of Boyle's influence? Is something a bit deeper not required? And is it really the case that the major figures who preceded Lavoisier were necessarily uninfluenced by Boyle and can even be counted among his opponents? All of this sounds suspiciously like a rational reconstruction of history rather than an empirically documented claim. Finally, and perhaps most importantly, Kuhn pays little attention throughout his article to the fact that Boyle postulated several distinct levels of corpuscular aggregation that were supposed to exist between the ultimate *prima naturalia* and the materials that account for most ordinary chemical reactions. Why would Boyle have bothered to create such a hierarchy if he really wished to deny or de-emphasize the reality of semipermanent combining corpuscles? And for that matter, what was the influence, if any, of Boyle's hierarchical corpuscular theory on seventeenth- and eighteenth-century chemists?

ANOTHER LOOK AT BOYLE'S MATTER THEORY AND ITS TERMINOLOGY

Before addressing the questions arising from Kuhn's claims, it will be necessary first to give a brief sketch of Boyle's hierarchical matter theory. Beginning with *The Sceptical Chymist* of 1661, Boyle distinguishes between the *prima naturalia* or *minima naturalia*—the first or smallest particles that cannot be divided further, and the composite corpuscles, which begin with the *prima mista* (i.e., *prima mixta*—first mixed bodies) and continue upward to form more and more complex particles, called "compounded" and "decompounded" bodies.[15] The individual corpuscles of lowest stage (*prima naturalia*) are endowed with only the essential characteristics of matter, which can be comprehended under the three generic categories of size, shape, and motion (or its absence). As Boyle moves up on the ladder of chymical composition, however, he places increasingly less emphasis on the primitive characteristics of mat-

---

[14] Kuhn, "Robert Boyle and Structural Chemistry" (cit. n. 8), 36.
[15] Boyle, *The Sceptical Chymist*, in *The Works of Robert Boyle,* ed. Michael Hunter and Edward Davis (London, 1999–2000), 2:296–7, 347; *The Origine of Formes and Qualities*, in ibid., 5:326; *The History of Particular Qualities*, in ibid., 6:270–1, 274–5; *Experiments and Notes about the Mechanical Origine and Production of Volatility*, in ibid., 8:425; *Producibleness*, in ibid., 9:114. The term "decompounded" had already been absorbed into discussions of matter theory by the time of Heinrich Cornelius Agrippa von Nettesheim's *De occulta philosophia*, if not earlier. See Agrippa, *De occulta philosophia libri tres*, ed. Vittoria Perrone Compagni (Leiden, 1992), 91.

ter and more on their "texture" or interparticular structure.[16] Following this tendency further, in fact, Boyle often leaves out any discussion of mechanical properties at all and speaks of what he calls "chymical qualities" such as volatility and fixity, corrosiveness and the ability to be corroded, susceptibility to precipitation, and ability to undergo amalgamation with mercury.[17]

A valuable example of this pattern can be seen in *The Sceptical Chymist*'s treatment of vitriol, which Boyle expresses in terms of his hierarchical theory of matter. As Boyle puts it there,

> it will be Difficult to evince that Nature her self does not make Decompound Bodies, I mean mingle together such mixt Bodies as are already Compounded of Elementary, or rather of more simple ones. For Vitriol (for instance) though I have sometimes taken it out of Minerall Earths, where Nature had without any assistance of Art prepar'd it to my Hand, is really, though Chymists are pleas'd to reckon it among Salts, a Decompounded body Consisting (as I shall have occasion to declare anon) of a Terrestrial Substance, of a Metal, and also of at least one Saline Body, of a peculiar and not Elementary Nature.[18]

Boyle claims here that vitriol is a decompounded body made of three simpler compounds or mixts—a terrestrial substance, a metal, and a saline body. His point is that the corpuscles making up each of these subsidiary compounds retain their own identity during the chymical combination that results in the production of vitriol: such corpuscles are semipermanent. As he continues, the same relative permanence sometimes occurs in particles of purging medicines. When a nurse has taken a purgative medicine, the nursing baby sometimes undergoes a purge because the robust corpuscles of the drug retain their identity within the milk. Again, Boyle's emphasis here is not on the putative ability of the vitriol or milk to be reduced to its uniform catholic matter and then built back up to form another substance—rather, he wishes to stress that the more primitive mixts retain their identity as robust corpuscles within the more complex corpuscles making up the vitriol or milk as such.

Boyle never devoted an entire treatise to his hierarchical theory of matter, but traces of it permeate *The Sceptical Chymist* and his later works. The most striking thing about Boyle's theory is not its hierarchical character per se—similar thoughts can already be found in the ancient atomists and in Boyle's immediate sources such as Pierre Gassendi and Daniel Sennert—but rather in Boyle's peculiar terminology. Neither Gassendi nor Sennert refer to the composite corpuscles of higher level as *decomposita* or "decompounded" particles. What then is the origin and sense of this peculiar term, so characteristic of Boyle's writings? Why would something that had

---

[16] As in Boyle, *About the Excellency and Grounds of the Mechanical Hypothesis,* in Hunter and Davis, *Works* (cit. n. 15), 8:111, where he speaks of "the union of insensible particles in a convenient Size, Shape, Motion or Rest, and Contexture; all which are but Mechanical Affections of convening Corpuscles." See also *Excellency of the Mechanical Hypothesis,* 8:105, as well as *Of the Imperfection of the Chymist's Doctrine of Qualities,* in ibid., 8:401–2; *Experiments and Notes about the Mechanical Origine and Production of Volatility,* in ibid., 8:431; *Experiments and Considerations Touching Colours,* in ibid., 4:99; *The Sceptical Chymist* (cit. n. 15), 2:356 (where Boyle explicitly equates "structure" and "texture"), and *Origine of Formes and Qualities,* in Hunter and Davis, *Works*, 5:306, 310, and throughout. See also Clericuzio, "Redefinition" (cit. n. 12), 580–1.

[17] See, e.g., Robert Boyle, "An Introduction to the History of Particular Qualities," in Hunter and Davis, *Works* (cit. n. 15), 6:268.

[18] Boyle, *The Sceptical Chymist* (cit. n. 15), 2:265.

been "decompounded" actually be equivalent to a supercompound? Is it not the case that a supercompound should have been re-compounded out of simpler corpuscles, while a *decompositum* should reasonably have been de-compounded into simpler components?

Despite the obvious Latin origin of the term "decompounded," the question can be answered easily by turning to the *Oxford English Dictionary* (*OED*). There we learn that the early modern English term "decomposite" means not uncompounded but rather further compounded. Moreover, the *OED* provides a convincing etymology for the Latin term *decompositum,* which is the origin of "decomposite."[19] *Decompositum* is a literal translation of the Greek grammarians' term *parasyntheton,* which referred to a word built on another word that is already a compound.[20] Conveniently, the word *parasyntheton* is itself precisely such a word, being formed by the addition of the Greek preposition *para* to the compound participle *syntheton* (meaning "compounded"), which is in turn formed from the preposition *syn* plus the participle *theton.* By Boyle's time, the term *decompositum* had already begun its migration from linguistic analysis into the realm of matter theory, for the famous magical writer of the sixteenth century, Heinrich Cornelius Agrippa von Nettesheim, used it to describe a second-order mixture or recombination of the four Aristotelian elements. These second-order mixed elements were not "elements *per se*, but decompounded (*decomposita*)" elements, which the magician had to reduce to a pure, simple state if he wished to perform his marvelous works.[21] The term only acquired a loosely corpuscularian sense when it was appropriated, probably from Agrippa, by Francis Bacon. In at least one passage of the *Novum organum* and occasionally in other works, Bacon uses *decompositum* to refer not only to "metaschematisms" or corpuscles that had undergone multiple combination but also to motions made up of combined simpler motions.[22] There can be little doubt that Boyle in turn acquired the term from his careful reading of Bacon. Unlike Bacon, however, Boyle transferred the term *decompositum* or "decompounded" fully into the realm of chymical theory and made it a key feature of his mechanical philosophy.

In the remainder of this essay, I will argue that Kuhn and his followers have missed a highly significant point—that Boyle, despite his belief in a uniform catholic matter, managed to exercise a considerable influence on the elemental and corpuscular underpinnings of what is often acknowledged as the most influential current in eighteenth-century chemistry before Lavoisier—namely, the school of Georg Ernst Stahl. In order to make this claim, I will focus here on the fortunes of Boyle's term "decompounded." "Decompounded" and its Latin equivalent *decompositum* were awkward terms even in the mid-eighteenth century, as a particularly prominent example reveals. Pierre-Joseph Macquer, while retaining the Boylean hierarchical theory of matter in his famous 1766 *Dictionnaire de chimie*, explicitly says that he

---

[19] See the online *Oxford English Dictionary*, s.v. "decomposite," http://www.oed.com.ezproxy.lib.indiana.edu/view/Entry/48354?redirectedFrom=decomposite#eid (accessed 16 November 2013).

[20] The word "decompositum" is not common in Latin, but it appears in Forcellini's *Totius latinitatis lexicon*, where the example of *parasyntheton* is also found. See the online *Database of Latin Dictionaries*, s.v. "decompositum," http://clt.brepolis.net.ezproxy.lib.indiana.edu/dld/pages/QuickSearch.aspx (accessed 16 November 2013).

[21] Agrippa, *De occulta philosophia* (cit. n. 15), 91.

[22] Francis Bacon, *The Instauratio magna Part II:* Novum organum *and Associated Texts*, ed. and trans. Graham Rees and Maria Wakely (Oxford, 2004), 383. Rees identifies Agrippa as Bacon's source in his commentary.

prefers a straightforward numerical ordering that does away with complicated Latinate terminology. Hence Macquer explicitly replaces *Composita, Decomposita,* and *Superdecomposita* with "*Composés* du premier, du second, du troisieme, du quatrieme dégré, &c."[23] Indeed, the primary sense of *décomposition* to Macquer is equivalent to the modern English "analysis"—the breaking down of a compound into its constituents; this had been the standard meaning of the term in French for at least half a century when Macquer composed his *Dictionnaire.*[24] It was only natural then for Macquer to restrict the term *décomposition* to analysis and to substitute a numerical terminology for increasingly complex products of synthesis.

Thanks to the archaic character of the term "decompounded" and its Latin original *decompositum* in the sense of "further compounded," we are provided with a marker that allows us to follow the influence of Boyle through various later authors. Let us begin here with our *terminus ad quem,* namely, Stahl himself, in order to show that the terminology of Boyle's hierarchical corpuscular theory still survived in the work of the German chemist. After that we will consider Stahl's immediate source, Johann Joachim Becher, and make the argument that Becher derived his terminology directly from Boyle.

Many readers will be familiar with the basics of Stahl's corpuscular theory from the presentation given by Hélène Metzger in her 1930 *Newton, Stahl, Boerhaave et la doctrine chimique.* Metzger's treatment, despite its merits, has the characteristic of obscuring the Stahlian terminology, since she worked with the publications of Stahl's students, and usually in French translation. As Metzger points out, the system of the famous chemist of Halle postulated a hierarchy of composition beginning with elemental earth and water. These materials, which Stahl sometimes calls atoms and sometimes *moleculae,* combined first to form mixts, which could in turn combine to form compounds or secondary mixts, and these in turn could combine to make up still more complex "supercompounds" (*surcomposés*) or tertiary mixts. To give an example of how the system worked, consider the example of antimony, which Metzger takes from the 1730 *Conspectus chymiae* of Stahl's follower Johann Juncker. Regulus of antimony, according to Juncker, is a compound made up of two primary mixts—an arsenical vitrifiable material and inflammable earth. The regulus in turn can combine with sulfur to form mineral antimony, which Juncker calls a supercompound.[25]

If we turn from Metzger's presentation to Stahl himself, the same hierarchical idea emerges, but with more complicated terminology. As Kevin Chang has noted, Stahl's early *Zymotechnia fundamentalis* of 1697 already contains his hierarchical matter theory, along with Stahl's first references to phlogiston. Here Stahl employs the following terminology in increasing order of material complexity—at the lowest level of matter are *Simplicia* or *Principia,* which combine to form *Mixta.* These in turn come together to make up *Composita,* which in turn recombine to make up

[23] Pierre Joseph Macquer, *Dictionnaire de chymie* (Paris, 1766), 1:277.

[24] Ibid., 1:325–7. In fact, this was already the primary sense of *décomposition* by the late seventeenth century, for the 1694 first edition of the *Dictionnaire de l'Académie française* gives the following definition of the word: "Décomposition. s. f. v. Terme de Chymie. Dissolution, resolution d'un corps mixte dans ses principes. La décomposition d'un corps mixte." See http://artflsrv01.uchicago.edu/cgi-bin/dicos/pubdico1look.pl?strippedhw=d%C3%A9composition&headword=&docyear=1600-1699&dicoid=ALL (accessed 6 November 2013).

[25] Hélène Metzger, *Newton, Stahl, Boerhaave et la doctrine chimique* (Paris, 1930), 121–3.

what Stahl here calls *Decomposita* and then, at an even higher level of complexity, *Superdecomposita*.[26] Comparing this scheme to the one given by Metzger, we find that Stahl's *Decomposita* and *Superdecomposita* correspond to Juncker's supercompounds, which Juncker has compressed into one category. With Stahl himself, therefore, we are back in the realm of Boyle's rather awkward terminology for supercompounds. As I have suggested, however, Stahl did not derive this terminology primarily from Boyle, but—and this should be no surprise—from the anointed hero of his younger years, Johann Joachim Becher.[27] As Stahl puts it in the *Zymotechnia fundamentalis*,

> In this affair, as for the most part, the distinction of that ingenious and excellent man Becher pleases me, which he proposes in his *Physica subterranea*, where he has divided bodies into *simplicia*, or *principia*, then *Mixta*, then *Composita*, finally *Decomposita*, even *Superdecomposita*.[28]

Thus Stahl not only pinpoints Becher as the source of his hierarchical matter theory but even gives us the very work that inspired him—the *Physica subterranea* that appeared first in 1669, followed by a succession of supplements (1671, 1675, 1680). The *Physica subterranea* is no doubt Becher's most famous chymical work, in no small part because Stahl himself published a posthumous edition with commentary in 1703; Stahl's edition was itself reprinted in 1738, four years after his own death.

THE EVOLUTION OF J. J. BECHER

Let us therefore turn to Becher himself. It is well known that in 1679 Becher moved to England, where he apparently entered into personal relations with Boyle, even publishing his 1680 *Magnalia naturae* to satisfy the curiosity of the English "naturalist."[29] But Becher's acquaintance with Boyle's writings predates his immigration by more than a decade, and indeed, the influence of Boyle in the 1660s seems to have been a decisive factor in Becher's transformation from a rather traditional early modern alchemist and projector to the much more critical and systematic theorist that one encounters in the *Physica subterranea*. The chymical works of this astonishing polymath reveal a striking evolution if one starts at his earliest publications and moves forward through his successive writings. In this section I will attempt to ascertain the first appearance in Becher's work of the hierarchical matter theory that Stahl took from the *Physica subterranea*. Fortunately, Becher left behind a *Catalogus librorum et manuscriptorum* documenting his writings and the dates of their publication. His major biographer, Herbert Hassinger, reproduces this as an appendix to his 1951

---

[26] Kevin Chang, "Fermentation, Phlogiston and Matter Theory: Chemistry and Natural Philosophy in Georg Ernst Stahl's *Zymotechnia Fundamentalis*," *Early Sci. Med.* 7 (2002): 31–64, on 50–1.

[27] Stahl's hero worship had declined by 1723, of course, when he published his *Billig Bedencken, Erinnerung und Erläuterung über D. J. Bechers Natur-Kündigung der Metallen,* for which see Kevin Chang's essay, "Communications of Chemical Knowledge: Georg Ernst Stahl and the Chemists at the French Academy of Sciences in the First Half of the Eighteenth Century," in this volume.

[28] G. E. Stahl, *Zymotechnia fundamentalis,* in *Georgii Ernesti Stahlii Opusculum chymico-physico-medicum* (Halle, 1715), 91: "Primo, arridet in hoc negotio, ut pleraque ingeniosi & exquisiti Viri, *Beccheri,* distinctio, quam in Physica sua Subterranea proponit, ubi Corpora dispecit in *simplicia* seu principia, deinde *Mixta,* inde *Composita,* tandem *Decomposita,* imo *Superdecomposita.*"

[29] Principe, *Aspiring Adept* (cit. n. 6), 112–3.

monograph, *Johann Joachim Becher.* At the very beginning of Becher's category "In Physicis, Medicis, et chymicis," we find the following entry:

> Solini Saltzthals Regiomontani de Lapide Trismegisto et Salinis Philosophicis. Germ. 1654. Reperitur Latine in Theatro chymico.[30]

Although Hassinger was unable to identify this work, it is clearly identical to the Latin *Discursus Solini Saltztal Regiomontani* found in volume 6 of the 1661 *Theatrum chemicum,* a large multivolume collection of medieval and early modern alchemy.[31] Composed by the nineteen-year-old Becher, this treatise comprises a valuable testament to the young prodigy's earliest recorded thoughts on chymistry. Perhaps we should not be surprised that the *Discursus* attributed by Becher to Solinus is entirely traditional in character. For further reading, Solinus recommends Paracelsus, Basilius Valentinus, Geber, and the *Turba philosophorum,* as well as a disciple of Chortolasseus, the German chymist Johann Grasseus.[32] Solinus claims to be the possessor of the philosophers' stone, which he oddly names *Trismegistus,* as if the name referred to the *lapis philosophorum* itself rather than to its most famous promoter, Hermes. This *Trismegistus* is to be made by combining the essences of man, the grape, and gold, since these three entities represent the highest and noblest beings in the animal, vegetable, and mineral world. Exactly how one is to carry this combination out Solinus does not reveal, instead taking the reader on a baroque dream-voyage accompanied by a little old man who appears to him in his sleep. The voyage carries Solinus from the city of *Ignorantia* to a mountain populated by the Rosicrucians. Solinus and his guide successively encounter three ornate fountains that disperse life-giving animal, vegetable, and mineral waters to the world, thus accounting for the perpetuation and growth of all species. Finally, Solinus learns that the constituents of the philosophers' stone, again gold, grapes or wine, and man, must be reduced to their first essence by means of a radical solution before they can combine as the wonder-working *Trismegistus.*

Needless to say, Becher's early *Discursus* contains nothing of the hierarchical matter theory to which Stahl alludes. The work is an elaborate allegory of the philosophers' stone that tells us more about Becher's youthful origins than about his mature theory. Let us therefore turn to the next work that Becher mentions in the chronological listing of his titles, the *Natur=Kündigung der Metallen,* published in Frankfurt in 1661. As Chang has pointed out, Stahl would explicitly distance himself from this work in his antichrysopoetic *Billig Bedencken, Erinnerung, und Erläuterung* of 1723. It is not difficult to see why—the *Natur=Kündigung* consists of an interesting mixture of metallurgical and mining observations partly derived from Basilius Valentinus's *Leztes Testament* and partly from other alchemical sources. Again, the work is strikingly traditional in character.

The first seventy-five pages of the *Natur=Kündigung* in the 1661 edition are dominated by Becher's interpretation of Aristotelian elemental theory. Like Aristotle,

---

[30] Herbert Hassinger, *Johann Joachim Becher (1635–1682)* (Vienna, 1951), 258.

[31] Anonymous, *Discursus Solini Saltztal Regiomontani,* in *Theatrum chemicum* (Strasbourg, 1661), 6:675–714.

[32] For Grasseus, see Thomas Lederer, "Leben, Werk und Wirkung des Stralsunder Fachschriftstellers Johann Grasse (nach 1560–1618)," in *Pommern in der Frühen Neuzeit,* ed. Wilhelm Kühlmann and Horst Langer (Tübingen, 1994), 227–37.

Becher stresses that the four elements have four qualities, hot, cold, wet, and dry, which account for qualitative change. Unlike Aristotle, however, Becher imports a rather specific theory ultimately going back to the Arabic corpus of Jābir ibn Ḥayyān, according to which two qualities dominate in the "center" of a body and two in the "circumference."[33] Possibly under the additional influence of J. B. Van Helmont, to whom he refers, Becher argues that a qualitative "inversion" of materials results in striking changes. Consider, for example, his treatment of solution, which I do not claim fully to understand. One cannot dissolve a fixed body by making its central qualities go outward and the peripheral qualities inward without their complete destruction, for such qualities, being fixed, can protect themselves against impinging accidental qualities. Only when a moisture that is warm in its center and cold in its circumference encounters a fixed body that is warm at its circumference and cold at its center can a solution happen. It is the sympathy between the internal heat of the solvent and the external heat of the fixed body that allows the latter to be inverted and hence to dissolve.[34]

After this complicated attempt to explain phenomenal change in terms of the elemental qualities, Becher then passes to a section on mineralogy and metallurgy that seems to have been influenced above all by Basilius Valentinus's *Letztes Testament.* Becher describes the growth and death of minerals within their mines at great length and ascribes to them a vital principle called *Ferch*, just as Basilius does. Also like Basilius, Becher spends a great deal of time discussing mineral exhalations or *Witterungen*. Traditional alchemical authorities such as Morienus, Geber, Maria the Hebrew, Raymond Lull, and Albertus Magnus figure very prominently alongside Basilius, and it is clear that Becher wishes to link his own authority to theirs.[35]

Finally, the second half of the book is devoted to a series of Maxims concerning the improvement of metals, which Becher calls *Sätze*. These too are quite interesting, since they display Becher's attempt to address scholastic philosophy as practiced in the schools of mid-seventeenth-century Germany. Consider Becher's attempt to deal with the issue of transmutation, for example. He is aware of the philosophical and theological problems that arise from saying that humans can transmute species.[36] Nonetheless, he also wants to do justice to the striking qualitative changes that occur in nature. For example, grass is indirectly converted to human flesh when a man eats a cow, so why should one metal not be able to turn into another? Since all metals share the name "metal," this would not necessarily involve the transmutation of species or specific forms. Becher illustrates this point with a poignant example taken from his experience of Europe in the Thirty Years War. He knows of starving children during the war who were forced to run into the woods and subsist upon grass; eventually they grew so famished that Becher assures us they even began consuming their own fingers. But meanwhile, as a result of having eaten too much grass, they also turned as green as frogs.[37] Since the green children were still children, they did not undergo a change

[33] See Paul Kraus, *Jābir ibn Ḥayyān: Contribution à l'histoire des idées scientifiques dans l'Islam*, vol. 44 of Mémoires présentés à l'Institut d'Egypte (Cairo, 1943), 2:2, 228, and throughout.

[34] Johann Joachim Becher, *Natur=Kündigung der Metallen. Mit vielen Curiosen/Beweißthumben/ natürlichen Gründen/Gleichnüssen/Erfahrenheiten/und bißhero ungemeinen Auffmerckungen vor Augen gestellet . . .* (Frankfurt am Main, 1661), 48.

[35] Ibid., 76–9.

[36] For a history of this long debate, see William R. Newman, *Promethean Ambitions* (Chicago, 2004), 34–114.

[37] Becher, *Natur=Kündigung* (cit. n. 34), 214–5.

of species—in short, to quote Becher, they did not experience a *Verwandlung* (transmutation) but rather a *Verenderung* (alteration). The same is true of metals that are matured and bettered to the point of becoming gold. They undergo no specific transmutation but only a *Verenderung* or alteration of qualities. In this way Becher sidesteps one of the principal scholastic objections often leveled against chrysopoeia, namely that it would require an impossible transmutation of species induced artificially by humans.

Given its highly traditional character, one should not be surprised that Becher's *Natur=Kündigung*, like the *Discursus* of Solinus Saltztal, contains no obvious trace of the Boylean hierarchy of corpuscles that Stahl found in the *Physica subterranea*. The same is true of Becher's *Oedipus chimicus*, a work published in Latin in 1664. Although the title page of the *Oedipus chimicus* bills the text as a "Little Work opening and resolving the Mysteries of More Obscure Chymical Terms and Principles," in reality very little of the book concerns obscure chymical language.[38] The bulk of the *Oedipus chimicus* is taken up with the attempt to distinguish "Hermetic Philosophy," as Becher calls it, from Peripatetic philosophy. As such, most of the book addresses the prime matter, principles, and elements of chymistry and reads like a continuation of the scholastic *Sätze* found in the *Natur=Kündigung*. Here too Stahl's language of *Decomposita* and *Superdecomposita* is largely if not completely lacking.

In a word, between the publication of his *Oedipus chimicus* in 1664 and *Physica subterranea* of 1669, Becher had begun employing the corpuscular hierarchy later alluded to by Stahl. It is during this transformative period, according to Becher's biographer, Herbert Hassinger, that the German polymath began a period (1666–70) during which he devoted himself to intensive study and writing.[39] Hence the *Physica subterranea* marks a new period for Becher in which we find most of the hierarchical matter theory later to be recapitulated in Stahl's *Zymotechnia* and elsewhere in his works. Becher tells us in the *Physica subterranea* that material things are composed ultimately of earth and water, which can be mixed together in different ratios. Sulfur, salt, and mercury, on the other hand, are mixed bodies, not ultimate principles. Indeed, Becher stresses that the Paracelsian principles are actually *decomposita* (in the Boylean sense) for no mercury, sulfur, or salt can be extracted from metals unless some other component is added, which results in a decompounding. The same is true of vitriols, *amausa* or metallic glasses, crocuses, butters, and flowers, which thereby have the same right to be called principles as do the Paracelsian *tria prima*. All of these materials are in reality *decomposita*, having been recombined from simpler components. If all of this sounds familiar to readers of *The Sceptical Chymist*, I suggest that there is good reason for the sense of déjà vu. It is clear that Becher had read Boyle by the time of composing the *Physica subterranea*, for on pages 203 and 204 he makes the following rather opaque comment:

> I would have granted the palm to Robert Boyle above all others of our age if he had set aside his *Elaterium* and continued further with his chymical experiments, and in expounding them had set himself both the task of concluding the matter and of doubting in individual cases.[40]

[38] Johann Joachim Becher, *Oedipus chemicus* (Frankfurt am Main, 1664); "Obscuriorum Terminorum & Principiorum Chemicorum, Mysteria Aperiens & resolvens. Opusculum."

[39] Hassinger, *Johann Joachim Becher* (cit. n. 30), 44, 51.

[40] Johann Joachim Becher, *Physica subterranea* (1669), 203–4: "*Roberto Boyle*, prae omnibus nostro seculo palmam concederem, si misso suo elaterio, Chymica experimenta ulteriùs continuâsset, &

Unfortunately, Hassinger makes too much of this passage, seeing it as an example of bitter sarcasm directed at the *dubia* set forth in the *Sceptical Chymist*.[41] His unduly negative reading allows Hassinger to eliminate Boyle as an influence on Becher's mixture theory and to imply that Daniel Sennert may have been Becher's source.[42] In all probability, however, the passage was meant merely as a mild reproof of Boyle's so-called skepticism, since Becher joins it immediately to a much stronger chastisement of Werner Rolfinck, author of a well-known debunking of chrysopoeia. Indeed, later in the text, Becher compares Boyle favorably to Rolfinck. At various points in the 1669 *Physica subterranea*, Boyle's pneumatic experiments, his *Sceptical Chymist*, and his *Experiments touching Colours* are explicitly invoked by Becher as authoritative sources.[43]

To sum up, it appears that Becher publicly adopted his hierarchical theory of matter, with the peculiar terminology later appropriated by Stahl, by the time of his fully mature *Physica subterranea* of 1669. Can it be coincidence that at this time Becher also begins to cite Robert Boyle as an authoritative source? Given the striking similarity between Boyle's comments about the formation of decompounded materials such as vitriol and Becher's extensive discussion of *decomposita* in the *Physica subterranea*, this seems quite unlikely. Again, the term *decomposita* and the related expression *superdecomposita* are conspicuously absent from other corpuscularian and atomist sources whom Becher read, such as Sennert and Descartes.

In short, Becher's *Physica subterranea* displays a striking contrast with the much more traditional series of chymical books that he composed between 1654 and 1664. His thinking clearly underwent a major transformation at some point between the composition of the *Natur=Kündigung* and the *Physica subterranea*, as Stahl himself would point out in his *Billig Bedencken, Erinnerung, und Erläuterung* of 1723. The easiest way to account for this change is by taking into account Becher's exposure to the works of Robert Boyle, whose corpuscular terminology underlay the language of Becher's maturity. No doubt there is a deeper story to be unearthed concerning Becher's Boylean "conversion," but this should form the subject of a detailed study based on the German chymist's biography, which we cannot present here. At the moment, we can only say that during the five years between 1664 and 1669, Becher came under the strong influence of Robert Boyle's chymical corpuscular theory.

CONCLUSION: TRANSMUTATION AND CHYMICAL ATOMISM

Let us now return to Kuhn's aggressive claims—first, that Boyle's transmutational views were fundamentally opposed to chemical matter theory as it developed in the eighteenth century, and, second, that "the true progenitors of Lavoisier's chemical revolution were necessarily among Boyle's opponents." Clearly the first of these statements cannot be true, since Becher managed to build upon Boyle's hierarchical corpuscular theory as expressed in *The Sceptical Chymist* despite Boyle's chryso-

---

in exponendis istis, non tàm materiam concludendi quàm in singulis dubitandi tractare sibi proposuisset." The term *elaterium* here may refer to Boyle's air pump, though the Latin word usually means a purging medicine.

[41] Hassinger, *Johann Joachim Becher* (cit. n. 30), 63–4.

[42] Ibid., 60.

[43] Becher, *Physica subterranea* (cit. n. 40), 477 for Boyle's "experimenta pneumatica"; 455, 479 for *The Sceptical Chymist*; and 631 for *Experiments touching Colours*.

poetic inclinations. Never mind the fact that Becher himself, the longtime hero of G. E. Stahl, was also a supporter of metallic transmutation, as Pamela Smith has underscored in her *Business of Alchemy*.[44] As for Kuhn's second claim, that the progenitors of the chemical revolution were necessarily Boyle's opponents, one can only accept this if the Becher-Stahl tradition is excluded from the lineage leading up to Lavoisier, a strategy that no responsible historian would accept, and that Kuhn himself did not advocate. Two lessons can be drawn from Kuhn's mistake. First, although we should not saddle Boyle with fatuous and hyperbolic titles such as the "founder of modern chemistry," it is also clear that he was not an insignificant player in the history of the discipline. Boyle was an important transmitter and reformulator of a chymical matter theory that had roots in medieval alchemy, as I have argued elsewhere.[45] Basing his empirical demonstrations of the corpuscular philosophy primarily on the chymical atomism of Daniel Sennert, Boyle was able to provide convincing experimental evidence for his theory. Second, the tradition of chymical atomism often stressed both the possibility of chrysopoeia *and* the existence of a hierarchy of robust combining corpuscles beneath the level of sense. Kuhn's fundamental error, and that of his followers, was to assume that transmutational aspirations necessarily ran counter to any belief in enduring chymical species. The case of Becher, the influence of Boyle, and the long tradition of alchemical corpuscular theory descending from Daniel Sennert and his sources reveal in reality that the opposite was more probably the case.[46]

How then do we resolve the seeming paradox that chymical atomism often coexisted with an active belief in chrysopoeia? This is not a trivial question, since the apparent if specious anomaly that it highlights underlies the common view that Boyle exercised little influence on subsequent chemistry. In order to address this puzzle it is important to recognize that most alchemists from the medieval period onward upheld some variation of the mercury-sulfur theory of metallic composition, which usually postulated that metals and minerals were composed of these two materials, and that these varied according to their purity, color, fixity, and relative quantity.[47] The fact that Paracelsus and his followers would later add the principle of salt to the preexistent two principles presented no serious incompatibility with this older theory. If one wanted to be a proper Aristotelian, of course, it was also necessary to think of the sulfur and mercury as being made up of the four elements, and there were additional philosophical considerations stemming ultimately from the corpuscular ruminations of Aristotle's *Meteorology*.[48] But alchemy had always been an art, not a disembodied system of pure speculation. When it came to transmuting metals, then, an obvious avenue for effecting this lay in the realm of laboratory operations that removed,

[44] Pamela H. Smith, *The Business of Alchemy: Science and Culture in the Holy Roman Empire* (Princeton, N.J., 1994), 173–227.

[45] Newman, *Atoms and Alchemy* (cit. n. 1), 157–225.

[46] See ibid., particularly 23–65, where the alchemical theory of *fortissima compositio* as an alternative to the Thomistic theory of the unity of the substantial form is discussed.

[47] For documentation concerning the sulfur-mercury theory, see John Norris's essay, "*Auß Queck-silber und Schwefel Rein*: Johann Mathesius (1504–65) and *Sulfur-Mercurius* in the Silver Mines of Joachimstal," in this volume.

[48] The reader who is accustomed to thinking of Aristotle as a purely continuist, antiatomist thinker should consult the fourth book of the *Meteorology* attributed to the Stagirite. For the role of this text and its corpuscular matter theory in the history of alchemy, see Newman, *Atoms and Alchemy* (cit. n. 1), 66–81.

replaced, or otherwise altered sulfur or mercury that was deemed to be excessive or deficient.[49] It was only natural, then, for alchemists to think of the different metals as retaining their sulfur and mercury within themselves *in actu*—as robust bits that could be manipulated ad libitum if one only had the proper set of hands-on skills. The easiest way to envision such compositional entities was in the form of small particles that in turn made up bigger, fully formed metallic corpuscles; in some cases the four elements were also thought to persist within the particles of sulfur and mercury. In this fashion the intellectual commitment to transmutation, coupled with corpuscularian ideas taken ultimately from Aristotle, encouraged a belief in the existence of mercury and sulfur as semipermanent particles that could be removed or introduced by operations such as sublimation, calcination, and dissolution in acids. This did not of course mean that the metals themselves were somehow seen as lacking in permanence or stability. To the contrary, once the habit of assuming the robust persistence of microstructural corpuscles had been acquired, it was a natural and obvious step to think of the particles of metals themselves as retaining their permanence when they combined as fully formed corpuscles with other particles of acid, salt, and so forth. And of course the fact that metals could be regained intact from their dissolution in acids reinforced the view that both acid and dissolved metal consisted of semipermanent corpuscles—in other words, chymical atoms.

Thus it was the operational necessity of the chrysopoetic laboratory that initially produced the hierarchical corpuscular theory of chymical atomism, and it was the hard reality of labor at the bench that led alchemists to the realization that metals can and usually do have their own stubborn permanence. The robust persistence of the metals is what led writers like Sennert, Van Helmont, and Boyle to speak of cinnabar, salts, vitriols, and other compounds as "wearing disguises" or "masks"—the ordinary laboratory processes known to "vulgar," unskilled chymists do not alter the fundamental nature of the metal hidden behind its new illusory appearance.[50] When the mask is removed, the same actor emerges unchanged. Unchanged, that is, unless one possesses a *menstruum* of such subtlety and penetrative power that it can break down the composite, metallic particles and operate on their own internal corpuscles. But in the very act of searching for such ultrasubtle analytical agents, such as Van Helmont's alkahest or the animated mercuries of the Philalethan tradition, chymists were tacitly acknowledging that normal, "vulgar" *menstrua* like the mineral acids left the metallic corpuscles intact, merely separating the gross mass of metal into imperceptibly small particles in solution. In his formative years, Boyle appropriated chymical atomism from writers with chrysopoetic interests such as Daniel Sennert;

---

[49] Excellent examples of such techniques for removing or adding mercury and sulfur can be found in book 3 of the *Summa perfectionis* of Geber, the classic work of transmutational alchemy in the High Middle Ages. See William R. Newman, ed., *The "Summa perfectionis" of Pseudo-Geber* (Leiden, 1991), 740–9.

[50] See, e.g., Robert Boyle, *Essay of the Holy Scriptures,* in Hunter and Davis, *Works* (cit. n. 15), 13:205: "But all this is Nothing to the numerous Disguizes that Proteus, Mercury will suffer himselfe to be mask'd in, without devesting his Nature. For tho Chymists have found wayes to disguise Quicksilver into almost as many formes as the Peripatetickes have ascrib'd to their Materia Prima, & seeme to have soe perfectly chang'd it into Oyles, Salts, Fumes, Butters, Plasters, Powders, & what not that they oftentimes believe themselves, when they boast those Appearing for Radicall Transformations; yet doth that Inscrutable Prodigie of Nature, (what ever Chymists are pleas'd to thinke to the Contrary) still remayne Reall & untransmuted Mercury; & but too often manifests himselfe to be so, to their Cost who fondly thinke, that for having forc'd him from his outward Scheme, they have robb'd him of his almost inaccessible Nature."

only later did he graft the uniform catholic matter taken from other sources onto the older trunk of this alchemical tree.

In a word, then, chymical atomism and chrysopoeia were anything but incompatible. The fact that metals stubbornly retained their own identity in the face of analytical assaults by chymists was a knowledge born of hard experience. Hierarchical theories of matter like those of Boyle, Becher, and Stahl originated out of the alchemical tradition's attempts to circumvent the obstinate refusal of nature to yield up the secret of transmutation. Since chymical atoms were not the solid and impenetrable units of Democritus and Leucippus, but structured composites made up of smaller particles, there was every reason to imagine that a sufficiently powerful solvent should be able to penetrate and break them into their components. The view that his belief in transmutation made Boyle unacceptable to the mainstream of eighteenth-century chemistry is wrong from an empirical perspective, since we now know that Stahl, following Becher, adopted the major features of Boyle's hierarchical matter theory, and also incorrect from the viewpoint of theory, now that we understand the complex motivations that underlay chymical atomism. It is time then to lay to rest the counterfactual, Kuhnian belief that Boyle's chrysopoetic inclinations doomed him to failure among the mainstream of chemists before Lavoisier, since Becher and Stahl, like the English "naturalist," accepted a matter theory that allowed for the possibility of transmutation by means of analysis and recombination of corpuscles.

TRANSITIONS FROM CHYMISTRY TO CHEMISTRY (1675–1750)

The Chymistry of "The Learned Dr Plot" (1640–96)

by Anna Marie Roos*

ABSTRACT

In the seventeenth century, there were developing norms of openness in the presentation of scientific knowledge that were at odds with traditions of secrecy among chymists, particularly practitioners of chrysopoeia, or the transmutation of metals. This chapter analyzes how Dr. Robert Plot, the first professor of chymistry at Oxford, negotiated these boundaries within an institutional context. I first delineate his chymical and experimental practice, which incorporated procedures from medieval alchemical sources, particularly the Lullian corpus, as well as more novel practices from seventeenth-century chymistry. Then, I analyze how personal and institutional ambitions and economic considerations shaped to what extent Plot negotiated the boundaries between secrecy and the public dissemination of chymical knowledge.

INTRODUCTION

In the seventeenth century, there were developing norms of openness in the dissemination and presentation of scientific knowledge that were at odds with traditions of secrecy among chymists,[1] particularly practitioners of chrysopoeia, or the transmutation of metals. Between these two standards a tension arose, evidenced by early modern writers' "vociferous criticisms" of chymical obscurity, with different strategies developed for negotiating the boundaries between secrecy and openness.[2] Newton and Boyle were both "in fearful awe" of the power of transmutation, concerned, for example, about its potential to disrupt economic and social stability, and thus kept their work secret.[3] Newton sequestered his more public research in physics and mathematics from his private chymical investigations.[4] Boyle, however, while keeping his chrysopoeia secret by using codes, ciphers, and the "principle of disper-

---

* Lincoln School of History and Heritage, University of Lincoln, Brayford Pool, Lincoln LN6 7TS, United Kingdom; aroos@lincoln.ac.uk.

[1] Use of the term "chymistry" is guided by Lawrence Principe and William Newman, "Alchemy vs Chemistry: The Etymological Origins of a Historiographic Mistake," *Early Sci. & Med.* 3 (1998): 32–65. States of secrecy in early modern science have recently served as the theme of the June 2012 issue of the *British Journal for the History of Science*, edited by Koen Vermeir and Dániel Margócsy.

[2] Lawrence Principe, "Robert Boyle's Alchemical Secrecy," *Ambix* 39 (July 1992): 63–74, on 63. See Principe, *The Aspiring Adept: Robert Boyle and His Alchemical Quest* (Princeton, N.J., 1998); William Newman, *Gehennical Fire: The Lives of George Starkey, an American Alchemist in the Scientific Revolution* (Chicago, 1994).

[3] Lawrence Principe, "The Alchemies of Robert Boyle and Isaac Newton," in *Rethinking the Scientific Revolution,* ed. Margaret Osler (Cambridge, 2000), 201–20, on 209.

[4] B. J. T. Dobbs, *The Foundations of Newton's Alchemy* (Cambridge, 1975), 94.

sion" or scattering logically connected parts of a procedure disconnectedly through printed text(s), published other forms of chymical knowledge widely.[5] Similarly, while Nicolas Lémery consciously and publicly excised material concerning metallic transmutation from the third edition of his enormously popular *Cours de Chimie*, he was simultaneously employing chrysopoetic "apparatus, practices, and skills" in his own work.[6] As William Newman has shown, the American alchemist George Starkey (1628–65) even went so far as to create an alter ego named Eirenaeus Philalethes. "Setting himself up as the middleman between his fictive adept and the Hartlib Circle, Starkey acquired a privileged position as the revealer of secret knowledge."[7] Not only did this mechanism enhance Starkey's prestige in learned circles, particularly in his communications with Boyle, but as a "metallurgist and inventor," Starkey could use his alter ego to protect his trade secrets.[8]

Less well studied, however, is to what extent other contemporaries negotiated these boundaries, particularly when they held a formal university appointment in chymistry. This is the case with Robert Plot, appointed as the first keeper of the Ashmolean, secretary of the Royal Society, and Oxford's first professor of chymistry. "The learned Dr Plot," as he was called, was a man of versatile talents, without whose flair for organization "the Ashmolean Museum might well never have come to Oxford."[9] Plot was educated at Magdalen Hall, Oxford, obtaining a BA (1661), an MA (1664), and a doctorate of civil law (1671). In 1667, Plot followed a course in practical chymistry given by William Wilden and was part of the group of natural philosophers who "congregated around Robert Boyle at Deep Hall until 1668."[10] Not just a chymist, Plot was also a natural historian, writing the *Natural History of Oxfordshire* (1677) and *Natural History of Staffordshire* (1686). He also founded the Oxford Philosophical Society (OPS), active from 1683 to 1688, the only university scientific society of this era for which we have extensive records. Plot directed its experiments, "occupying a position roughly comparable to that of Hooke in the Royal Society," and edited the *Philosophical Transactions of the Royal Society* from 1683 to 1684.[11]

Despite his importance to the development of early chymistry and natural history, there has been little historical analysis of his chymical work and practice, much less how he negotiated the boundaries between secrecy and dissemination of knowledge within and outside of the university and scholarly societies.[12] Some of this state of affairs may be because unlike Boyle, Plot did not publish any books about chymistry, his oeuvre confined to books concerning natural history and chorography (regional

---

[5] Principe, "Secrecy" (cit. n. 2), 63.

[6] William Newman, "From Alchemy to Chymistry," in *The Cambridge History of Science: Early Modern Science*, ed. Katharine Park and Lorraine Daston (Cambridge, 2006), 497–517, on 511 and 515.

[7] Newman, *Gehennical Fire* (cit. n. 2), 62.

[8] Ibid., 62–3.

[9] Ken Arnold, *Cabinets for the Curious* (Aldershot, 1988), 50.

[10] Anthony Turner, "Robert Plot (1640–1696)," *Sphaera* 4 (1996), http://www.mhs.ox.ac.uk/sphaera (accessed 23 October 2011).

[11] Stanley Mendyk, "Robert Plot: Britain's Genial Father of County Natural Histories," *Notes Rec. Roy. Soc. Lond.* 39 (1985): 159–77, on 164.

[12] Exceptions are Marcos Martinón-Torres, "Inside Solomon's House," *Ambix* 59 (2012): 22–48; F. Sherwood Taylor, "Alchemical Papers of Dr Robert Plot," *Ambix* 4 (1949): 67–79; K. Hoppen, "The Nature of the Early Royal Society Part II," *Brit. J. Hist. Sci.* 9 (1976): 243–73.

landscape studies). Unlike Newton, Boyle, or Starkey, Plot also did not leave behind a large corpus of chymical manuscripts. Material is confined to his correspondence,[13] MS Sloane 3646 in the British Library, which consists of a considerable amount of papers in his hand concerning his own experiments, and OPS meeting minutes.[14] By close analysis of these sources, we will remedy omissions in the literature about Plot and analyze the role of chymical secrecy in an institutional context.

In his "rules of Praticall Chimistry," most likely composed for his Oxford students, Plot wrote, "chymistry and Alchemy cannot be separated."[15] This chapter will first delineate his chymical and his experimental practice, which indeed incorporated procedures from medieval alchemical sources, particularly the corpus traditionally though wrongly ascribed to Raymond Lull (1232–1316), as well as more novel practices from seventeenth-century chymistry that, on the surface, had little to do with metallic transmutation but were intimately inspired by it. Then I will analyze how personal and institutional ambitions and economic considerations shaped to what extent Plot negotiated the boundaries between secrecy and the public dissemination of chymical knowledge. We will see that for Plot (as in the case of Starkey), the tension between secret and shared and private and public knowledge was informed by ambition and self-fashioning. However, Plot's strategy was also influenced by his legal knowledge of contracts of tender and his financial needs specific to his position as an Oxford professor. As chymistry was not yet part of the regular university curriculum at Oxford, Plot was expected to provide running laboratory expenses and his salary by privately selling his inventions, chymical medicaments, and secrets, which he did on contract. In the public sphere, Plot demonstrated his expertise by performing experiments concerning less lucrative though chymically interesting procedures for the Royal Society and OPS; he also intimated to their members that he possessed valuable secret knowledge but did not reveal it.

PLOT, CHYMICAL MENSTRUA, AND THE SPIRIT OF WINE

An analysis of MS Sloane 3646, written by Plot in the late 1670s and 1680s, shows that he was familiar with several chymical techniques and authors.[16] He included material concerning iatrochemistry and *chymiatria* (the preparation of medicines) including a defense of Tachenius's *Hippocrates chymicus* (1666), which combined Johannes Baptista Van Helmont's iatrochemistry with Sylvius de la Boë's (1614–72) theory of acids and alkalis in medical treatments, as well as "an example of the disease of the stone in Van Helmont."[17] In the OPS meeting minutes, Plot appended his own list of "chemical arcana and desiderata" featuring the synthesis of several Helmontian substances, including "Metallus Masculus" (a type of alkahest or

---

[13] Plot's correspondence is in the Bodleian Library in MSS Ashmole, Aubrey, Ballard, Lister, Tanner, and Wood, indexed in Early Modern Letters Online (http://emlo.bodleian.ox.ac.uk) and partially transcribed in R. T. Gunther, *Dr. Plot and the Correspondence of the Philosophical Society of Oxford: Early Science in Oxford* (Oxford, 1939). The OPS minutes are in Bodl. MS Ashmole 1810–2 and reprinted in Gunther, *The Philosophical Society: Early Science in Oxford* (Oxford, 1925).

[14] For a catalogue description of MS Sloane 3646, see Taylor, "Alchemical Papers" (cit. n. 12).

[15] British Library, MS Sloane 3646, 44r.

[16] Most of the manuscript is in Plot's writing, although there are chymical excerpts in other hands.

[17] Sloane 3646 (cit. n. 15), 58r–61v.

universal dissolvent), the "Asoph" (the ultimate arcanum, which contained the universal spirit of God), and his chemical arcanum against gallstones.[18] Overall, however, Plot's chymical manuscripts have more of the character of Boyle's than of Newton's; whereas Newton emphasized the written traditions of chymistry, "favoring an approach based on textual analysis and the drawing of compendia . . . thereafter complemented with laboratory experimentation," Boyle "favored a more empirical approach to learning" about chymistry, employing experimentation more immediately.[19] So, apparently, did Plot.

Most of Sloane 3646 is dedicated to Plot's own observations with an emphasis on praxis, which may be due to his role as a chymistry professor; in his "rules of Practical Chemistry," he stated, "because the Philosophers haue set forth theoreticall and practicall bookes we are to chuse the praticall because they are to be understood literally."[20] On the other hand, he noted that when chymical language was deliberately obscured by adepti to preserve secrets, that "the designe of the process will easily show whether these words or any of them are to be taken literally or no"; he advised that it was, for instance, easy to tell the differences between the wet way (*via humida*) to the Philosophers' Stone, which employs watery solvents, and the other major route of chrysopoeia, the dry way (*via sicca*), which uses no such watery corrosives.[21]

Plot also noted that he deciphered obscure published procedures to create his own secret chymical menstrua or solvents that were both medical and involved in transmutation of matter.[22] He defined a menstruum as "not a thing dissolving bodies superficially, or dividing them into the most minute parts by corrosion, but a liquor, remaining with the thing dissolved in it, inseparable and altering . . . the constitution of the thing dissolved by the addition of itselfe into a new thing which neither is nor can be again that which it was."[23] Menstrua were usually classified as corrosive dissolvents, especially those for "dissolving metal in the attempt to convert base metals into gold."[24] Plot's definition of the menstruum was more akin to that of an alkahest, a substance that could aid the practitioner to break down matter into its smallest parts to enable not only transmutation but also the creation of medicines from constitutive matter. Plot then asserted in a section "On the use of secret menstruum," that "In Chymistry unless a man know exactly the use of Menstruums, hee will without doubt never performe anything worthy of prayse."[25] "On the contrary if a man know nothing but that," he would be able to tell true medicine from false and understand the books of chymical adepti and "the progresses of our art."[26]

Plot thought this secret menstruum was to be found in spirit of wine "actuated and rectified to a preternatural subtlety" to create a substance described as the philosophical spirit of wine.[27] In particular, Plot's manuscript notes displayed experi-

[18] The desiderata are in the OPS minutes (17 March 1684/5) and reprinted in Gunther, *The Philosophical Society* (cit. n. 13), 130–2.
[19] Principe, "Alchemies" (cit. n. 3), 210.
[20] Sloane 3646 (cit. n. 15), 44r.
[21] Ibid., 44r–44v.
[22] Taylor, "Alchemical Papers" (cit. n. 12), 69.
[23] Sloane 3646 (cit. n. 15), 55r.
[24] *Oxford English Dictionary,* s.v. "menstruum," www.oed.com (accessed 27 October 2013).
[25] Sloane 3646 (cit. n. 15), 9r.
[26] Ibid.
[27] Taylor, "Alchemical Papers" (cit. n. 12), 69.

ments with menstrua described in the oeuvre attributed to the Majorcan philosopher and mystic Raymond Lull.[28] Although the historical Lull denied the possibility of transmutation, over 120 posthumous texts composed on the subject between the fourteenth and seventeenth centuries were attributed to him.[29] Michela Pereira has demonstrated that some of these works were circulating in England in the fourteenth century.[30] By the late seventeenth century, Johann Seger von Weidenfeld, the Lithuanian chymist and acquaintance of Boyle, had composed an English edition of *Four Books of Johannes Segerus Weidenfeld, Concerning the Secrets of the Adepts, or, Of the Use of Lully's Spirit of Wine* (London, 1685), which was disseminated widely and used extensively by American chymists such as Gershom Bulkeley (1636/7–1713).[31] Plot's manuscripts indicate that he, like his colleagues, was also interested in the "philosophical spirit of wine" (akin to ethanol), which was identified in the corpus of pseudo-Lull as the source of the universal quintessence or fifth element, the essence of all metals, and an incorruptible, pure, and original substance of the world.[32] It was believed this vital spirit of the cosmos could be extracted by means of distillation and other techniques.[33]

Plot wrote, "the spirit of wine being now prepared, wee have indeed the root and basis of all secret menstruums, yea also medicines"; for example, he noted that as a menstruum par excellence, it would dissolve volatile alkaline salts, such as salt of tartar, and when the mixture was distilled, the salt of tartar would be volatilized.[34] This was especially important to chymistry, as Van Helmont saw the ability to make the salt of tartar volatile and thereby achieve complete dissolutions as a means to produce a solvent with powers akin to the alkahest, the universal dissolvent of great use in transmutation. Because alkalis were not only caustic but also cleansing and purgative, they were thought to have dissolving qualities; if they were volatilized, their corpuscles had been reduced in size (volatility and particle size had been related in chymistry since the Middle Ages), and they would approach corpuscles of the alkahest in size.[35] But, "the practical problem is that alkalis . . . such as salt of tartar (potassium carbonate) are steadfastly non volatile. Salt of tartar can withstand hours of

[28] Sloane 3646 indicates that Plot's interest in Lull went beyond derivative transcription to active experimentation.

[29] Jennifer Rampling, "Establishing the Canon: George Ripley and His Alchemical Sources," *Ambix* 53 (2006): 189–208, on 191.

[30] Michela Pereira, *The Alchemical Corpus Attributed to Raymond Lull* (London, 1989), 3:3.

[31] Ibid., 3:22–3; Johannes Weidenfeld, *Four Books . . . Of the Use of Lully's Spirit of Wine* (London, 1685). For Bulkeley, see Newman, *Gehennical Fire* (cit. n. 2), 44–6, and Thomas W. Jodziewicz, "A Stranger in the Land: Gershom Bulkeley of Connecticut," *Trans. Amer. Phil. Soc.* 78 (1988): 1–106, on 19.

[32] Rampling has discussed a chymical practice advocated by George Ripley, using sericon, a lead compound that when dissolved in spirit of wine (or vinegar) formed sugar of lead. This sugar of lead was thought to combine the mineral and vegetable qualities of materials and was thought could be used to prepare elixirs or to produce a composite water that could transmute base metals into gold. Although Plot wished to achieve a similar objective, his procedure uses strong acids and spirit of wine and is chymically distinct from Ripley's; both authors interpreted the Lullian corpus a bit differently, Plot's interpretation emphasizing the medieval interest in corrosives (discussed below). Rampling, "The Catalogue of the Ripley Corpus: Alchemical Writings Attributed to George Ripley (d. ca. 1490)," *Ambix* 57 (2010): 125–201, on 129–30.

[33] William Newman and Anthony Grafton, "Introduction: The Problematic Status of Astrology and Alchemy in Premodern Europe," in *Secrets of Nature: Astrology and Alchemy in Early Modern Europe*, ed. William Newman and Anthony Grafton (Cambridge, Mass., 2001), 1–37, on 24.

[34] Sloane 3646 (cit. n. 15), 17r, 20r.

[35] William Newman and Lawrence Principe, *Alchemy Tried in the Fire* (Chicago, 2002), 138.

red heat without evaporating in the least."[36] Treating salt of tartar with philosophical spirit of wine was widely seen by chymists as a means of volatilizing salt of tartar. Plot went on to assert that the true philosophical spirit of wine could "dissolve miner-alls . . . joyne them to itselfe, make them volatile," creating a menstruum akin to the alkahest.[37] With such an alkahest, it was possible not only to carry out chymical analy-sis but, by "stopping the process at the right point, and distilling off the alkahest, the 'first essence' of the dissolved substance would be left behind as a crystalline salt."[38] This salt contained the concentrated medicinal properties of the substance. So the philosophical spirit of wine could create a powerful tool for transmutation and *chymiatria* alike.

So how did Plot define the regular and the philosophical spirit of wine? He first describes regular spirit of wine, or ethanol, the twice-distilled product of vinous fer-mentation. Regular vinous distillation produced what Plot called "common brandy," which would then be mixed with different herbs to produce compound liqueurs.[39] However, this was different from the philosophical spirit of wine, the menstruum that would reveal the true spirit of metals. Ever sensitive to the obscurity of chymical writing, Plot warned, "Every thing in the practick writings are to be taken secundum litteram except the word wine or some word synonimous with it as Lunaria."[40]

In the pseudo-Lullian corpus, the *lunaria* is described as follows:

> our Menstruum rectified and actuated, or the heavenly lunaria, which among the Phi-losophers is called Vegetable Mercury, produced from Wine red or white . . . it behooves us to draw out our Menstruum . . . from the Impurities and Phlegm of Wine, through the agency of an Alembic, and to actuate it in distillation with appropriate Vegetables; such as are Apium sylvestre, [Wild parsley], Squilla [sea onion]. . . . Then the Menstruum it-self must be circulated continually in a vessel intended for this purpose for the space of ten days in hot Dung steeped in wine in a water bath [bain marie].[41]

We have seen why the philosophical spirit of wine was important to *chymiatria* in that it could be harnessed to make the alkahest, which in turn could make medicine. Why was the philosophical spirit of wine specifically important to transmutation or chrysopoeia? Pseudo-Lull's work *Liber de secretis naturae seu de quinta essentia* (ca. 1350–1400) was concerned with the question of how perfect metals such as gold and silver could unite in a new individual substance together with the philosophical spirit of wine, called *menstruum vegetabile*.[42] Pseudo-Lull also revealed that he saw the purpose of alchemy as *vegetare seu transmutare* (to vegetate or transmute) and

[36] Ibid.

[37] Sloane 3646 (cit. n. 15), 20r.

[38] Lawrence M. Principe, *The Secrets of Alchemy* (Chicago, 2013), 134.

[39] Sloane 3646 (cit. n. 15), 3r.

[40] Ibid., 44r.

[41] Raymond Lull, "Compendium Animae Transmutationis," in the *Theatrum Chemicum* (Stras-bourg, 1659), 4:172–3; "menstruum nostrum rectificatum et acuatum, seu lunaria coelica quae apud philosophos vocatur Mercurius vegetabilis ortus à vino rubeo vel albo. . . . Sed tamen oportet prin-ceps serenissime prius nostrum menstruum per . . . sordibus vini et phlegmate extrahere per officium alembici, et acuatur in destillatione cum vegetabilibus pertinentibus, quae sunt, Apium sylvestre, Squilla . . . Ex altera parte ipsum menstruum in vase circulationis rotetur continue spacio decem di-erum in fimo calido, vinatico, balneo Mariae."

[42] Michela Pereira, "'Vegetare seu Transmutare': Vegetable Soul and Pseudo-Lullian Alchemy," in *Arbor Scientiae,* ed. Fernando Reboiras, Pere Varneda, and Peter Walter (Turnhout, 2002), 93–119, on 97.

the Philosophers' Stone as *vegetatum* (a living substance).[43] To harness this transformative principle to make the Philosophers' Stone, the use of animated substances, particularly oils from vegetative substances, was important.

Plot thus followed suit. As pseudo-Lull advised in the *Compendium*, Plot described in his notes doing circulatory distillation of ethanol or spirit of wine with such "hott and oily vegetables" as black pepper; the vegetables were thought to actuate or enliven the distilled wine with their properties.[44] Indeed, all the vegetables pseudo-Lull listed either have a distinctive smell or had potent medical properties. Most of the vegetables were "hot" in quality, pungent and spicy in taste; in Galenic medicine, hot herbs and medicines stimulate and warm the metabolism and disperse chills. These qualities may have led to the belief that they were especially efficacious chymically; it did not hurt matters that many of the plants, including the resin of euphorbia, are soluble in alcohol. This distillation process of the spirit of wine with the vegetables would then produce the "lunaria." Weidenfeld, a commentator on the Lullian corpus who was contemporary to Plot, advised "the Oyl drawn out of Oyly Vegetables is by distillation together with the Spirit of Philosophical Wine, circulated into a Magistery by which the Spirit of wine is multiplied."[45]

The *lunaria* or philosophical spirit of wine was widely considered among chymists to consist not only of aqueous principles but also oily ones; Weidenfeld advised the spirit of wine distilled with vegetables "doth by simple digestion divide into two distinct parts, two Oils or Fats, whereof one is the Essence of the thing, the other the Body."[46] And indeed Plot does create something that he refers to in his notes as his *Magisterium V[ini]*, as "oleose perfectly dissolved . . . the use of it is to be a more powerful Menstruum than [ordinary] Spirit of Wine."[47] For Pseudo-Lull and his followers, "this oil or oily moisture," called *unctuositas,* was a basic principle, "the unifying principle of cohesion of living and non living beings, an equivalent of the medieval concept of *humidum radicale* as the support of life."[48] In Pseudo-Lull's *Testamentum*, there was an identification of the oil distilled from vegetables with chymical transmutation.[49]

However, Plot did not stop with the oleose qualities of the *lunaria*: he had a way to the preparation of his own *magisterium vini* that was "more secret," concluding from his experiments that "this process ought to be managed with manifest Acids."[50] In his idea of using acids to make his own menstruum of the spirit of wine, Plot was clearly influenced again by pseudo-Lull. Lull's corpus was heavily indebted to the corpuscular matter theory in Geber's *Summa perfectionis*, and Geber's theory was "immensely appealing to the self-styled atomists and corpuscularians of the seventeenth century."[51] While Plot's manuscripts do not indicate that he speculated about corpuscularianism in his use of an acid to make philosophical spirit of wine, Plot inherited a clear conviction "from medieval alchemy, that processes such as . . . dissolution in

[43] Ibid., 97–8.
[44] Sloane 3646 (cit. n. 15), 17r.
[45] Weidenfeld, *Four Books* (cit. n. 31), 17.
[46] Ibid.
[47] Sloane 3646 (cit. n. 15), 3r.
[48] Pereira, "'Vegetare'" (cit. n. 42), 105.
[49] Ibid., 105–6.
[50] Sloane 3646 (cit. n. 15), 20r.
[51] William Newman, *Atoms and Alchemy: Chymistry and the Experimental Origins of the Scientific Revolution* (Chicago, 2006), 43–4.

corrosives provided ocular testimony to the analysis of matter" into its most pure and basic form.[52] Hermann Kopp claimed that Lull supposedly described the heating of ethanol with sulfuric acid or vitriol to produce *oleum vitrioli dulce verum* (diethyl ether; Claus Priesner, however, stated that "only the residue, and not the distillate, is described"; there is thus doubt whether ether was formed or escaped unnoticed).[53]

Nevertheless, oily or unctuous substances combined with acids were considered important for chrysopoeia. The *Epistola accurtacionis lapidis,* attributed to Lull, provided a further explanation:

> there exists a great affinity between the Spirit of Vitriol, and the Nature of Gold. . . . For this reason the Spirit of Vitriol being joined with the Spirit of Aqua ardens [spirit of wine] inspissates [thickens or condenses] it, and makes it quickly adhere to Gold, so as to be fixed with it; this is a very excellent way of Abbreviation.[54]

The product of the distilling of the oil of vitriol with spirit of wine was an unctuous substance that could combine with gold, creating a substance that could abbreviate transmutation.

Plot adapted the Lullian process, his notes showing that, like Lull, he used spirit or wine or ethanol but mixed it instead with double *aqua fortis,* a stronger version of nitric acid.[55] Although the theoretical basis of his work was taken from Lull, Plot's innovative use of nitric acid reflected a trend in seventeenth-century chymistry in which acids became more important as chemical reactions in solution began to predominate over older distillation processes; Newton himself made frequent use of mineral acids.[56] Using nitric acid, Plot created the sweet spirit of niter, or spirit of nitrous ether (ethyl nitrite).[57] Johannes Kunkel, in his *Epistola contra spiritum vini sine acido* (Leipzig, 1681), has generally been considered to be the first to observe that a liquid that swims on the surface of water may be obtained from this mixture, but this observation "remained unnoticed, because the nitric ether, largely used as a medicine, was obtained by distilling a considerable quantity of alcohol with a small quantity of nitric acid, and was, therefore, only obtained in dilute alcoholic solution."[58] In MS Sloane 3646, Plot listed several recipes for the process, including "spirit of wine: part 1. of Aqua fortis parts 4, destill nine times, or till the matter remains in the bottom like an oyle, which destills by an Alembick with more aqua fortis."[59] Further

[52] Ibid., 44.

[53] H. Kopp, *Geschichte der Chemie* (Brunswick, 1843–7), 4:299, cited by Claus Priesner, "Spiritus Aethereus—Formation of Ether and Theories on Etherification from Valerius Cordus to Alexander Williamson," *Ambix* 33 (1986): 129–52, on 129, n. 2.

[54] Raymond Lull, "Epistola accurtatoria ad Regem Neopolitanum," in Raymond Lull and Michael Toxites, *Raimundi Lullii Majoricani philosophi sui temporis doctissimi . . .* (Basel, 1600), 327–8: "et etiá[m] magna est concordia inter spiritum vitrioli et naturá[m] auri . . . Ideo spiritus vitrioli coniunctus spiritui aquae ardentis inspissat eum, et adhaerere facit eum citò auro, ut secum fixetur. Et crede mihi, quòd haec curtatio est praeexcellens in arte quantum ad aurum albricum."

[55] *Aqua fortis* was made by distilling niter and oil of vitriol together.

[56] Norma Emerton, *The Scientific Reinterpretation of Form* (Ithaca, N.Y., 1984), 184; William Newman, "Experiments in Mineral Acids," The Chymistry of Isaac Newton, http://webapp1.dlib.indiana.edu/newton/reference/mineral.do (accessed 5 March 2012).

[57] H. E. Roscoe and C. Schlorlemmer, *Treatise on Chemistry* (New York, 1884), vol. 3, pt. 1, 357, noted: "The compound formed by this action of nitric acid on alcohol is . . . not ethyl nitrate . . . but ethyl nitrite, one part of the alcohol being oxidized, and the nitrogen trioxide, thus formed, combining with another part of the alcohol in the following way: $2C_2H_5OH + N_2O_3 = 2C_2H_5NO_2 + H_2O$."

[58] Ibid., 357. Plot's experiments with the sweet spirit of niter were performed at the same time as Kunkel's work was published, making it difficult to attribute priority of discovery.

[59] Sloane 3646 (cit. n. 15), 21r.

down the page, he indicated to "destill aqua fortis from Vitriol and [salt] peter into the spirit of wine, afterwards digest and destill."[60]

Plot's experiment was dangerous, as a reaction of ethanol and nitric acid is exothermic and can become explosive if the concentration of nitric acid reaches over 10%.[61] Plot even noted that in his reading about making the philosophical spirit of wine, "they have almost in all places for pitty and charity sake admonished the unwary and ignorant to be cautious of themselves, yea alltogether abstaine from these ignivomous dragons, lightenings, and thunders, unless they know how to handle these beasts ingeniously and gently."[62] Plot then put the "distilled oleose" in a "Circulatory glasse"[63] or pelican to concentrate and refine it. He stated, "by circulating it in Balneo [in a water bath or bain marie] so long till you see in the glasse a separation that is a cleare oyle swimming on a Muddy water: which clear Oyle being taken by a separating funnell is our spirit or essence of wine."[64] This is the same reaction noted by Kunkel. Boyle, who was looking for a similar menstruum that contained the subtle essence and efficacy of its constituent parts but that would also be safe to use for medicines, made the sweet spirit of niter, utilizing a similar process:

> Take one ounce of strong Spirit of Nitre . . . and put to it by little and little (which caution if you neglect, you may soon repent it), and another ounce of such rectified Spirit of Wine . . . you may . . . unite them exquisitely into one liquour [of] . . . a Vinous tast[e], very pleasing, as if it belonged to some new or unknown Spice.[65]

PLOT'S SECRET CHYMISTRY

Boyle, unlike Plot, never indicated the sweet spirit of niter had anything to do with chrysopoeia and in 1676 freely published the preparation for it. Plot, however, apparently thought otherwise and was particularly concerned to keep the details of his search for the philosophical spirit of wine and other menstrua under cover, saving his "most secret secret of the more abstruse Chymistry or of the first and onely matter of Menstruums."[66] Plot's secrecy had largely to do with the economic necessities of his position. As no one formally read chemistry at Oxford before the late nineteenth century, Plot's chymistry chair was in keeping with the "seventeenth-century University's encouragement of what one might call 'mind-broadening' extracurricular subjects."[67] In this spirit, as part of his appointment in 1683, Oxford defrayed the initial cost of equipping his "public" laboratory in the basement of the Ashmolean (accessible at least to all members of the University) by providing the tin, copper, and iron vessels (e.g., an alembic, a reverberatory furnace, and an iron digester).[68] Plot lived, taught, and collected in the Ashmolean, from the time that it opened in 1683

---

[60] Ibid.

[61] Pierre Macquer warned to add the acid gradually to the ethanol. Pierre Macquer and Andrew Reid, *Elements of the Theory and Practice of Chymistry* (London, 1758), 2:255–6.

[62] Sloane 3646 (cit. n. 15), 20r.

[63] A retort having the neck bent back to reenter its lower part.

[64] Sloane 3646 (cit. n. 15), 20r.

[65] Robert Boyle, *Experiments, Notes, &c. about the Mechanical Origine or production of divers qualities* (London, 1676), 23–4.

[66] Sloane 3646 (cit. n. 15), 55r.

[67] Robert Williams, Allan Chapman, and John Rowlinson, *Chemistry at Oxford* (London, 2009), 40.

[68] Plot to Arthur Charlet, 5 October 1695, Bodleian MS Ballard 14, 16r, printed in Gunther, *Dr. Plot* (cit. n. 13), 404–6.

until 1690. However, for running expenses and his own salary, he was expected to rely on instructional fees and upon the sale of chymical preparations and medicines that he could manufacture in the laboratory, assisted by an operator and chymical demonstrator named Christopher White, who also was expected to supply his own laboratory equipment.[69] Simply put, Plot needed funds and hoped that the manufacture of menstrua as well as instructional fees could support his academic post. His philosophical spirit of wine was equivalent to a valuable trade secret that he could not only use to make medicines but also to engage in *chrysopoeia*. Unlike Boyle, who was independently wealthy and publicly denied any personal interest in financial by-products of such processes, and Newton, whose position as Lucasian Professor at Cambridge was in no way dependent on his chymical research, Plot needed to make a living.[70]

Plot was also ruthlessly ambitious in pursuing gain. Although he had a reputation for being outwardly genial, he was thought quite grasping in character; "a willingness even to embrace Catholicism in order to obtain advancement was imputed to him."[71] In 1689, James Bobart, who directed Oxford's botanical gardens, remarked that Plot's ambitions meant that he was not entirely dedicated to his post: "he does noe good in his station, but totally neglects it, wandring abroad where he pleases."[72] Edward Lhwyd, who served as Plot's deputy keeper of the Ashmolean and was considered by Hans Sloane to be "the best naturalist in Europe," was continually doubtful that Plot would treat him fairly; in particular, Lhwyd worried Plot would not give him the amount of the museum's takings promised to him during his seven years of service as deputy keeper.[73] Lhwyd wrote to his friend Martin Lister, "And then I hope the Dr will let me have the one half of the money receiv'd (for as to the whole which he promisd us at London, I know him better than to expect any such matter).[74] After Plot resigned his post as Professor and Keeper in 1690 to marry a Kent heiress, Lhwyd remarked to Lister, "I think he's a man of as bad Morals as ever took a Doctors degree. I wish his wife a good bargain of him; & to my self that I may never meet with the like again."[75] Lhwyd went on to complain that Plot had been reimbursed for an "Arabick Monument" that Plot had presented as his supposed charitable donation to the University.

Papers in MS Sloane 3646 also give us a sense of Plot's lofty and grasping ambitions for his laboratory and chymical work. Plot wrote a draft petition to the King seeking assistance to create an endowed Hermetic "Colledge" that made direct reference to Sir Francis Bacon's Solomon's House in Bensalem as described in the *New Atlantis*. This institution would be partially based upon Plot's experiences in the Royal Society and involve members of the "Colledge of physicians" as well as those who belonged to the "Society of the Sophic" or chymical adepti.[76] In his petition, Plot then mentioned a menstruum recently discovered as "being the same or equivalent to . . . the grand liquor Alkahest . . . an inestimable dissolvent"; Plot compared it to

---

[69] Gunther, *Dr. Plot* (cit. n. 13), 355.

[70] Principe, *Aspiring Adept* (cit. n. 2), 185.

[71] *Oxford Dictionary of National Biography*, s.v. "Plot, Robert (*bap.* 1640, *d.* 1696)," by A. J. Turner, http://www.oxforddnb.com/view/article/22385 (accessed 8 February 2014).

[72] Edward Lhwyd to Martin Lister, 15 August 1689, Bodleian MS Lister 35, 134–5.

[73] Thomas Hearne, *Hearne's Collections*, ed. Thomas Doble (Oxford, 1701–7), 1:244.

[74] Lhwyd to Lister, September 1689, Bodleian MS Lister 3, 165–6.

[75] Lhwyd to Lister, 17 January 1691, Bodleian MS Lister 36, 8r.

[76] Hoppen, "Nature" (cit. n. 12), 260.

that used by Paracelsus to prepare "noble and generous medicines."[77] Plot claimed the "Sophi" valued it at "thousands of pounds the pint"; whether this was a vain boast, or he really was in contact with other trusted adepti, is unclear. Although we do not know precisely this substance's composition—perhaps it was even his own sweet spirit of niter—clearly Plot identified it as a menstruum of great importance. He indicated to the King that the proposed hermetic institution could manufacture his menstruum, thus restoring "the true legitimate physicians (who are not like farryers to practice with a bundle of receipts, but are taught in the experimentall schoole of nature) . . . to their . . . due repute."[78]

Apart from royal patronage, Plot realized his chymical secrets could be valuable currency to those less exalted than the King. MS Sloane 3646 shows that he entered into several chymical contracts with unnamed individuals, before and during his time in post in Oxford.[79] First, dated 1677, there is a tripartite agreement in blank, which is a template contract related to the making and sale of "Dropps and Chymicall Medicines," in which all parties agree to provide equal shares in setting up a laboratory and split the profits accordingly.[80] As Plot had a doctorate in civil law, it is not surprising he was using legal safeguards for his secret processes. In another undated agreement in Plot's hand, it was proposed that for the "competent sum of money" of £350, his partners would be given the recipe for Plot's secret menstruum, which they would use to prepare the alkahest and the Grand Arcanum to use in chrysopoeia, subsequently sharing the proceeds with Plot.[81] To sweeten the deal, for a mere £50 more, Plot's partner would be shown how to prepare "noble Medicines"; Plot continued, "and if we consider what estates some ordinary fellows from trifleing things have purchased, witness Russel with his powder, Daffy with his elixir, Bromfield with his pills, then how much more probable will it be to raise considerable advantage from some of the choycest medicines."[82] (Plot here was referring to a number of popular patent medicines.) He noted, "doubtless in a little time great profitts may be made" and then indicated that if his partner would "manage the Laboratory he will have (over and besides the share in the advantage aforesaid) £20 a year allowed him by the rest of the partners for the management thereof."[83] As a further inducement, Plot indicated that the King had promised him a patent for a preparation that secured ships "from the worme."[84] Plot then concluded, "Then what would any rational man have more."[85]

In this contract, Plot was careful to distinguish himself from "vain boasting chymists," saying any fraud perpetrated on his part would be a "blott in his Eschution," an appropriate metaphor, as Plot would later become registrar of the College of Heralds.[86] A note below the agreement indicated that the other party promised "that not any Aqua fortis, nor any other corrosive liquor is us'd through out the whole work of the Elixir."

---

[77] Sloane 3646 (cit. n. 15), 76r.
[78] Ibid., 76v.
[79] Taylor, "Alchemical Papers" (cit. n. 12), 75.
[80] Sloane 3646 (cit. n. 15), 81–2.
[81] Ibid., 77–8.
[82] Ibid., 78v.
[83] Ibid.
[84] Shipworms are mollusks, usually *Teredo navalis*, notorious for boring into and destroying wood immersed in seawater. *Encyclopedia of Life*, s.v. "*Teredo navalis*," http://eol.org/pages/439957 /details (accessed 28 June 2013).
[85] Sloane 3646 (cit. n. 15), 78v.
[86] Ibid., 77v.

Corrosive liquors like *aqua regia* (a mixture of nitric and hydrochloric acids) was well known as a "vulgar menstruum" that would dissolve gold, and it was often fraudulently passed off as a more exalted substance in the path to the Philosophers' Stone.

F. Sherwood Taylor speculated that Plot's use of legal safeguards in his contracts was due to the fact that Plot only possessed a bare chymical secret and that he thus was unsure of success.[87] However, it seems rather that Plot's caution was due to the fact that he saw it merely as a commercial, albeit risky, venture. Plot first acknowledged that "no man upon a bare supposition" should hazard the money, especially "if it were upon a person who had no other way of livelyhood."[88] He then compared the chymical quest to investing in the voyages of ships that could come back laden with treasure from the New World. He wrote, "How many vessels richly fraught, loaden with the Eastern and Western treasures whose owners and venturers have like their sayles swelled with expectation of great advantage, have yet by storms been driven upon rocks?"[89] It is interesting to speculate whether Plot had in mind Francis Bacon's emblem in the frontispiece of his *Instauratio magna* (1620), in which a ship sailed out between the "Pillars of Hercules" at the western end of the Mediterranean Sea, serving as a metaphor for discovery with its rewards and risks. As Tara Nummedal has shown in her analysis of chymistry and fraud in the Holy Roman Empire, such contracts were common. Adepti could "point to their contracts as proof that they were . . . genuine practitioners," and by addressing the concerns of patrons and practitioners, contracts like Plot's "played a central role in facilitating chymical practice as not just private study, but as a collaborative entrepreneurial endeavor."[90]

In MS Sloane 3646, there followed another draft agreement, dated ca. 1683, the year of Plot's appointment as professor of chymistry at Oxford. Here Plot promised to "shew" another individual the secrets of chrysopoeia in exchange for his usual fee of £350 and "half the charges of the worke," though he noted "no great varietie of furnaces nor much charge is required."[91] Plot was apparently also not only using corrosives in his work with philosophical wine but engaging in mercurialist chrysopoeia, in which the key to attaining the Philosophers' Stone is preparing a philosophical mercury from ordinary mercury via purification and animation. Plot indicated in the agreement that the person he was dealing with in making "this great Arcanum should be willing to worke some part at my laboratory here, and the rest in his Laboratory in the country when he shall go down, which perhaps may be shortly" and "should be enjoyned to faithful secrecy."[92] As Taylor has indicated, the dating of the contract to 1683, the fact that Plot wrote of "my laboratory here" and the other party was about to "go down," may suggest Plot wrote from his Oxford Laboratory at the Ashmolean.[93]

There is other evidence that suggests this agreement was made when Plot was in his official capacity as professor of chymistry. Plot promised in his agreement to show his fellow adept the meaning of the alchemical "wolfe," a common symbol for the antimony ore, stibnite (Sb_2S_3).[94] As I have shown elsewhere, Plot speculated in

[87] Taylor, "Alchemical Papers" (cit. n. 12), 76.
[88] Sloane 3646 (cit. n. 15), 77v.
[89] Ibid.
[90] Tara Nummedal, *Alchemy and Authority in the Holy Roman Empire* (Chicago, 2007), 116.
[91] Sloane 3646 (cit. n. 15), 80r; Taylor, "Alchemical Papers" (cit. n. 12), 75.
[92] Sloane 3646 (cit. n. 15), 80r.
[93] Taylor, "Alchemical Papers" (cit. n. 12), 69.
[94] Sloane 3646 (cit. n. 15), 79r.

his *Natural History of Oxfordshire* (1677) that fossils were not the result of animal or plant remains but were formed by the interaction of salt crystals to form their different shapes.[95] He then postulated that the regulus or stellate form of antimony could be the source of star-shaped fossils, such as pointed sea lily fossils and sea urchins, so he clearly was au fait with antimonial chymistry.[96] In a recent archaeological analysis of early modern crucibles found in Plot's laboratory, Marcos Martinón-Torres has also found a thin corroded layer of antimony oxide (Sb_2O_3) and lime (CaO), together with silica (SiO_2) and small amounts of potash (K_2O).[97] This could have been due to the making of the emetic medicament of the glass of antimony, or "this antimony could also be linked with the roasting or calcination of the mineral stibnite, with a view to producing antimonial mercury," a form of "incalescent" or philosophical mercury that also interested Boyle and Newton.[98] Martinón-Torres argues that the crucibles are "likely to date" from the late seventeenth century and are thus "connected" to Plot's work; Plot's manuscript contract and references to antimony in his *Natural History* support this archaeological evidence.[99]

PLOT'S PUBLIC CHYMISTRY

Although a professor at Oxford, Plot was also clearly a chymical entrepreneur both before and after his appointment, the terms of his university post requiring that he make the Ashmolean laboratory profitable. Concerned as he was to keep his trade secrets, part of the chymical work he did at Oxford was very public, involving not only his lectures to students but also research collaboration between the OPS and the Royal Society. While the Ashmolean was being built from 1679 to 1683 and Oxford was making provisions for a meeting room for the OPS, Plot exchanged correspondence with natural philosophers in London. He worked in earnest to link the OPS with the Royal Society and regularly gave the Royal Society accounts of the scientific proceedings of Oxford's "ingenious assembly," which discussed problems of meteorology, physiology, chymistry, and mineralogy. Plot's colleague Martin Lister told him in a letter of October 1683, "I can assure you the best of our Entertainment [in the Royal Society] . . . was your Letter. Your new Oxford Society will be of great use, it will excite this other here, and emulation is the great promoter of learning."[100] In this manner, Plot became familiar with experiments being done at the Royal Society, experiments that he himself was also performing.

This collaborative work included experiments with the spirit of wine. We recall that Kunkel had written a treatise on this subject in 1681, and in 1684, he was in correspondence with the Royal Society concerning the reaction of spirit of wine with syrup of violets (an indicator of acidity/alkalinity), milk, and water. The OPS received a letter from Francis Aston, the Royal Society secretary, on 18 November 1684, "giving an account of severall experiments mentioned in a book lately writ-

---

[95] Robert Plot, *The Natural History of Oxfordshire* (Oxford, 1677), 111; A. M. Roos, "Salient Theories in the Fossil Debate in the Early Royal Society," in *Controversies within the Scientific Revolution,* ed. Victor Boantza, Marcelo Dascal, and Adelino Cattani (Amsterdam, 2011), 151–70.

[96] Plot, *Oxfordshire* (cit. n. 95), 122–3.

[97] Martinón-Torres, "Inside Solomon's House" (cit. n. 12), 46.

[98] Ibid.

[99] Ibid., 22.

[100] Bodleian MS Ashmole 1813, 64r.

ten by Kunckell."[101] Frederic Slare, the chemistry curator at the Royal Society, performed several of the experiments to test Kunkel's results. As Birch reported in the meeting minutes, the spirit of wine and milk indeed curdled, syrup of violets and spirit of wine turned green, and spirit of wine and water mixed together grew hot, an accurate observation as the reaction is exothermic.[102] On 9 December 1684 and 12 May 1685, the OPS repeated the experiments but found that the syrup of violets and spirit of wine turned red rather than green, indicating an acid component to the mixture rather than an alkali. (In Boyle's *Experiments and Considerations Touching Colours* [1664], he established an early pH indicator, in which liquids of a deep blue or purple color would turn red when an acid was added or green at the addition of an alkaline.) The OPS minutes reported, "but spirit of wine was so farr from turning it [syrup of violets] green, as Kunkel saies . . . that it made it immediately become as red, as when Acids were mingled with it, or that of Vinegar."[103] These experiments may have indicated to Plot an affinity between spirit of wine and its ability to be turned into a philosophical spirit of wine with acidic substances.

There were also several reported attempts made by the OPS to volatilize alkali salts such as salt of tartar, sal ammoniac, and hartshorn by adding them to spirit of wine "without any visible motion of the mingled liquors."[104] We recall that the ability to make salt of tartar volatile was seen as a means to produce a solvent akin to the alkahest. In the case of the society's experiments, they had little luck, as salt of tartar is insoluble in ethanol, sal ammoniac [ammonium chloride] only sparingly soluble, and ammonium bicarbonate [salt of hartshorn] is not. Although this may have led Plot to question whether spirit of wine could be involved in the volatilization of alkali salts, he may have reasoned that the regular spirit of wine was utilized in the trials at the OPS, not the secret philosophical spirit of wine or other chymical menstrua that he had created and was trying to sell.

While Plot would not reveal the particular secrets of his chrysopeia, it was certainly not a secret to his colleagues that he was interested in such processes. On 17 March 1684/5, Plot presented to the OPS a "wish list" of "arcana and desiderata in chymistry."[105] And, in its meeting of 18 November 1685, the Royal Society meeting minutes reveal that the "secretary was desired to procure a copy of Dr Plot's Desiderata in chemistry," and the list was also presented to the Dublin Philosophical Society.[106] Much of the list was devoted to aspirational processes and products of chrysopoeia and *chymiatria*, including:

> to make a Universall Medicine
> a Mercury which . . . may be melted into Gold
> to make Aurum Potabile (potable gold, a panacea).

Plot's desiderata list was nothing new. Bacon had already printed the idea of the wish list in his *Advancement of Learning*.[107] As Vera Keller has argued, Bacon in-

[101] Gunther, *The Philosophical Society* (cit. n. 13), 104.
[102] Thomas Birch, *History of the Royal Society* (London, 1756–7), 3:328.
[103] Gunther, *The Philosophical Society* (cit. n. 13), 148.
[104] Ibid.
[105] Ibid., 130–2.
[106] Birch, *History* (cit. n. 102), 4:409.
[107] Vera Keller, "The New World of Sciences," *Isis* 103 (2012): 727–34, on 730.

tended his desiderata to be fulfilled gradually and collaboratively for collective advancement, rather than presenting claims about what any self-interested individual could do.[108] Boyle also wrote a desiderata list not limited to chymistry but devoted to natural philosophy as a whole. Like Bacon's, Boyle's list was not predictive but a tool of skepticism concerning our ability to know the "limits of the real," suspending "objects in a state of continued doubt that invited further investigation."[109] Whilst Plot's desiderata could, in Keller's terms, be seen as a communal wish list presented to three scholarly societies, allowing them to attempt seemingly impossible feats without claiming ability to achieve any results, Plot's item 32 is "to make Lully's lunaria."[110] As we have seen, Plot did not think this process impossible. Plot also had on his wish list items that were not *chrysopoetic*, for instance, "staining marble black," and he met with artisans about how to accomplish this goal. In a Royal Society meeting of 25 May 1681, Plot related that one Mr. Bird, an Oxford mason, "had very much perfected the invention of staining . . . but he could not yet find out a way of staining a perfect black."[111] Undaunted, Plot later sent a letter to the Royal Society on 10 November 1683 that contained an account of him having tinged white marble a quarter of an inch deep.[112] Unlike Boyle, who was testing limits of knowledge in a philosophical sense for Baconian "experiments of light," Plot's motivation was largely economic and practical: Bacon's "experiments of fruit."

In the seventeenth century, despite the fact that chymical procedures were increasingly made available to larger audiences, the secrecy of the adepti still survived. But this secrecy was maintained for different reasons, depending on the chymist's circumstances. In the case of Plot, his maintenance of the boundaries between public and private knowledge was informed by his desire to preserve trade secrets that not only would fund his livelihood as Professor of Chymistry, but that also would enhance his scholarly prestige. Plot's contemporary, Thomas Hearne, recalled that Plot had failed to return a fossil that he borrowed from Magdalen College, claiming it "was a Rule among Antiquaries to receive and never restore."[113] As a chymist, Plot also was content to receive knowledge, but generally he would only restore it for a price.

[108] Ibid., 728.
[109] Ibid., 732.
[110] Ibid.
[111] Birch, *History* (cit. n. 102), 4:88.
[112] Ibid., 4:225.
[113] Hearne, *Hearne's Collections* (cit. n. 73), 1:67.

The End of Alchemy?

The Repudiation and Persistence of Chrysopoeia at the Académie Royale des Sciences in the Eighteenth Century

*by Lawrence M. Principe**

ABSTRACT

The general abandonment of serious endeavor toward metallic transmutation represents a major development in the history of chemistry, yet its exact causes and timing remain unclear. This essay examines the fate of chrysopoeia at the eighteenth-century Académie Royale des Sciences. It reveals a long-standing tension between Académie chemists, who pursued transmutation, and administrators, who tried to suppress it. This tension provides background for Etienne-François Geoffroy's 1722 paper describing fraudulent practices around transmutation. Although transmutation seems to disappear after Geoffroy's paper, manuscripts reveal that most of the institution's chemists continued to pursue it privately until at least the 1760s, long after widely accepted dates for the "demise of alchemy" in learned circles.

The transmutation of metals has provided considerable challenges not only for hopeful chrysopoeians endeavoring to produce gold but also for historians of science endeavoring to understand this crucial part of the history of chemistry. Thanks to groundbreaking work carried out during the past forty years or so, we now understand many more of the dimensions of the chrysopoetic quest, perhaps most significantly its nature as a serious and rational endeavor, undergirded by coherent theoretical and observational foundations, to understand the natural world and to make use of its powers. In short, transmutational alchemy's important position in the history of science is no longer questioned. Pursued for a millennium and a half in various cultural, intellectual, and theoretical contexts, chrysopoeia represents one of the central endeavors of the chymical tradition.[1] But now that the important role of chrysopoeia for the history of science has been established, this reembrace of a formerly ostracized subject underscores another challenge, namely, how then to explain the disappearance of metallic transmutation from the normal operations and goals of

* Department of the History of Science and Technology, 301 Gilman Hall, Johns Hopkins University, 3400 N. Charles Street, Baltimore, MD 21218; Lmafp@jhu.edu.

[1] For a survey of chrysopoeia in the history of alchemy, see Lawrence M. Principe, *The Secrets of Alchemy* (Chicago, 2013); for an overview of the new historiographical understanding of alchemy, see Principe, "Alchemy Restored," *Isis* 102 (2011): 305–12.

chymists. This issue proves to be complex and resists simple and universal answers. This essay cannot hope to address all the questions and local contexts involved and so limits itself to investigating this issue in one locale—albeit a central and highly influential one for chemistry as a whole—namely, the Parisian Académie Royale des Sciences.[2]

In most general histories of chemistry, the date offered for the disappearance of chrysopoeia from serious consideration ranges from the late seventeenth to the early eighteenth century. At one time, the 1661 publication of Robert Boyle's *Sceptical Chymist* was routinely seen, at least by English-language authors, as a watershed moment marking the repudiation and subsequent demise of "alchemy" (which in this context meant primarily metallic transmutation). Yet we now know that the primary targets of the *Sceptical Chymist* were not chrysopoeians but Paracelsian pharmacists, and far from repudiating alchemical ideas and the Philosophers' Stone, Boyle himself made use of the former and pursued the latter throughout his adult life.[3] Another putative landmark, this time in the French context, is the 1679 edition of Nicolas Lémery's *Cours de chymie*, where, in the revised chapter on gold, Lémery launches a broad attack on transmutation.[4] Lémery's objections cite a select few chrysopoetic notions—by no means the most prevalent ones—and ridicule them as foolish and primitive. He then moves on to fraud, extending the well-known and long-term connection of cheating practices to transmutation to claim that all chrysopoeia is simply fraud. The exact reasons for Lémery's sudden attack remain unclear, but his views were nothing new in 1679. Accusations of fraud and vigorous debates over the possibility of transmutation and the reality of the Philosophers' Stone did not appear suddenly in the seventeenth century; they had been the constant companion of chrysopoeia since the Middle Ages.[5] What is striking is that theory-based arguments against transmutation changed very little during all that time—some of the same arguments against transmutation cited by Geber in the thirteenth century were voiced by Thomas Erastus in the sixteenth and by various other antichrysopoeians in the seventeenth.

The fact that Lémery's critique depends primarily upon ridicule and accusations of fraud highlights a key point: one searches in vain for contemporaneous theoretical innovations or new experiments that were used—or even could be used—to discredit chrysopoeia.[6] Some historians, following the lead of Hélène Metzger, have thought of Lémery as a key break in the history of chemistry due to his supposed embrace of "mechanical" or Cartesian theories, which served to doom a supposedly vitalistic

[2] For relevant scholarship on the early Académie, see, for example, Roger Hahn, *The Anatomy of a Scientific Institution: The Paris Academy of Sciences, 1666–1803* (Berkeley and Los Angeles, 1971); Alice Stroup, *A Company of Scientists: Botany, Patronage, and Community at the Seventeenth Century Parisian Academy of Sciences* (Berkeley and Los Angeles, 1990); David J. Sturdy, *Science and Social Status: The Members of the Académie des Sciences, 1666–1750* (Woodbridge, 1995); for chemistry at the early Académie, see Frederic L. Holmes, *Eighteenth-Century Chemistry as an Investigative Enterprise* (Berkeley, Calif., 1989).

[3] Lawrence M. Principe, *The Aspiring Adept: Robert Boyle and His Alchemical Quest* (Princeton, N.J., 1998).

[4] Nicolas Lémery, *Cours de chymie*, 3rd ed. (Paris, 1679), 57–60; on Lémery, see Michel Bougard, *La chimie de Nicolas Lemery* (Turnhout, 1999).

[5] On fraud, see Tara Nummedal, *Alchemy and Authority in the Holy Roman Empire* (Chicago, 2007), 48–72.

[6] The one exception in regard to experiments occurs over fifty years later in Boerhaave's 1730s processes with mercury (see below). In any event, these seem to have had little if any effect.

"alchemy" with its pretensions to metallic transmutation.[7] Yet Lémery's deployment of mechanism, which can hardly be called Cartesian save in a very superficial sense, is fitful at best and only sporadically used to undergird a text that is primarily a recipe book concerned with the preparation of pharmaceuticals, not with the advancement or development of chemical theories.[8] Moreover, mechanistic notions could have little impact upon belief in transmutation. Boyle, whose devotion to mechanical explanations was far more profound and coherent than Lémery's, found no conflict of this system with his quest for transmutation and the Stone. Quite the opposite, seventeenth-century mechanical views of matter and its transformations, far from weakening belief in transmutation, actually strengthened it. For if all substances are composed of the same universal matter it would then be possible to transform anything into anything—indeed, to accomplish chymical changes far beyond what most chrysopoeians considered possible. Time and time again, chrysopoeia proved itself extraordinarily flexible in terms of accommodating itself to, and actually benefiting from, new theoretical conceptions. Thus we must look elsewhere for more substantial evidence of chrysopoeia's decline and the causes thereof.

The early eighteenth century holds more indications of a serious decline of transmutation than the late seventeenth. For example, the flood of books on chrysopoeia that appeared throughout the seventeenth century rapidly dwindled to a trickle in the first decades of the eighteenth. At the same time, repudiations of gold making appeared from a range of prominent authorities: in France, in 1722, Etienne-François Geoffroy's paper "Des supercheries concernant la pierre philosophale" ("Some Cheats Concerning the Philosophers' Stone"); in Germany, in 1726, Georg Ernst Stahl's "Bedencken von der Gold-Macherey" ("Thoughts on Gold-Making"); and in the Netherlands, although published in England's *Philosophical Transactions* in the 1730s, Herman Boerhaave's long-term experiments with quicksilver that steadfastly refused to produce any gold.[9] Of these three important publications, it is Geoffroy's paper that is most frequently cited by historians as marking the end of "alchemy." Yet despite the attention paid to Geoffroy's paper, it tends to be used rather iconically— the way the *Sceptical Chymist* was often used—and has not hitherto been fully contextualized. Let me now begin by trying to illuminate the paper's background and the foregoing position of chrysopoeia within the Académie.

Etienne-François Geoffroy (1672–1731) was interested in chymistry from a young

---

[7] Hélène Metzger, *Les doctrines chimiques en France du début du XVIIe à la fin du XVIIIe siècle* (Paris, 1923; repr., Paris, 1969), esp. 229–338.

[8] On this point, see Lawrence M. Principe, "A Revolution Nobody Noticed?" in *New Narratives in Eighteenth-Century Chemistry,* ed. Lawrence M. Principe (Dordrecht, 2007), 1–22; and Bernard Joly, "L'anti-Newtonianisme dans la chimie française au début du XVIIIe siècle," *Archives internationales d'histoire des sciences* 53 (2003): 213–24, esp. 215–6. On Descartes and chemistry, see Bernard Joly, *Descartes et la chimie* (Paris, 2011).

[9] Etienne-François Geoffroy, "Des supercheries concernant la pierre philosophale," *Mémoires de l'Académie Royale des Sciences* (hereafter *MARS*) 24 (1722): 61–70. Georg Ernst Stahl's "Bedencken von der Gold-Macherey," in *Chymischer Glückshafen* (Halle, 1726); on Stahl and transmutation, see Kevin Chang, "Georg Ernst Stahl's Alchemical Publications: Anachronism, Reading Market, and a Scientific Lineage Redefined," in Principe, *New Narratives* (cit. n. 8), 23–43; on Boerhaave, see John Powers, *Inventing Chemistry: Herman Boerhaave and the Reform of the Chemical Arts* (Chicago, 2012), esp. 170–91, and "Scrutinizing the Alchemists: Herman Boerhaave and the Testing of Chymistry," in *Chymists and Chymistry: Studies in the History of Alchemy and Early Modern Chemistry*, ed. Lawrence M. Principe (Canton, Mass., 2007), 227–38.

age. In the 1690s, his father—a Parisian apothecary—had him tutored informally by the Académie's chief chymist Wilhelm Homberg. In 1699, Geoffroy was admitted to the Académie as Homberg's student, was quickly promoted to associate chemist, and worked often in concert with Homberg until Homberg's death in 1715, whereupon Geoffroy received Homberg's pensionnaire place in the Académie. Geoffroy is perhaps best known for his *Table des rapports* published in 1718, a chart of the relative "affinities" between various chemical substances, which would prove highly influential for much of the rest of the eighteenth century.[10]

Geoffroy's mentor, Wilhelm Homberg (1653–1715), was in his day the Académie's most prominent and productive chymist, and perhaps the most intriguing. He was also a firm believer and active experimenter in chrysopoeia. In the 1690s, he began writing a textbook of chymistry where he declared that alchemy is "a particular science within chemistry that teaches the transmutation of metals" and investigated in detail the method for preparing the Philosophers' Stone proposed by Eirenaeus Philalethes.[11] Although he concluded that section of his text by saying that he had not been successful with the process, a few years later in 1704, he published—in the Académie's memoirs—an account of his successful transmutation of a small quantity of mercury into true gold using a predominantly Philalethean process and based much of his comprehensive and influential theory of chemistry (expounded serially in his *Essais de chimie*) upon these transmutational results.[12] Thereafter, Homberg carried out extensive experiments to determine the composition of metals, always in terms of the traditional Mercury-Sulphur theory. Many of these experiments were done in concert with Philippe II, Duc d'Orléans, the nephew of Louis XIV, who had engaged Homberg as his physician and chemistry tutor, given him an apartment in the Palais Royal, and built there a magnificent laboratory in which they worked together on chemical experiments, including transmutational ones.

Geoffroy assisted Homberg in many experiments and carried out ancillary ones exploring the same or related issues. In 1704, for example, Geoffroy published a paper claiming that iron could be synthesized from nonmetallic ingredients, thus launching a vitriolic dispute with Louis Lémery (1677–1743, son of the more famous Nicolas) that would continue for the next four years.[13] In 1713, Geoffroy presented a memoir to the Académie on the tinctures, or metallic sulphurs, extracted from various metals

---

[10] On Geoffroy, see Bernard Joly, "Etienne-François Geoffroy (1672–1731): A Chemist on the Frontiers," in this volume. On the *Table,* see Frederic L. Holmes, "The Communal Context for Etienne-François Geoffroy's 'Table des rapports,'" *Sci. Context* 9 (1996): 289–311, and Ursula Klein, "The Chemical Workshop Tradition and the Experimental Practice: Discontinuities within Continuities," *Sci. Context* 9 (1996): 251–87.

[11] Wilhelm Homberg, Voenno-meditsinskoi Akademii, MS 130, fols. 119v–27v; "une science particuliere dans la chimie qui enseigne la transmutation des metaux," on fols. 119v–20r.

[12] Lawrence M. Principe, "Wilhelm Homberg: Chymical Corpuscularianism and Chrysopoeia in the Early Eighteenth Century," in *Late Medieval and Early Modern Corpuscular Matter Theories*, ed. C. Lüthy, J. E. Murdoch, and W. R. Newman (Leiden, 2001), 535–56, and "Wilhelm Homberg et la chimie de la lumière," in "Chimie et mécanisme à l'âge classique," special issue, *Methodos* 8 (2008), http://methodos.revues.org/1223 (accessed 2 November 2013).

[13] Bernard Joly, "Quarrels between Etienne-François Geoffroy and Louis Lémery," in Principe, *Chymists and Chymistry* (cit. n. 9), 203–14; Joly, "Geoffroy on Frontiers" (cit. n. 10); and Joly, "Le mécanisme et la chimie dans la nouvelle Académie royale des sciences: les débats entre Louis Lémery et Étienne-François Geoffroy," in "Chimie et mécanisme à l'âge classique," special issue, *Methodos* 8 (2008), http://methodos.revues.org/1403 (accessed 2 November 2013).

"especially gold."[14] Indeed, much of his paper is devoted to the preparation of potable gold, the panacea long sought after by generations of chymists. (Geoffroy himself avoids making any strong claims for the medicinal virtue of such a substance.) The chief recipe that Geoffroy cites actually comes from Homberg, who had set down the very same recipe in the 1690s in an unpublished manuscript and where he claims to have brought it to a successful conclusion.[15] Geoffroy's report indicates that further work and study had been expended—by Geoffroy himself, or by Homberg, or by both—on this preparation. The recipe witnesses the close relationship between Homberg and Geoffroy: the elder chymist shared prized information with the younger, and the two collaborated on chymical projects, including traditionally alchemical ones such as potable gold, metallic transmutation, and the composition and decomposition of metals. Thus, in the first decade and a half of the eighteenth century, chrysopoetic endeavors were alive and well within the Académie under the guidance of Homberg, who had Geoffroy as protégé and the Duc d'Orléans as patron.

TENSIONS OVER CHRYSOPOEIA WITHIN THE ACADÉMIE

Such interest in transmutation did not have a comfortable existence within the Académie. In fact, we can trace a continuing tension within the Académie between those who pursued chrysopoeia and those who wanted to ban any discussion or exploration of it. When Louis XIV's minister, Jean-Baptiste Colbert, founded the Académie in 1666, he forbade only two topics of study: astrological prognostication and the Philosophers' Stone.[16] Today one could too easily interpret this regulation as a forward-looking rejection of so-called pseudosciences. But this ban probably reflects instead the fact that these two topics were potentially the most politically subversive and controversial; astrological prognostications (for example, about the king's health or a coming war or famine) could threaten political stability, while transmutation could threaten economic stability. Both topics provoked their share of vigorous debate, and abuses of both were the targets of public ridicule. For such reasons, these subjects were just the sort of things in which an agent of the Crown like the Académie should not be involved. Yet despite Colbert's prohibition, the Académie's first chymist and one of its most active members, Samuel Cottereau Duclos (1598–1685), busied himself with traditional chymical pursuits that included the Philosophers' Stone and metallic transmutation and did so in the Académie's laboratory, which he had designed.[17]

---

[14] Etienne-François Geoffroy, "Des teintures des métaux & particuliérement des teintures d'or," Archives de l'Académie des Sciences (hereafter AdS), Paris, Procès-verbaux (15 March 1713), vol. 32 (the foliation has been trimmed off of much of this volume).

[15] Homberg, MS 130 (cit. n. 11), "De l'or potable," fols. 151v–153v.

[16] "Notes et desseins de Claude Perrault, August 1667," in Lettres, instructions, et mémoires de Colbert, ed. Pierre Clément, 8 vols. (Paris, 1861–70), 5:515.

[17] On Duclos, see Alice Stroup, "Censure ou querelles savantes: L'Affaire Duclos (1666–1685)," in Règlement, usages et science dans la France de l'absolutisme, ed. Christiane Demeulenaere-Douyère and Éric Brian (Paris, 2002), 435–52; Doru Todériciu, "Sur la vraie biographie de Samuel Duclos (Du Clos) Cotreau," Rev. Hist. Sci. 27 (1974): 64–7; Sturdy, Science and Social Status (cit. n. 2), 107–9; and Victor Boantza, "Reflections on Matter and Manner: Duclos Reads Boyle, 1668–69," in Principe, Chymists and Chymistry (cit. n. 9), 181–92. For Duclos's active participation in alchemical circles in Paris and in correspondence abroad, see Lawrence M. Principe, "Sir Kenelm Digby and His Alchemical Circle in 1650s Paris: Newly Discovered Manuscripts," Ambix 60 (2013): 3–24, esp. 17–9.

Shortly after Duclos's death in 1685, Pierre Bayle's *Nouvelles de la République des Lettres* published a deathbed repudiation of the Philosophers' Stone that Duclos supposedly gave to Nicolas Clément (1647–1712), the keeper of the Bibliothèque du Roi.[18] In general, one should be suspicious of deathbed conversions. If it is a true repudiation, it does nothing to diminish the fact that Duclos spent so much of his life working on the problem of transmutation within the Académie. But one cannot discount the possibility that this repudiation is at least partly a show for public consumption, akin to Duclos's publicized conversion to Catholicism that took place at about the same time.[19] Indeed, Clément's account tells how he requested that Duclos make such an avowal "for the public and the service of the King" in order to "restrain those who would too easily engage themselves with the unhappy passion of puffery [*la malheureuse passion de soufflerie*]."[20]

It is conceivable—although it cannot be confirmed—that the statement is not really Duclos's. The account does seem exaggerated in some details. For example, it states that Duclos burned all his papers on alchemy, yet many volumes survive, scattered in several archives in France, and many more existed at least until the late 1750s.[21] It is clear that the Académie was already uncomfortable with Duclos's work or, perhaps more accurately, with public knowledge of it. During Duclos's life the Académie refused him permission to publish his major work on chymistry, the *Dissertation sur les principes des mixtes naturels*, which he had to send out of the country to be published in Amsterdam. His massive work on salts in the format of a series of letters—ready for the press—and including large sections on such things as the Helmontian alkahest, remains unpublished to this day.[22]

In January 1686, just a few months after Duclos's death, François-Michel le Tellier, Marquis de Louvois (1641–91), Colbert's successor as Louis XIV's minister overseeing the Académie, intervened and sent a letter to the assembly ordering them to stay away from any work dealing with "the extraction of the mercuries of the metals, the transmutation of metals, and their multiplication, about which

---

[18] *Nouvelles de la république des lettres,* October 1685, 1139–43.

[19] *Mercure galant,* August 1685, 136–7.

[20] *Nouvelles* (cit. n. 18), October 1685, 1141–2.

[21] Several copies of an *Abregé de la transmutation perfective des métaux* survive: Bibliothèque de l'Arsenal, Paris (hereafter Arsenal), MS 2517; Université de Bordeaux, Lettres, MS 3, and Bibliothèque interuniversitaire de la Sorbonne, MS 1884. Since these are all eighteenth-century documents, they must have been copied from an original that survived Duclos. The surviving notebooks of Jean Hellot cited later in this paper contain substantial material, most of it dealing with metallic transmutation, that was made directly from Duclos's papers in the mid-eighteenth century, indicating their survival until at least that date (see, e.g., Bibliothèque municipale de Caen [hereafter Caen], MS in-4° 171, vol. 3, fols. 25r–27v, 148r–149v, 174r–182v, and passim in 267r–300v.) A work entitled *Extraits des recherches curieuses de physique et de chymie de M. Duclos* was preserved as Université de Bordeaux MS 10, but upon the division of the library into separate faculties in the second half of the twentieth century, this manuscript and all the other fifteen manuscripts of the *anciens fonds* that were assigned to "Sciences et Techniques" were lost. ("Lettres" has, fortunately, proven a better guardian of the *patrimoine*, but the persons responsible for arbitrarily dividing the *anciens fonds* and then "losing" a substantial part of it have much to answer for.)

[22] See Stroup, "Censure ou querelles" (cit. n. 17), where the work on salts (Bibliothèque Nationale, MS fr. 12309) is identified as Duclos's for the first time, on 439. The *Dissertations* were finally published by the Académie some fifty years later in the collection of "anciens mémoires," namely, writings composed by academicians prior to 1699; *Historie et Memoires de l'Académie Royale des Sciences*, 11 vols. (Paris, 1728–33); Duclos's *Dissertation* appears in vol. 4 (1731), 1–41. On this collected publication see Robert Halleux, James McClellan, Daniela Berariu, and Geneviève Xhayet, *Les publications de l'Académie Royale des Sciences de Paris (1666–1793)*, 2 vols. (Turnhout, 2001), 1:57–91.

Mr. de Louvois does not want to hear anything spoken."[23] Given the timing, this injunction was surely an administrative response to what Duclos had been doing in the Académie's laboratory. But this second ministerial warning had no more effect than the first: Duclos's replacement, although delayed by several years until 1691, was Homberg—who would spend all of his twenty-four years as an academician involved in chrysopoetic pursuits.

The unease expressed by both Colbert and Louvois continued under their successor Louis Phélypeaux, comte de Pontchartrain (1643–1727), who took up ministerial administration of the Académie upon Louvois' death in 1691. While I have found no ruling from Pontchartrain's pen akin to those promulgated earlier by Colbert and Louvois, a letter written by a visitor to the Académie in 1692 provides remarkable testimony about the institution's "official" attitude regarding the study of chrysopoeia at this time. In the early 1690s, the Swedish chymist Erich Odhelius embarked on a tour of England and France for the purpose of gathering information on scientific and technological matters. In June 1692, Odhelius was in Paris and wrote a letter back to Urban Hiärne (1641–1724), head of the Royal Laboratory in Stockholm, about his visit to the Académie Royale. Odhelius recounts how he met Homberg, then fairly recently made a member "in the place of Duclos" on account of "his curiosity and experience in chymical matters," and reported that the Académie was in the midst of a reinvigoration thanks to Pontchartrain. He further reported that there was new work in progress on minerals but, most interestingly, that the academicians were "expressly forbidden by the directors of the Academy and by the King to work on alchemical matters, for the King does not wish it to be thought that his money is produced by goldmaking [*per aurifactionem*]."[24] This remarkable statement clearly places the issue of chrysopoeia, at least for the Académie's administrators and royal patron, not in the scientific realm but rather in the political and the economic. It is not a question of the reality or possibility of metallic transmutation but rather a serious concern over broader effects and reactions should the king's academicians be seen working on the matter. The exact nature of these feared reactions is not spelled out, but one can imagine very practical worries over rumors that French coinage was debased with artificial metal, or ones more related to public image and royal *fama*, namely, that Louis XIV's might and glory came not from military prowess or wise governance but rather from chymical labors in smoky laboratories.

These concerns over chrysopoeia were carried forward and intensified by the institution's official public spokesman, Bernard le Bovier de Fontenelle (1657–1757), appointed the Académie's perpetual secretary in 1697. Fontenelle held a generally low opinion of chymistry—seemingly because it could not be, or at least had not yet been, reduced to deductive axioms—to "l'esprit géometrique" of Descartes—like mathematics and physics.[25] In his lengthy essay on the utility of the sciences, Fon-

[23] AdS, Procès-verbaux (cit. n. 14; 30 January 1686), vol. 11, fols. 157r–158r (on fol. 157r).

[24] Erich Odhelius to Urban Hiärne, 6 June 1692, in Carl Christoffer Gjörwell, *Det Swenska Biblioteket*, 2 vols. (Stockholm, 1757–62), 1:337–9: "warande expres förbudna af *Directoribus Academiae* och Konungen at laborera *in Alchymicis*, at icke Konungen må säjas hafwa sina penningar *per aurifactionem*." I thank Hjalmar Fors for bringing this letter to my attention. For more on Odhelius's travels and his relations with Urban Hiärne, see Fors, *The Limits of Matter: Chemistry, Mining, and Enlightenment* (Chicago, 2014).

[25] On Fontenelle and the Académie, see Leonard M. Marsak, *Bernard de Fontenelle: The Idea of Science in the French Enlightenment*, *Transactions of the American Philosophical Society* 49 (Philadelphia, 1957); Simone Mazauric, *Fontenelle et l'invention de l'histoire des sciences à l'aube des*

tenelle mentions chymistry in only a single sentence, and then only as an adjunct to medicine.[26] This attitude reappears frequently. For example, in 1700 Homberg published a sophisticated paper that literally sets the foundations for the standardization of reagents and analysis, yet Fontenelle misses—or ignores—the paper's whole point and instead picks out a trivial side comment made by Homberg about using ground oyster shells as an antacid, and so concludes pompously in his summary of the paper that "it is principally to these sorts of [medical] uses that all chymical discoveries ought to be turned."[27] The search for hidden arcana tainted with a reputation for fraud—like transmutation and the alkahest—only made things worse. Indeed, one of Fontenelle's popular *Dialogues of the Dead* summons up the ghost of Raymond Lull, supposed author of numerous chrysopoetic works, who admits that after his death he finally realized (too late!) that the Philosophers' Stone was a lie. But Lull concludes happily that "though I was not able to make the Stone, at least I was able to fool other people into believing I had."[28]

To be fair to Fontenelle, it must be said that chymistry had long suffered an ambiguous status and a dubious reputation. Unlike astronomy and physics, it could boast neither a classical pedigree nor a place in the university curriculum. The figure of the chymist was often ridiculed in literature, on the stage, and in the visual arts.[29] The memory of the celebrated "affaire des poisons" which caused the chymist Christophle Glaser to be cast into the Bastille (where he may have died) in 1672 was kept alive by the continuing royal prohibition against owning chymical furnaces in Paris without a license.[30] Throughout the period, many other chymists, accused of poisoning or fraud (or simple failure) with respect to transmutation, ended up in the Bastille.[31] Chymists were widely associated in the public mind with poisoners, counterfeiters, frauds, and fools. When the Dauphin, Dauphine, and petit Dauphin all died within a short space of time in 1711–2, fingers pointed instinctively at the nearest chymist, namely, the Duc d'Orléans's tutor, and Homberg narrowly escaped being sent to the Bastille.[32] A high-profile state-sponsored institution like the Académie Royale des Sciences required an image free from such blemishes and associations, and it was in large measure Fontenelle's job to develop and maintain that image.

Lumières, (Paris, 2007); and Alain Niderst, ed., *Fontenelle* (Paris, 1991); for his attitude toward chemistry, see Luc Peterschmitt, "Fontenelle et la chimie: la recherche d'un 'loi fondamentale' pour la chimie," in "Un siècle de chimie à l'Académie royale des sciences," special issue, *Methodos* 12 (2012), http://methodos.revues.org/2873 (accessed 2 November 2013).

[26] *Oeuvres diverses de M. de Fontenelle*, 3 vols. (Paris, 1724), 1:1–35 (not paginated), "Sur l'utilité des mathematiques et de la physique," on sig. Aiiiv.

[27] Wilhelm Homberg, "Observations sur la quantité d'acides absorbés par les alcalis terreux," *MARS* 2 (1700): 64–71; Bernard de Fontenelle, *Histoire de l'Académie Royale des Sciences* (hereafter *HARS*) 2 (1700): 50.

[28] *Oeuvres diverses* (cit. n. 26), 1:117–20.

[29] Principe, *Secrets of Alchemy* (cit. n. 1), 182–90.

[30] Jean-Christian Petitfils, *L'affaire des poisons: alchimistes et sorciers sous Louis XIV* (Paris, 1977; rev. ed., Paris, 2010); Arlette Lebigre, *L'affaire des poisons, 1679–1682* (Brussels, 1989); and Lynn Wood Mollenauer, *Strange Revelations: Magic, Poison, and Sacrilege in Louis XIV's France* (State College, Pa., 2007).

[31] *Archives de la Bastille*, 19 vols, (Paris, 1866–1904), esp. vol. 12 (1881): *Règnes de Louis XIV et de Louis XV* (1709 à 1772), 1–5, 52–4; Clara de Milt, "Christophle Glaser," *J. Chem. Educ.* 19 (1942): 53–60.

[32] Louis de Rouvroy, duc de Saint-Simon, *Mémoires*, ed. Yves Coirault, 8 vols. (Paris, 1983–88), 4:459–66; *Aus der Briefe der Herzogin Elisabeth Charlotte von Orléans an die Kurfürstin Sophie von Hannover*, ed. Eduard Bodemann, 2 vols. (Hannover, 1891), 2:302–3, 307.

Accordingly, Fontenelle's biographical *Éloges* of deceased academicians invariably "sanitize" the chymists. Fontenelle claims that both Homberg and Lémery (both of whom died in 1715) literally fled from practitioners of the "old disreputable" chymistry. Lémery purportedly fled from Glaser, whom Fontenelle describes as "a true chemist, full of obscure ideas, greedy of such ideas, and unsociable."[33] For Homberg, his alleged fear over association with a chrysopoeian supposedly compelled him to leave Paris and flee to Italy. At this point, Fontenelle declares loudly that "Homberg was too capable to aspire to the Philosophers' Stone and too sincere to put such a vain idea into anyone's head." But Fontenelle protests too much, for Homberg himself unabashedly described in print how at just this time he was trying to transmute mercury into silver using an oil distilled from human feces.[34] Fontenelle's attempt to enhance the status of chymists and chymistry involved not only denying relationships with the supposedly disreputable but also inventing relationships with the reputable. Thus Fontenelle's eloge of Homberg provides him with university training in medicine he never had and apprentices him to respectable people in respectable sciences like physics and anatomy—even when those people would have been dead before Homberg could possibly have met them.[35] Fontenelle is correct to say that Homberg met Boyle, although my own research shows that it is impossible that he stayed with him for a year to study, as Fontenelle claims, in "one of the most learned schools of physics."[36] Homberg was in England for only a few weeks in late 1679 or early 1680, and I note with delicious irony that the only thing I can confidently assert that Homberg did learn from Boyle was the secret preparation of the prized Philosophical Mercury needed to make the Philosophers' Stone.[37] Presumably the only thing that prevented Fontenelle from opposing Homberg's own explicit claims to transmutation was fear of offending the Duc d'Orléans, who had himself claimed to have made gold before witnesses with a powder he and Homberg had prepared. Yet Fontenelle nevertheless engaged in a campaign to redefine the boundaries of acceptable chymistry, as a utilitarian chemistry purged of "alchemy."

Fontenelle, or perhaps the Académie in a more corporate sense, might possibly have played a role in preventing Homberg's life-work from being published. Upon his death, Homberg left behind a completed version of his *Essais de chimie*, on which he had been working for over a decade. The manuscript was entrusted to Geoffroy, with the request to publish it as soon as possible.[38] But nothing ever appeared. Given the "alchemical" origins of the experiments upon which so much of the text was based, and its claim successfully to have produced gold from mercury, the publication of this work—bearing Homberg's name and his title as academician—may well

[33] Fontenelle, "Éloge de M. Lemery," *HARS* 17 (1715): 73–82, on 73. On Lémery's relationship with Glaser, see Bougard, *Nicolas Lemery* (cit. n. 4), 24–26.

[34] Wilhelm Homberg, "Observations sur la matiere fecale," *MARS* 13 (1711): 39–47.

[35] Alice Stroup, "Wilhelm Homberg and the Search for the Constituents of Plants at the Seventeenth-Century Académie Royale des Sciences," *Ambix* 26 (1979): 184–202, on 185–6.

[36] Fontenelle, "Éloge de M. Homberg," *HARS* 17 (1715): 82–93, on 85.

[37] Principe, "Homberg: Chymical Corpuscularianism and Chrysopoeia" (cit. n. 12), and William R. Newman and Lawrence M. Principe, *Alchemy Tried in the Fire: Starkey, Boyle, and the Fate of Helmontian Chymistry* (Chicago, 2004), 304–6; my *Transmutations of Chymistry* (forthcoming), chap. 1, completely rewrites Fontenelle's biographical account of Homberg.

[38] Nicolas Remond to Gottfried Wilhelm Leibniz, 23 December 1715, Niedersächsische Landesbibliothek Hannover, Leibniz Briefe 768, fols. 53r–54v, on fol. 54r.

have been unwelcome for the image of the Académie and of chemistry being constructed in 1716. If some kind of suppression did take place, it would correlate with the suppression of Duclos's treatises a generation earlier.

There is thus clear evidence of long-standing tension between the nonchemist administrators of the Académie (Colbert, Louvois, Fontenelle), who tried to suppress transmutation, and the chemists themselves (Duclos, Homberg, Geoffroy), who were busily working on the problem despite repeated prohibitions against such endeavors. This tension provides crucial background to Geoffroy's 1722 paper. It is now time to turn to its contents and impact.

GEOFFROY'S 1722 "SUPERCHERIES CONCERNANT LA PIERRE PHILOSOPHALE"

Geoffroy's "Some Cheats Concerning the Philosophers' Stone" was presented orally in 1722 and published in 1724. It relates methods used by fraudulent would-be transmuters of metals to trick people into believing that they have witnessed a transmutation—for example, using crucibles that contain gold hidden under a false bottom, or stirring a molten mixture with a hollow rod that contains gold hidden inside. It has been shown that the majority of Geoffroy's paper is cribbed from the *Examen fucorum pseudochymicorum*, a well-known work published in 1617 by Michael Maier and intended to defend transmutational alchemy and to help his fellow chrysopoeians distinguish true from false transmutations. Much of Maier's work is in turn borrowed from Heinrich Khunrath's *Trewhertzige Warnungs-Vermahnung* of 1597.[39] Thus Geoffroy's paper contained nothing new; it was merely a restatement of material already over a century old, and written originally by defenders of metallic transmutation. This is immediately strange, since it renders Geoffroy's paper virtually unique among memoirs presented to the Académie in that it does not contain any research results whatsoever. Moreover, it would be incorrect to conclude that Geoffroy was necessarily himself utterly opposed to the possibility of transmutation, for although he describes several cheating practices he nowhere claims that all chrysopoeia is fraudulent. Indeed, the catalogue of Geoffroy's library shows that at the time of his death he owned more than seventy books on transmutational alchemy, including classic works by Philalethes, Valentine, and others, as well as Manget's huge 1702 compendium *Bibliotheca chemica curiosa*.[40] Nevertheless, Fontenelle used Geoffroy's paper as the platform from which to launch a lengthy commentary containing his most vitriolic and sarcastic blanket condemnations of "les Alchimistes." Fontenelle also used this opportunity explicitly to distinguish "alchemical" claims from the work done previously by Homberg. He further asserts that alchemists have never made a single grain even of an imperfect metal, much less silver or gold—perhaps in reference to Geoffroy's earlier claims to have synthesized iron.[41]

---

[39] Wolfgang Beck, "Michael Maiers Examen Fucorum Pseudo-chymicorum: eine Schrift wider die falschen Alchemisten" (PhD diss., Technische Univ. München, 1992); Robert Halleux, "L'alchimiste et l'essayeur," in *Die Alchemie in der europäischen Kultur- und Wissenschaftsgeschichte*, ed. Christoph Meinel (Wiesbaden, 1986); Michael Maier, *Examen fucorum pseudo-chymicorum detectorum et in gratiam veritatis amantium succincte refutatorum* (Frankfurt, 1617); Heinrich Khunrath, *Trewhertzige Warnungs-Vermahnung* (Magdeburg, 1597).

[40] *Catalogus librorum Stephani-Francisci Geoffroy* (Paris, 1731); I thank Brigitte Van Tiggelen for kindly bringing the existence of this source to my attention.

[41] Fontenelle, *HARS* 24 (1722): 37–9.

A crucial, but hitherto overlooked, feature of Geoffroy's paper is that it was presented not at a private séance attended only by members of the Académie but rather at one of the institution's semiannual public assemblies (on 15 April 1722). Papers given at these special assemblies were carefully chosen by committee to showcase the Académie and its work; thus, Geoffroy's memoir was certainly intended for a wider audience than the Académie. Indeed, the popular press got involved too. The monthly periodical *Mercure galant* routinely mentioned the Académie's public meetings in one or two paragraphs, as it did for the April 1722 meeting. But in this case, anomalously, the following issue of May 1722 carried a five-page reprise of Geoffroy's paper—a unique treatment for a paper given at an *assemblée publique* in this period.[42] Fontenelle, who had personal ties to the *Mercure*, may have orchestrated this enhanced coverage.

I therefore suggest that Geoffroy's paper was primarily an act of public relations. It functioned both as a statement of the Académie's official views and as a tacit renunciation of the work Geoffroy and his master Homberg had performed previously. It may also have been intended to deflect contemporaneous rumors. At the time of Geoffroy's paper, a rumor was circulating that the Duc d'Orléans—now the Regent of France—had ordered the Académie's chemists to direct their attention to the problem of chrysopoeia. It is difficult to assess how widespread this rumor became, but it certainly traveled as far as Germany, where Johann Thomas Hensing (1683–1726) reported in the course of lectures he gave in 1722 that "the current Regent has himself requested the members of the Académie Royale who apply themselves to chemistry to seek out the Philosophers' Stone."[43] The rumor might also explain why the agent of the English ambassador to France sent home a special account of the Regent's abilities in chemistry and his work with Homberg.[44] Indeed, the broader situation in France would have served to support the rumor's claim. In 1720, the banking scheme organized by John Law with the backing of the Regent had begun a spectacular collapse. Too many bank shares had been sold, and there was not enough gold to back them up. Insufficient gold is of course the traditional problem for alchemists. Might the Regent have sought to stave off economic troubles through the chymical manufacture of gold? Philippe had after all claimed firmly that he and Homberg had succeeded in making gold a decade earlier. If the rumor of Philippe's orders to the Académie is true, then it would mark a complete reversal of the earlier orders of Colbert, Louvois, and Louis XIV, who wanted the Académie to steer well clear of chrysopoeia. Whether or not it is true, such a rumor ran entirely counter to the image of the Académie's respectable, domesticated chemistry that Fontenelle had been struggling to craft.

[42] *Mercure galant*, April 1722, 97; May 1722, 121–5.

[43] Hensing's "Discurs von dem Stein der Weisen," which he prefixed to his November 1722 chemistry lectures delivered at Giessen, was published in Georg von Welling, *Opus mago-cabbalisticum et theosophicum* (Homburg, 1735), 518: "hat der jetzige *Regent*, diejenigen Glieder der Königlichen *Socie*tät, so die *Chemie excoli*ren, selbst angehalten, den beruffenen *Lap. Phil.* zu untersuchen." A French version was published later as Johann Thomas Hensing, "Dissertation sur la pierre philosophale," in *Mémoires littéraires*, ed. Marc Antoine Eidous (Paris, 1750), 121–54, on 122–3: "M. le Régent a voulu que les Membres de l'Académie qui s'appliquent à la Chymie, travaillassent de tout leur pouvoir à découvrir la Pierre Philosophale."

[44] National Archives, Kew, State Papers 78/166, fols. 339–41; the text is excerpted from Fontenelle's eloge of Homberg.

CHRYSOPOEIA AT THE ACADÉMIE AFTER GEOFFROY'S "SUPERCHERIES"

What effect did Geoffroy's paper have? No other papers even suggesting the possibility of transmutation were thereafter published by the Académie.[45] Thus the paper did in this sense signal a "point of no return" for chrysopoeia. Attempts at transmutation were now "officially" considered fraudulent, and transmuters as socially and intellectually unacceptable frauds. But did chrysopocia actually evaporate from learned French circles? Despite the continuing decrease in French publications on transmutational alchemy from the 1720s to the 1730s and further on into the eighteenth century, some experimenters nevertheless continued to work on the problem of chrysopoeia, and this persistence was reflected in continuing concerns on the part of Fontenelle. For example, in summarizing Boerhaave's 1734 memoir on his experiments with mercury, Fontenelle writes of the continuing importance of refuting entirely the possibility of metallic transmutation. This "former hope . . . abused by so many imposters to engage gullible and greedy persons in infinite labors and ruinous expenses . . . has up to the present day deceived everyone who has given himself up to it." He concludes that "the true chemists shall leave the alchemists nothing but the refuge of invincible stubbornness, a refuge always open to those who can profit from it, where in fact an infinity of people boldly take their stand."[46]

Fontenelle's concluding remark sets up an unbridgeable division between *les vrais chimistes* and *les alchimistes*, two groups that remained imperfectly or not at all distinguished a generation earlier, and places them in active opposition to one another. Given the presence of this declaration in the Académie's official publication, it would be reasonable enough to assume that the academicians exemplify the "true chemists." *Pace* Fontenelle, what were the actual views of Académie chemists toward the old problem of metallic transmutation and the endeavors of *les alchimistes*?

The difficulty in answering this question has been that so few personal papers of early academicians have survived, making it difficult to assess privately held opinions and privately conducted investigations. Fortunately, the Académie's chemist Jean Hellot (1685–1766) comes to the rescue. Hellot studied chemistry during the 1720s under Claude-Joseph Geoffroy (1685–1752), the younger brother of Etienne-François. (Claude-Joseph Geoffroy later married Hellot's niece.) In 1735, Hellot became *adjoint* chemist in the Académie Royale, and *pensionnaire* three years later (skipping over the intermediate rank of *associé*), a post he held until his death in 1766. He also fulfilled several official functions, such as overseeing dyeworks, examining mining and assaying operations, improving the porcelain production at Sèvres, and reforming weights and measures. Crucially, Hellot compiled substantial notebooks of material copied from the then-surviving papers of earlier Académie chemists or acquired orally from his contemporaries; indeed, his eloge mentions the "many memoirs and manuscripts on the arts and sciences" he left behind at his death.[47] Fortunately, some of these documents have not only survived but have also

---

[45] The only later *Mémoires* paper of which I am aware that deals at all with transmutation is Herman Boerhaave's "Sur le mercure," *MARS* 36 (1734): 539–52, the original French version of the paper that, translated into Latin by the Secretary of the Royal Society, would appear in the 1736 *Philosophical Transactions* as "Part II" of the three-part paper "De mercurio experimenta," describing attempts to alter the nature of mercury; see Powers, "Scrutinizing the Alchemists" (cit. n. 9).

[46] Fontenelle, *HARS* 36 (1734): 55–7, quotations on 55 and 57.

[47] Jean-Paul Grandjean de Fouchy, "Éloge de M. Hellot," *HARS* 68 (1766): 167–79, on 179.

been recovered in recent years. In the 1960s, nine substantial notebooks representing about three thousand pages of material in Hellot's minuscule hand were discovered in Caen.[48] Two more notebooks of "chemical recipes taken from the literary remains of the late Mr. Hellot of the Académie des Sciences and written in his hand" are preserved at the Bibliothèque de l'Arsenal in Paris.[49] In 2010, I discovered a further three, previously unidentified, autograph notebooks of Hellot at the same library.[50] Fortunately, Hellot was a thorough and conscientious note taker; he copied out materials in extenso and always noted his sources explicitly. He became familiar enough with these documents to be able to recognize autograph hands and noted this fact where appropriate. Hellot's surviving notebooks provide an otherwise unobtainable glimpse of the private work of Académie chemists.

In regard to Geoffroy, Hellot's notebooks record his work on the mercurification of metals, the "animal stone," the use of lead ore as a starting material for the Philosophers' Stone, and the interpretation of allegorical chrysopoetic texts, and preserve a lengthy section on "The Great Work . . . from the hand of Mr. Geoffroy" that follows the general principles of the Philalethes treatises.[51] Other notebooks contain processes on gold, methods for making animated mercury (a crucial starting material for the Philosophers' Stone) with bismuth, and even a transmutation of copper into silver labeled as "épreuve veritable," all explicitly noted as having been copied from Geoffroy's autograph manuscripts.[52] Hellot also owned a two-volume manuscript translation of Alchymia denudata prepared by Geoffroy himself.[53] Some of this material certainly dates from Geoffroy's younger years, when he was working on transmutational endeavors alongside Homberg. But some entries are explicitly dated after 1722, that is, after his famous "Supercheries" paper. For example, Hellot's notes preserve a lengthy account of producing a transmuting powder able to turn one hundred times its weight of mercury into gold, copied from "a manuscript that Geoffroy gave me in 1730" as well as an "augmentation of gold" using iron and a process for potable gold both dating from the same year.[54] It seems unlikely that Geoffroy shared these manuscripts with Hellot out of some mutual antiquarian interest, but rather because Geoffroy remained involved in the problem of chrysopoeia—even after his association with Homberg and well after 1722.

But the Hellot documents tell a much wider story about the fate of chrysopoeia at the Académie in the eighteenth century. Arsenal MSS 3006, 3007, and 3008 comprise a coherent three-volume compilation of materials dealing almost exclusively

[48] Arthur Birembaut and Guy Thuiller, "Une source inédite: Les cahiers du chimiste Jean Hellot (1685–1766)," Ann. Econ. Soc. Civilisations 21 (1966): 357–64; Doru Todériciu, "La bibliothèque d'un savant chimiste et technologue parisien du XVIIIe siècle: Livres et manuscrits de Jean Hellot (1685–1766)," Physis 18 (1976): 198–216; see also n. 69. Bibliothèque municipale de Caen MS in-4° 171 (cit. n. 21), 9 vols.

[49] Arsenal MSS 2824 and 2825.

[50] Arsenal MSS 3006, 3007, and 3008. These three volumes exist within a series of alchemical manuscripts; how many of the others may have likewise belonged to Hellot remains undetermined at present. A flyleaf note on Arsenal MS 2520 indicates that that manuscript, which contains the Interruption du sommeil cabalisticque, was owned by Hellot; this work was of particular interest to the Parisian alchemical circle gathered around Kenelm Digby and Samuel Cottereau Duclos in the mid-seventeenth century, see Principe, "Sir Kenelm Digby" (cit. n. 17), 7, 10.

[51] Caen MS in-4° 171 (cit. n. 21), vol. 3, fols. 146v–147r, 270r–271v, and 276v–277r.

[52] Arsenal MS 3006, 25–6, 126–7, 138; MS 3008, 53.

[53] Catalogue des livres de feu M. Hellot (Paris, 1766), 101, no. 1805. The Alchymia denudata was first published in 1708 and reissued in extended form in 1716, 1723, and 1728. Unfortunately, since Geoffroy's MS is now lost, we cannot determine which edition he translated.

[54] Caen MS in-4° 171 (cit. n. 21), vol. 2, 86, 97, and 98.

with achieving metallic transmutation. Hellot organized these notebooks into sections, each dealing with a particular starting material. The first notebook is devoted exclusively to processes using gold and silver and approaches to the Philosophers' Stone; the second notebook records processes beginning with salts, sulphurs, and nostoc; and the third contains those employing the base metals, semimetals (bismuth, zinc, antimony, etc.), minerals, and organic materials.[55] Significantly, about half of the entries these volumes contain come from chemists within the Académie, thus indicating the interests of these other chemists. Some sections were copied from the surviving papers of chemists before Hellot's days at the Académie, such as Homberg and Duclos, but a larger number are listed as coming from Hellot's contemporaries, collected by direct contact or copied from their papers.[56]

A large number of processes and accounts are listed as originating from Antoine-Tristan Danty d'Isnard. Danty d'Isnard (1663–1743), a member of the Académie since 1716, held positions both as a botanist and as a chemist. Hellot's notebooks record Danty d'Isnard's processes for such things as the extraction of the sulphur of gold and the mercury of antimony, for the transmutation of tin into gold and of mercury into silver, for the alkahest, and for preparing several different transmuting powders from gold.[57] Some entries are in the first person and give circumstantial details of working through the process—such as a case where the digesting vessel broke three times in a row such that "we did not obtain any profit"—indicating that Danty d'Isnard was actively experimenting with transmutational processes.[58] Moreover, there are accounts and processes that Danty d'Isnard collected from other experimenters outside the Académie, indicating his participation in a network of exchanges regarding transmutation with other hopeful chrysopoeians. One process, for example, describing how to "exalt" gold into a transmuting agent able to convert silver and mercury into gold, he obtained from "Mr. de la Motte"; Hellot highlighted this process by drawing a hand pointing to it in the margin.[59] Other receipts acquired by Danty d'Isnard include one for potable gold from "Mr. Clecy," another for a powder to turn silver into gold from "Mr. de Rasé," a transmutation of mercury into silver from "Mr. Lambert," a "magnet of the philosophers for the celestial mercury" from "Mr. Jacques," and two for making mercury of antimony—the first from a "Sieur Denis" and the second from "Benato."[60] Thus it is clear that Danty d'Isnard maintained an active interest in the topic of transmutation—collecting processes and experimenting with them.

There are also dozens of chrysopoetic processes attributed to Charles François de Cisternay Dufay (1698–1739), pensionnaire chemist of the Académie during the 1730s, Intendant of the Jardin du Roi, and a close associate of Hellot.[61] Materials

---

[55] Hellot signaled these sections throughout by writing the symbol for the featured material in the upper right corner of the first page and then on every other folio of each section.

[56] Dating these three manuscripts is difficult; however, Etienne-François Geoffroy is referred to as "le feu Mr Geoffroy" (MS 3008, 53 and 185–7), indicating a date after 1731, and MS 3006 contains the dates 1732 (20) and 1734 (24).

[57] Arsenal MS 3006, 95–6, 101–7; MS 3007, 33–37 (this material on the alkahest is cited as "tiré du manuscrit que ma presté Mr Danty d'Isnard") and 51; MS 3008, 154, 157, 188–9.

[58] Arsenal MS 3007, 111–2; "n'y trouvâmes aucun profit."

[59] Arsenal MS 3006, 96–8.

[60] Ibid., 99, 110, 111–5; MS 3007, 74; and MS 3008, 189 and 190–1.

[61] On Dufay, see Bernard de Fontenelle, "Éloge de M. Dufay," HARS 41 (1739): 73–83. Dufay's sharing of secrets with Hellot is mentioned in Hellot's eloge (cit. n. 47, 174), in regard to a process for making gold bas-reliefs not to be communicated until after Dufay's death. Hellot was also Dufay's executor, further testifying to their close relationship, "Éloge de M. Dufay," 82.

from Dufay record, for example, the transmutation of silver into gold and the preparation of a red oil of vitriol able to convert silver into gold.[62] A process for preparing a salt capable of extracting the "soul of silver" contains first-person narrative, indicating that Dufay carried out this operation himself.[63] Hellot also records that while Dufay was in Strasbourg in 1732, a skilled chymist "who appeared to know the [Great] Work" gave him the first steps of a process for making the Philosophers' Stone. Dufay and Hellot discussed the progress of this experiment, then under way, in May 1739, while Dufay was awaiting further instructions from the Alsatian chymist. Unfortunately, such instructions did not arrive by the time of Dufay's premature death two months later in July, and Hellot subsequently found four flasks still containing the slowly reacting materials when he sorted Dufay's effects.[64] Dufay's own labor on such transmutational processes is equally clear in another account, where Hellot records that "this operation is very certain, for Mr. Dufay used one *marc* of silver to do it, from which he himself separated the gold by quartation." The same entry also contains Dufay's theoretical explanation of the operation, which he then used to improve the procedure so that from one ounce of fine silver "he found after quartation four grains of very beautiful gold."[65] Here and elsewhere, Hellot's expressions make it clear that he was not only copying out information from written remains but rather that he and Dufay discussed these procedures and their results.

Both the Arsenal and the Caen notebooks further record Dufay's association with Francesco Maria Pompeo Colonna (1646–1726), one of the most famous of the "late period" chrysopoeians in France. Colonna authored several books about chymical medicine and metallic transmutation in the decade prior to his sudden death in a fire in the early morning hours of 6 March 1726 (ironically enough, Ash Wednesday).[66] One of the best known of these books, *Les secrets les plus cachés de la philosophie des anciens*, published in 1722 under the pseudonym of Crosset de la Haumerie, received a generally positive review in the *Journal des sçavans*. Although the reviewer remarked that the book was "very badly written in terms of its French," he concluded that the book was worthy of careful consideration because it "does not consist of vain speculations" but rather "is founded upon the operations of nature."[67] Colonna was on familiar terms with several persons connected with the Académie Royale; he had collaborated with Mathieu-François Geoffroy (1644–1708)—Etienne-François's father—and was a personal friend of the astronomer Gian Domenico Cassini (1625–1712) and of the brother of Jean-Paul Bignon (1662–1743), president of the Académie and head of the Bibliothèque du Roi.[68]

[62] Arsenal MS 3006, 12–14, 128–31; MS 3007, 80

[63] Arsenal MS 3006, 132–4, "Jai pris dit Mr Dufay, la matiere blanche resteé dans la matras," on 133–4.

[64] Caen MS in-4° 171 (cit. n. 21), vol. 1, fol. 85: "Feu Mr. dufai etant a Strasbourg en 1732 y fit connoissance avec un tres galand homme . . . qui s'amusoit depuis 30 ans a la chimie [et] parroissoit scavoir l'oeuvre." My thanks to Michael Bycroft, who kindly supplied me with a transcript of this entry.

[65] Arsenal MS 3006, pp. 128–30; on 129: "cette operation est tres certaine, car Mr Dufay s'est servi pour la faire d'un marc d'argent dont il a separé luy meme l'or par le depart"; 130: "par le Depart il a trouvé 4 graines de tres bel or." Since previously he had obtained four grains from one marc of silver, that is, eight ounces, the improved process gave eight times as much transmuted gold.

[66] *Mercure de France*, March 1726, 603–7.

[67] *Journal des sçavans*, March 1723, 147–51. Colonna also published under the pseudonym Alexandre le Crom, and additional books under his own name appeared posthumously.

[68] Gustavo Costa, "Un Collaboratore italiano del Conte di Boulainviller: Francesco Maria Pompeo Colonna (1644–1726)," *Atti e memorie dell'Accademia Toscano di scienze e lettere* 29 (1964):

Hellot records that Dufay was present at two transmutations performed by Colonna. One case involved the conversion of some silver into gold "seen by Mr. Dufay."[69] The other event was more significant—a witness of the transmutation of lead into gold using the Philosophers' Stone itself. Hellot underscored this account's importance with a marginal note reading "NB: Proof of the Great Work."

> It is claimed that Mr. Colonna succeeded in the Great Work using this philosophical mer-cury [described in the previous entry]. Mr. Dufay has assured me that he was present at the Hôtel de Richelieu for the conversion of one ounce of molten lead, from which four *gros* [half an ounce] and a few grains of gold were obtained by throwing into it a little ball of wax into which had been introduced, in his presence, in Mr. Colonna's chamber, a very small quantity of a powder of a very beautiful red.[70]

Another process gathered from Dufay, for a transmutation of silver into gold using "fixed sulfur," suddenly switches in mid-sentence from French to Italian, suggesting that it comes ultimately from Colonna. It is preceded in fact by an account of "silver tinged into gold with orpiment and fusible salt," which Hellot attributes simply to "Colonne," and an Italian recipe exists elsewhere in the manuscript for a "solfo fisso" drawn from Dufay's papers.[71]

Clearly then, Dufay, like Danty d'Isnard, not only had a serious interest in chryso-poeia but also worked on the topic himself and associated with others who were like-wise searching for solutions to the problem. And in Dufay's case, Fontenelle once again "protests too much" in the *Éloge* he wrote for him. The perpetual secretary re-marks that Dufay's grandfather "got chemistry into his head, with the design, truth be told, to achieve the Great Work. He worked a great deal and spent a great deal, with the usual degree of success." Fontenelle then notes that "perhaps the blood of that grandfather of whom we have just spoken acted in" the grandson, but Fontenelle im-mediately reassures his readers that "it was corrected in that grandson, who never as-pired to the Great Work."[72] If Fontenelle felt it important in 1739 to stress that Dufay had no interest in chrysopoeia, we can be certain that he *did* have such an interest—as Hellot's notes clearly bear witness.

Hellot himself had a grandfather interested in chymistry. His grandfather, another Jean Hellot, translated William Davisson's 1640 *Philosophia pyrotechnica* into French.[73] Indeed, the official *Éloge* for the younger Hellot mentions that he was di-

---

207–95, on 218–9. Interestingly, the copy of *Les secrets* preserved at Johns Hopkins University bears the ownership stamp of the Bibliothèque du Roi.

[69] Caen MS in-4° 171 (cit. n. 21), vol. 9, fol. 105v: "Transmutation de Mr. Colonne, vuë par Mr duFay de l'arg[en]t en Or." Here Hellot cites "Chym. p. 281," presumably a lost notebook with more details; Todériciu suggests it was an unfinished treatise, "Le traité de chimie inachevé de Jean Hellot," *Physis* 19 (1977): 355–75,

[70] Caen MS in-4° 171 (cit. n. 21), vol. 5, fol. 189: "On a pretendu que Mr. Colonne avoit réussi par ce mercure philosophique à la grande opération. Mr du Fay m'a assuré qu'il avoit esté présent à l'hotel de richelieu à une conversion d'une once de plomb fondu, dont on tira 4 gros quelques grains d'or en y jettant une petite boule de cire, où on avoit introduit, devant lui, dans la chambre de Mr Colonne une tres petite quantité d'une poudre d'un tres beau rouge."

[71] Arsenal MS 3006, 130–1; MS 3007, 91. There is another transmutative process, using liver of antimony and mercury precipitate, attributed to Colonna at MS 3008, 18–19.

[72] Fontenelle, "Éloge de M. Dufay" (cit. n. 61), 73–4.

[73] *Les elemens de la philosophie de l'art du feu ou chemie . . . traduit du Latin du Sieur Davis-sone . . . par Jean Hellot* (Paris, 1651). A second edition appeared in Paris in 1657, and a third under the title *"Le cours de chymie"* at Amiens in 1675.

verted from his initial interest in pursuing theology by discovering his grandfather's chymical papers.[74] In accord with this information, Hellot's notebooks occasionally cite entries taken from "an old manuscript in the hand of my grandfather." In one case, for example, he copies out a process for making the mercuries of antimony and of lead.[75] It would be significant (and somewhat amusing) if one of the most productive and prominent chemists of mid-eighteenth-century France actually took up the study of chemistry by becoming fascinated with transmutational processes. While there is currently no direct evidence that Hellot himself engaged in practical attempts at gold making, his notes on chrysopoeia are so extensive and so well organized that I find it difficult to believe that his interest was purely historical. Could it possibly even be the case that Hellot—forced to act as editor of the *Gazette de France* throughout the 1720s in order to maintain a livelihood due to the loss of his family fortune in the collapse of John Law's banking schemes—saw in transmutational alchemy a way to restore his former financial security?[76] However that might be, we certainly have yet another example in Hellot of a highly active and important Académie chemist who maintained a serious interest in chrysopoeia, now into the 1760s. Nor was Hellot alone, even at that late date.

Another influential chemist of the Académie, Pierre-Joseph Macquer (1718–84), actually wrote a short essay entitled "Sur la pierre philosophale" sometime between 1753 and 1766 as a response to a dismissive essay on the subject written by his colleague Pierre-Louis Moreau de Maupertuis (1698–1759). While Macquer did not publish this text, its manuscript—displaying a degree of rewriting that implies he intended to publish it—survives in the Bibliothèque Nationale and has recently been studied by Christine Lehman.[77] While Maupertuis admitted that one could not absolutely prove the impossibility of the stone, he nevertheless characterized those who seek it as fools. Macquer, for his part, responds by asking, "Can one consider insane a person who labors for the Philosophers' Stone only for the pleasure of resolving one of the most beautiful and most difficult problems in natural philosophy?"[78] Macquer recounts the frauds with which chrysopoeia is associated, rather like Geoffroy's "Supercheries," as well as famous accounts of successful transmutations—in regard to the latter he "suspends judgement" in terms of their veracity. While he stresses the difficulty of finding the means of transmutation, he nonetheless calls upon the current state of chemical knowledge to propose more fruitful ways in which it might be pursued. He begins with the Stahlian idea that "the metals are all formed from a more or less vitrifiable earth united and combined with the principle of inflammability that the chemists call phlogiston." He then notes that in gold and silver, the union between the earth and the phlogiston is "more intimate and more perfect" (which results in

[74] Grandjean de Fouchy, "Éloge de M. Hellot" (cit. n. 47), 167.

[75] Arsenal MS 3008, 186–7; "ancien MSC de la main de mon ayeul."

[76] Grandjean de Fouchy, "Éloge de M. Hellot" (cit. n. 47), 168.

[77] Bibliothèque Nationale, Paris, MS français 9132, fols. 98r–106v; for a description of this manuscript, see William Smeaton, "Macquer on the Composition of Metals and the Artificial Production of Gold and Silver," *Chymia* 11 (1966): 81–8, and for more complete analysis, see Christine Lehman, "Alchemy Revisited by the Mid-Eighteenth Century Chemists in France: An Unpublished Manuscript by Pierre-Joseph Macquer," *Nuncius* 28 (2013): 165–216, which includes a full transcription of the manuscript on 202–16. Macquer's essay responds to Maupertuis's "Sur la pierre philosophale," published in his *Lettres de M. Maupertuis* (Dresden, 1752), 81–5; much of Macquer's rewriting tones down his criticism of Maupertuis and deletes all references to him by name. For more on Macquer, see Lehman, "Pierre-Joseph Macquer: Chemistry in the French Enlightenment," in this volume.

[78] Lehman, "Alchemy Revisited" (cit. n. 77), 204 (Macquer, BN MS 9132, fol. 99v).

the more stable properties of these metals). Thus, one approach would be "to perfect all the metallic substances that are not gold or silver by giving their principles the degree of union that they lack."[79] It is possible, however, he cautions, that each metal might contain an earth that is proper and unique to it, in which case "true transmutations would be impossible." If that is the case, one could still undertake a project of attempting to combine phlogiston with every sort of earth available—"for nothing prevents us from believing that nature has not converted into gold and silver all the species of earths proper for entering into the composition of these metals; one could hope to find them after a more or less large number of experiments."[80] Macquer underlined how difficult and costly such a program of research would be, and therefore doing so for the sole purpose of becoming rich would indeed be foolhardy. Yet those willing to be satisfied with gaining knowledge—without monetary profit—might well undertake the project as something worthwhile.

While there is no evidence that Macquer himself dabbled in the aurific art, the very fact that he wrote such an essay indicates that transmutation was by no means a dead issue in the mid-eighteenth century, even among members of the Académie Royale. Lehman has in fact connected this manuscript with Macquer's interest in determining more about the composition of metals, and his later replication—in collaboration with Antoine-Laurent Lavoisier—of the experiments on the decomposition of metals that Homberg conducted earlier in the century with the Tschirnhaus burning lens.[81]

More impressive testimony about the persistence of transmutational alchemy comes from the long-lived Académie chemist Michel-Eugène Chevreul (1786–1889). He recalled that the brother of the chemist Joseph-Louis Proust (1754–1826) "many times affirmed" to him that no less a figure than Guillaume-François Rouelle (1703–70) worked on transmutation secretly in a special laboratory he kept for this purpose in Paris. In this laboratory on the rue Copeau (now rue Lacépède), Rouelle "delivered himself over, in the greatest secrecy, to his alchemical labors."[82] Located not far from the Jardin du Roi, this special workspace was clearly a different locale than Rouelle's well-known private laboratory on the rue Jacob—over a mile away in St. Germain—where he customarily worked and where he taught his private chemistry course (as opposed to the well-attended public one he taught at the Jardin du Roi).[83] Chevreul's report of Rouelle's interest in transmutational alchemy is corroborated by the testimony of Diderot, who attended Rouelle's course of chemistry at the Jardin du Roi in the 1750s. Writing in 1771, Diderot recalled that Rouelle "believed in alchemy; he used the last two lectures of a course that lasted eight months to demonstrate its reality by facts and principles." Although he urged his students not to

[79] Ibid., 212–3 (Macquer, BN MS 9132, fols. 104v–r).

[80] Ibid., 213 (Macquer, BN MS 9132, fol. 105r).

[81] Ibid., 190–200.

[82] Chevreul's comments appear in the context of his four-part essay on alchemy in the guise of a review of L. P. François Cambriel's *Cours de philosophie hermétique* (Paris, 1843): Michel-François Chevreul, *Journal des savants*, May 1851, 284–98; June 1851, 337–52; August 1851, 492–506; and December 1851, 752–68. Quotation on 293: "s'il est vrai, comme le frère du célèbre Proust nous l'a affirmé plusieurs fois, que Guillaume François Rouelle fut alchimiste, cela prouverait qu'au XVIIIe siècle l'alchimie n'était pas une chimère pour tous les hommes vraiment distingués." Proust was a student of Hilaire-Marin Rouelle, Guillaume-François's younger brother.

[83] For more on Rouelle's laboratory on rue Jacob, see Marco Beretta, "Rinman, Diderot, and Lavoisier: New Evidence Regarding Guillaume-François Rouelle's Private Laboratory and Chemistry Course," *Nuncius* 26 (2011): 355–79, esp. 362–70.

pursue chrysopoeia, "nevertheless," continues Diderot, "he often confided to me that it would be the object of his work in his later years."[84]

Diderot's claim is fully borne out by surviving student notes from Rouelle's lectures; the chemist did in fact end his course with an investigation of chrysopoeia.[85] Rouelle's statements reveal deep familiarity with traditional alchemical sources and principles, as well as belief in metallic transmutation and the Philosophers' Stone's powers. In regard to the topic of transmutational alchemy has a whole, Rouelle asserts that "the generality of natural philosophers [*commun des phisiciens*] doubt the truth of this science, but they are not able to be judges of a subject that is entirely unknown to them; the most knowledgeable chemists, even those who do not possess its principles, do not call it into doubt." When discussing the possibility of the Philosophers' Stone's power to transmute, Rouelle declares emphatically that "the most sensible and knowledgeable chemists have believed it; the ignorant ones and uneducated people have denied it."[86] Rather than merely perpetuating the ideas and explanatory systems of earlier alchemical authors, Rouelle—like Macquer—"updates" chrysopoeia into the theoretical context of his contemporaneous eighteenth-century chemistry. He applies Stahlian principles to the constitution of metals and remarks that "the Philosophers' Stone is nothing other than the result of a fermentation of gold with mercury—not common mercury but a special mercury oversaturated with phlogiston."[87] This special overphlogisticated mercury is identical to the "animated" Philosophical Mercury described by chrysopoetic writers. Unfortunately, Rouelle continues, the written accounts left by the adepts—such as Philalethes—are enigmatic and ambiguous, and it is possible that no one to date has been able to interpret them correctly. Thus, just as Diderot recounts, Rouelle discouraged his students from pursuing the Stone, but only because success in so expensive an undertaking was very uncertain "without a sure guide to lead one through an operation that is preserved only in oral transmission."[88]

Rouelle taught chemistry to a vast number of students both at the Jardin du Roi

---

[84] Diderot, *Oeuvres complètes*, ed. Herbert Dieckmann et al. (Paris, 1995), 20:630; on Diderot and Rouelle's course, see Jean Jacques, "Le Cours de chimie de G.- F. Rouelle recuelli par Diderot," *Rev. Hist. Sci.* 38 (1985): 43–53, and Beretta, "Rinman, Diderot, and Lavoisier" (cit. n. 83), 370–79. On Rouelle's course, see Christine Lehman, "Innovation in Chemistry Courses in France in the Mid-Eighteenth Century," *Ambix* 57 (2010): 3–26, and, on Rouelle more generally, see Rhoda Rappaport, "G.-F. Rouelle: An Eighteenth Century Teacher and Chemist," *Chymia* 6 (1960): 68–101.

[85] There are numerous copies of notes from Rouelle's course preserving his comments on transmutational alchemy. I have used the Clifton College Science Library (Bristol, UK) manuscript "Cours de chimie, ou Leçons de Monsieur Rouelle, recueilles pendant les années 1754, 1755, et rédigées en 1756, revües, et corrigées, en 1757 et 1758" (alchemy section on 894–903) with comparisons to Bibliothèque Interuniversitaire de Santé, Paris, Pharmacie MS 19 (fols. 394v–401r) and Bibliothèque Interuniversitaire de Santé, Paris, Médecine MS 5022 (126–40). The Clifton College MS is attributed to "Messrs. Roux et Darcet," who were preparing Rouelle's lectures for publication. For more on these and other MSS of Rouelle's course, see Rappaport, "G.-F. Rouelle" (cit. n. 84), 86–92, and references therein.

[86] Guillaume-François Rouelle, Clifton College MS 894: "le commun des phisiciens doute de la verité de cette science; mais ils ne peuvent pas être juges dans une matiere qui leur est entierement inconnüe; les plus sçavants d'entre les Chimistes, ceux même qui n'ont pas possedé ces principes ne le revoquent pas en doute." And on 898: "Les plus sensés et les plus sçavants chimistes l'ont crû, les ignorans et les gens peu instruits l'ont nié."

[87] Ibid., 898: "M. Rouelle pense que la pierre philosophale, n'est autre chose que le résultat d'une fermentation de l'or, avec le mercure; non pas le mercure ordinaire, mais un mercure particulier surchargé de phlogistique."

[88] Ibid., 903: "faute d'un guide seur pour se conduire dans une operation qui n'est conservée que par tradition."

and in his private course on rue Jacob. These students included not only Diderot but also most of the important chemists who gained prominence in the second half of the eighteenth century, including Lavoisier. How much longer after 1770 interest in and pursuit of chrysopoeia continued on the part of Académie chemists remains an open question, one I cannot treat at this time. Yet it is clear from the evidence presented above that the pursuit of transmutational alchemy persisted far longer than has commonly been believed. It should be underscored that I have here documented the persistence of chrysopoeia only among the high-profile chemists of the Parisian Académie and said nothing about what was going on in wider French circles, save for noting the collaborative connections between some non-Académie practitioners and Académie chemists. The significant point is that virtually every chemist—Geoffroy, Danty d'Isnard, Dufay, Hellot, Rouelle—within the Académie Royale des Sciences up to the 1770s continued to study, and often to experiment on, the problem of transmutation, or at least, like Macquer, defended its possibility and the study of it. This extension of serious work on chrysopoeia bridges what had previously seemed a "gap" in the pursuit of transmutational alchemy between its putative "demise" in the 1720s and its revival in the 1770s and 1780s in German circles; meaningful links between the French and the German communities, however, remain to be identified.[89]

Strikingly, we are left with a stark divide between the official face of the Académie and the private practices of its chemists. Governmental overseers like Colbert and Louvois tried to restrict "alchemical" activity within the Académie, and this attitude continued under Pontchartrain. The Académie's perpetual secretary and publicist Fontenelle tried to deny the existence of chrysopoeians within the institution's august membership. Pharmacists who wished to restrict chemistry's domain to their own practices—like Nicolas and Louis Lémery—joined the assault on transmutation. Given chymistry's poor public image in the seventeenth century—as portrayed in literature, art, and on the stage, and made worse by the stories of trickery and fraud that clustered around transmutation in particular—it is not surprising that those concerned about the reputation of the Académie wanted to dissociate that institution from anything that savored of chrysopoeia. Yet throughout it all, the chemists continued to find the topic of interest and continued to conduct experiments and exchange information in pursuit of it. The manuscript materials cited here indicate that a majority of the chemists within the Académie for its entire first century of existence, from 1666 until 1770, were seriously involved in chrysopoeia—despite administrative directives and public rhetoric to the contrary. Geoffroy's 1722 "Supercheries" does not, therefore, mark the "end of alchemy" at the Académie, much less in the wider ambit of France, but signals only a rather superficial victory for administrative naysayers. In the earlier years, Duclos, Homberg, and Geoffroy remained more or less open about their activities—Homberg in particular being protected by his close relationship and collaboration with the Duc d'Orléans. But after the death of Homberg in 1715 and Philippe in 1723, and after Geoffroy's 1722 paper, members interested in chrysopoeia kept their continuing scientific interests in transmutation quiet. Geoffroy's paper, and Fontenelle's trumpeting of it, effectively linked metallic transmutation to immoral frauds, thus making it virtually impossible for respectable chemists to be seen as actively engaged in the subject, much less defending it or making open

[89] On the alchemical revival in late eighteenth-century Germany, see Principe, *Secrets of Alchemy* (cit. n. 1), 89–92 and references therein.

claims of their success (however limited) in it. But, as Hellot's notebooks show perfectly clearly, the chemists' persistent, serious interest in gold making continued to drive their private practices. Without private papers to tell us differently, one might well acquiesce in a more or less triumphalist narrative of chrysopoeia's demise. But now we can no longer say that chrysopoeia ended in the 1720s in respectable scientific circles in France—much less that it was defeated by the experimental or theoretical work of prominent chemists—only that it went "underground," thereby leaving chrysopoeia without open defenders and thereby surrendering debate on the topic to those who ridiculed transmutation for their own, generally nonscientific, purposes. Chrysopoeia may indeed have vanished from public view in France in the 1720s, but not from the minds and hands of the chemists.

Etienne-François Geoffroy
(1672–1731), a Chemist
on the Frontiers

by Bernard Joly*

ABSTRACT

Etienne-François Geoffroy is certainly the most representative chemist of the Paris Académie Royale des Sciences in the early eighteenth century. Interested in Newtonian ideas, he did not reject Cartesian mechanism. He is the inventor of the "Table des rapports entre les substances chimiques," which remained in use throughout the eighteenth century, but he drew from the alchemical tradition. He readily theorized about the composition of metals or the laws of chemical affinities, but he practiced a chemistry that was rooted in laboratory work and the search for substances useful to craftsmen.

Etienne-François Geoffroy, chemist at the Paris Académie Royale des Sciences, presented a memoir to his colleagues on 15 April 1722 titled *Des supercheries concernant la pierre philosophale.*[1] Observing that "the Philosophers' stone opens a very large field for deception in chemistry," he explained the diverse methods and recipes by which alchemists attempted to prove that they had succeeded in the transmutation of metals. For a long time the scholarly world regarded Geoffroy as a champion in the fight against alchemy because of this memoir. As Partington remarked in his celebrated *History of Chemistry,* "He was an opponent of alchemy and exposed several alchemical tricks."[2] One could then consider, as Robert Halleux did, "that alchemy did not survive the memory of Geoffroy"[3] and that his work showed that chemistry separated itself definitively and completely from alchemy at the beginning of the eighteenth century. Yet, as Halleux noted,[4] Geoffroy's arguments were heavily indebted to a work published a century earlier, in 1617, by the physician alchemist

* UMR 8163 "Savoirs, textes, langage," Université de Lille 3, BP 60149, 59653 Villeneuve d'Ascq Cedex, France; bernard.joly@univ-lille3.fr.

This translation was a team effort, for which I am exceedingly grateful to the editors of the present volume. I want to specially thank Dr. Brenna K. Heitzman, Visiting Assistant Professor of French, Department of Romance Studies, Duke University.

[1] *Mémoires de l'Académie Royale des Sciences* (hereafter cited as *MARS*) 1722 (1724), 62–70.
[2] J. R. Partington, *A History of Chemistry* (London, 1961–70), 3:50.
[3] Robert Halleux, "L'alchimiste et l'essayeur," in *Die Alchemie in der europäischen Kultur- und Wissenschaftsgeschichte*, Wolfenbütteler Forschungen, vol. 32, ed. Christoph Meinel (Wiesbaden, 1986), 277.
[4] Ibid.

Michael Maier, who denounced the cheating of counterfeiters in order to defend authentic alchemy from the unjust attacks to which it was subject.[5] This surprising connection leads to a reinterpretation of Geoffroy's position, namely, that he could only have been criticizing the fraudulent claims of certain alchemists the better to defend a tradition on which his works were heavily based. I wish to show here that Geoffroy, while being one of the most innovative chemists of his day, did not renounce the alchemical background of chemistry. While certain colleagues at the Académie Royale des Sciences, such as Louis Lémery, rejected alchemy as outdated chemistry of centuries past and embraced the scientific novelty of Cartesian mechanical philosophy, Geoffroy to the contrary rejected the boundaries between past and present, between Cartesianism and alchemy, and between French chemistry and that practiced in the neighboring countries of England and Germany. Rather than pursuing chemistry within the narrow frame of Cartesian mechanism that was in style at the time in France, he preferred to explore the legacy of the alchemical tradition, and to draw inspiration from the many experiments transmitted by the alchemists. Without doubt this is what enabled him to be the pioneer and inventor of the "Table des différents rapports observés en chimie entre les différentes substances," which became one of the pillars of eighteenth-century chemistry under the name "Tables of affinity." We shall thus see that characterizing French science by debates between Cartesians and Newtonians is no more suitable for understanding the transformation of chemistry at the beginning of the eighteenth century than a consummate rupture with the alchemical tradition. The analysis of Geoffroy's works is at once a part of the reevaluation of the place of alchemy in the history of sciences that has been developing in recent years,[6] and of a new approach to the history of chemistry at the start of the eighteenth century. If Geoffroy was one of the great chemists of his time, this is not despite his connections to alchemy but rather because of them.

A CHEMIST'S EDUCATION

Etienne-François Geoffroy (1672–1731), who should not be confused with his younger brother Claude-Joseph (1685–1752) who had a similar career, received, according to Fontenelle, a genuine "éducation d'un fils de ministre." In the *éloge* that he gave as secretary of the Académie Royale des Sciences,[7] Fontenelle describes the meetings that Geoffroy's father, an apothecary, organized at his home where *physiciens*, doctors, and chemists gathered, including Wilhelm Homberg, member of the Académie

[5] Michael Maier, *Examen fucorum pseudo-chymicorum detectorum et in gratiam veritatis amantium succincte refutatum* (Frankfurt am Main, 1617). This work has been reedited and commented on in the framework of a thesis defended in 1991 at the University of Munich by Wolfgang Beck under the title "Michael Maiers *Examen Fucorum Pseudo-Chymicorum*—Eine Schrift wider die falschen Alchemisten." Also see Ulrich Neumann, "Michel Maier (1569–1622) philosophe et médecin," in *Alchimie et philosophie à la renaissance*, ed. J.-C. Margolin and S. Matton (Paris, 1993), 307–26.

[6] Bernard Joly, *La rationalité de l'alchimie au XVIIe siècle* (Paris, 1992); Joly, "Quelle place reconnaître à l'alchimie dans l'histoire de la chimie?" *Sci. Tech. Persp.* 25 (1993): 111–21; William Newman and Lawrence Principe, "Alchemy vs. Chemistry: The Etymological Origins of a Historiographical Mistake," *Early Sci. & Med.* 3 (1998): 32–65; Principe and Newman, "Some Problems with the Historiography of Alchemy," in *Secrets of Nature: Astrology and Alchemy in Early Modern Europe*, ed. William Newman and Anthony Grafton (Cambridge, Mass., 2001), 385–431; Joly, "A propos d'une prétendue distinction entre la chimie et l'alchimie au XVIIe siècle: questions d'histoire et de méthode," *Rev. Hist. Sci.* 60–1 (2007): 167–83.

[7] Fontenelle, "Eloge de M. Geoffroy," *Histoire de l'Académie Royale des Sciences* (hereafter cited as *HARS*) 1734 (1731), 93–100.

Royale des Sciences starting in 1691, who later became Geoffroy's teacher of chemistry.[8] Mathieu-François Geoffroy, the father, wanted his son to become an apothecary and provided opportunities for him to gain the extensive knowledge required for the profession.

In his twenties, Geoffroy traveled extensively. First, he went to Montpellier to complete the pharmacy studies that he had begun with his father in Paris. The following year, in 1693, he journeyed along the French scacoasts, and in 1698, at age twenty-six, he went to England, where, writes Fontenelle, "he earned the friendship of the majority of illustrious men in a country that produces many," then to Holland, "where he visited other savants, made other observations, and gained new knowledge,"[9] and finally to Italy in 1700. During his stay in London, where he accompanied the new French ambassador, the count of Tallard, in order to take care of his health even though he was not yet a doctor, he formed a strong friendship with Hans Sloane, secretary of the Royal Society, and maintained a correspondence for many years.[10] Thanks to these privileged ties, he quickly became a member of the Royal Society and presented communications there.[11]

When he returned to Paris and joined the Académie Royale des Sciences as Homberg's student in January 1699, he was charged with establishing and developing links between the two institutions.[12] So, on 25 April 1699, Geoffroy wrote to Sloane in English that

> Monsr. Cassini and I understanding how great advantage should be to the Royal Academy the mutual commerce of Letters with you, we have proposed it to the Academy, and promised to intertain it for our part, imparting you with all things worthy of notice in our meetings, viz. Mr. Cassini concerning the mathematical and I the Physical Knowledges. We hope you will do the favour of requiting us, as we have bond it to the Academy.[13]

Geoffroy shared several of the works he presented at the Académie Royale des Sciences with his English colleagues. At the same time, he informed his French colleagues of the works of knowledgeable British scientists. Thus, the minutes of the

[8] On Homberg, see in particular Lawrence Principe, "Wilhelm Homberg: Chymical Corpuscularism and Chrysopoeia in the Early Eighteenth Century," in *Late Medieval and Early Modern Corpuscular Matter Theories*, ed. Christoph Lüthy, John E. Murdoch, and William R. Newman (Leiden, 2001), 535–56; Principe, "Wilhelm Homberg et la chimie de la lumière," in *Chimie et mécanisme au tournant XVIIe-XVIIIe siècle, Methodos*, no. 8 (2008), http://methodos.revues.org/1223 (accessed 25 May 2014).

[9] I am drawing from Fontenelle's expressions.

[10] This correspondence, preserved in the archives of the Royal Society, was partially published by I. Bernard Cohen, "Isaac Newton, Hans Sloane and the Académie royale des sciences," in *L'aventure de la science: Mélanges Alexandre Koyré*, 2 vols. (Paris, 1964), 1:60–116. It was also studied by Jean Jacquot, *Le naturaliste Sir Hans Sloane (1660–1753) et les échanges scientifiques entre la France et l'Angleterre*, Conférences du Palais de la Découverte, installment 25 (Paris, 1954); Jacquot, "Sir Hans Sloane and French Men of Science," *Notes Rec. Roy. Soc. Lond.* 10 (1953): 87; Gavin R. De Beer, "The Relations between Fellows of the Royal Society and French Men of Science when France and Britain Were at War," *Notes Rec. Roy. Soc. Lond.* 9 (1952): 244–99.

[11] The *Philosophical Transactions* of February 1699 reports the communication that he presented under the title, "A way to make two clear spirituous inflammable liquors, which differ very little in taste and smell, and being mixted together, do give a fine carnation colour, without either sensible fermentation or alteration. Communicated by Monsieur Geoffroy F. R. S.," *Phil. Trans. Roy. Soc. Lond.* 21 (1699): 43–4.

[12] *Procès-verbaux de l'Académie Royale des Sciences* 18 (1699), fols.143v–144r; British Museum, London, Sloane MSS 4037, fol. 222; Cohen, "Isaac Newton" (cit. n. 10), 83–4.

[13] British Museum, Sloane MSS 4037, fol. 258; Cohen, "Isaac Newton" (cit. n. 10), 86.

Académie Royale des Sciences from 24 July 1703 reported that "M. Geoffroy read an extract from the *Philosophical Transactions* containing comparisons between various degrees of heat made according to a certain method using monsieur Amontons's thermometer."[14]

It is not surprising, under these conditions, that Geoffroy had read extracts from Newton's *Opticks*, which he translated from English, to his colleagues at the Académie Royale des Sciences from June 1706 to the middle of 1707.[15] But he also used his relationships with England in a more personal way, sending communications to London that he had given in Paris; translated into English, they were then published in the *Philosophical Transactions*. Thus, the first statement that he gave before the Académie Royale des Sciences on 12 May 1700,[16] was published in the *Philosophical Transactions* in September 1701 under the title "Observations upon the dissolutions and fermentations which we may call cold."[17] Several years later, an English translation of a paper he presented to the Académie on 2 May 1709, entitled "Experiments upon metals, made with the burning-glass of the Duke of Orleans,"[18] appeared in the *Philosophical Transactions* in July 1709.[19]

In Paris, meanwhile, Geoffroy had an excellent career as a chemist. *Associé* of the Académie Royale des Sciences since 1699, he became a *pensionnaire* in 1715. He was also named demonstrator of chemistry at the Jardin Royal des Plantes in 1707, and then professor in 1712. The lectures he gave supplied a good part of the material for the *Nouveau cours de chymie suivant les principes de Newton & de Stahl*. Published anonymously in Paris in 1724, this *cours* was afterwards attributed to a physician named Senac, who was twenty-nine at the time. But the role of the latter was no doubt limited to collecting the lessons given by Geoffroy, together with his colleague Boulduc.[20] The work is uneven, but it adequately represents the teachings of

---

[14] *Procès-verbaux de l'Académie royale des sciences* 22 (1703), fol. 266v. Amontons indeed presented a comparison between the measures performed by the English author and his own (*MARS* 1703 [1720], 200–212).

[15] Henry Guerlac points out that the copy from *Opticks* sent by Sloane to Geoffroy, which contains the handwritten quote "Pour Mr Geoffroy de l'académie des sciences" as well as Geoffroy's signature, is found in the Cornell Library at Cornell University; see Guerlac, *Newton on the Continent* (Ithaca, N.Y., 1981), 76. Bernard Cohen thinks he has found the translation from Geoffroy at the Arsenal Library in Paris in the form of an anonymous manuscript titled *Optique ou traité des réflexions, réfractions, inflexions et couleurs de la lumière par Mr Neuton 1704*; see Cohen, "Isaac Newton" (cit. n. 10).

[16] Geoffroy, "Observations sur les dissolutions et sur les fermentations que l'on peut appeler froides parce qu'elles sont accompagnées du refroidissement des liqueurs dans lesquelles elles se passent," *MARS* 1700 (1742), 110–21.

[17] *Phil. Trans. Roy. Soc. Lond.* 22 (1701): 951–62: "Observations upon the dissolutions and fermentations which we may call cold, because they are accompanied with a coolness of the liquors into which they pass. And of a new thermometer. Extracted out of a discourse, which Mr Geoffroy, FRS, made in the public meeting of the Académie royale des sciences the 21st of April" (the date is incorrect).

[18] *Phil. Trans. Roy. Soc. Lond.* 26 (1709): 378–86. As we will see later, it is about the giant lens constructed by Tschirnhaus at the request of Homberg, financed by the Duke of Orléans and installed in Paris in the gardens of the Palais Royal.

[19] Geoffroy, "Expériences sur les métaux faites avec le verre ardent du palais Royal," *MARS* 1700 (1742), 162–76. By burning off the metals with the burning glass, Geoffroy thought he was engaging in their analysis and making the two principal components of all metal appear: a sulfur or oily substance and a vitrifiable earth or calx, which seemed to confirm his position in the quarrel that had gone on for four years with Louis Lémery, as we will soon see. However, in the English version, Geoffroy took care to erase the references to his previous publications that held traces of the argument with Lémery.

[20] This is what Baron asserts in the preface of his republished *Cours de chymie* of Nicolas Lémery (Paris, 1757). I thank Christine Lehman for having brought this text to my attention.

a chemist who incorporated the doctrines of English chemistry, inspired by Newtonian physics as well as the chemistry of Becher and Stahl, without breaking with the French tradition of the "Cours de chymie."[21] The first part recounts the theory of three earths developed by the German alchemist Johann-Joachim Becher, then the Stahlian doctrine of phlogiston; it ends with a resume of the chemistry that Freind and Keil developed in England "according to the principles of monsieur Newton." The second part, by far the longest (more than 500 pages), mixes references to the chemistry of Stahl with numerous formulas borrowed from (without citing) the *Cours de chymie* of Nicolas Lémery.

With the exception of his *Tractatus de materia medica* to which we will soon return, Geoffroy's work was made up of the thirty or so memoirs that he presented before the Académie Royale des Sciences from 1700 to 1727. To these personal works were added the reports of works of other scientists as well as the numerous expert assessments that he was commissioned by the Académie to write, often in collaboration with other academics, about machines, inventions, and new processes. To only take a few examples, he gave his opinion on a machine for desalinating sea water (1 September 1717), on "two machines from England to laminate and mold lead pipes" (24 January 1728), or again on the "method to make tin harder" (31 August 1729).[22]

DOCTOR GEOFFROY

Like Boerhaave and Stahl at this time, Geoffroy was also a physician.[23] Alongside his commitments as a chemist at the Académie Royale des Sciences and at the Jardin Royal, he was named professor of medicine in 1709 at the Collège Royal, then at the Faculté de Médecine de Paris, where he became dean in 1726. But his medical publications did not stray far from chemistry, because his medical courses at the Collège Royal, which were gathered together after his death and published in Paris in 1741 as the *Tractatus de materia medica*, related to the preparation of remedies using chemical operations.

Geoffroy was not satisfied just teaching medicine; he practiced it and acquired the reputation among his contemporaries of being an excellent doctor. Fontenelle, in his *éloge*, states that "convinced that a doctor belonged equally to all the sick, he made no distinction between good and evil patients, or between illustrious and obscure ones."[24] A recent study by Laurence Brockliss focusing on the treatments that Geoffroy dispensed through his correspondence reveals him to be careful and attentive, distrusting all doctrinal approaches and concerned about the specifics of each sick person.[25] Certainly, he prescribed chemical medications, but his preference went to thermal cures and the consumption of mineral waters. He also, as Fontenelle

[21] On this tradition, see Bernard Joly, "L'édition des *Cours de chymie* aux XVIIᵉ et XVIIIᵉ siècles: obscurités et lumières d'une nouvelle discipline scientifique," in *La diffusion du savoir scientifique: XVIᵉ–XIXᵉ siècle*, Archives et bibliothèques de Belgique, no. 51 (Brussels, 1996), 57–81.

[22] Christian Fabre, memoir, "Etienne-François Geoffroy entre chimie et alchimie: tentative de démarcation" (MA thesis, Univ. of Lille 3, 1994).

[23] Here I am summarizing the analyses that I developed in "Le médecin, philosophe mécanico-chimiste selon Etienne-François Geoffroy (1672–1731)," in *Machine and Wife*, ed. Nunzio Alloca (Münster, forthcoming).

[24] Fontenelle, "Eloge de M. Geoffroy" (cit. n. 7), 97.

[25] Laurence Brockliss, "Consultation by Letter in Early Eighteenth-Century Paris: The Medical Practice of Etienne-François Geoffroy," in *French Medical Culture in the Nineteenth Century*, ed. Ann La Berge and Mordechai Feingold (Amsterdam, 1994), 79–117.

points out, "very carefully listened to nature and did not disturb nature by using remedies under the pretext of helping."[26] The image of a doctor who comes from the Hippocratic tradition takes shape, Geoffroy thus falling within the trend of "reinventing Hippocrates" that is characteristic of his time.[27] Brockliss summarizes Geoffroy's position well when he writes: "Geoffroy, then, was very much the Hippocratic doctor: he treated an individual, not a disease entity."[28]

Indeed, Geoffroy referred to Hippocrates in his medical thesis that he defended on 31 May 1703 under the title *An medicus, Philosophus Mechanico-Chymicus?* The first part is an introduction intended to justify the necessity for doctors to avoid the two pitfalls of dogmatism and empiricism by means of natural philosophy. The doctor only arrives at useful knowledge of his art by exploring multiple aspects of the economy of the human body. The second part is a description of the solid parts of the "humana machina"; these are the small conduits that make it up, as well as the diverse fluids that travel along them and constitute its wet parts. The doctor who claimed to have cured sicknesses without possessing the understanding of the working of these parts would be comparable to an artisan who repaired clocks without understanding their clockwork. However, Geoffroy was not satisfied with this mechanistic approach. Admittedly, he recalls these principles in the third part of his thesis, by referring notably to the *Projet d'une nouvelle mechanique* by Pierre Varignon published in 1687. But he immediately emphasized that this necessary usage of mechanics in medicine is insufficient, because only chemistry is capable of penetrating the most intimate parts of the body to understand the nature of the internal forces that govern these movements. The critique of Cartesianism is explicit here: "It is in vain for one to attempt to explain [these interior forces] by using the terms *motion, shape,* and *bulk,* if one does not have recourse to this art, namely chemistry, which opens the hidden qualities of bodies and which reveals their *motion, shape,* and *bulk.*"[29]

The knowledge of chemistry that resolves mixed bodies into simple substances and composes new mixtures based on these substances[30] is then as necessary to the doctor as is the knowledge of mechanics. This is explained in the fourth part of the thesis that applies the teachings that were just mentioned to the functioning of the human machine. Certainly, movement of muscles and nerves, the circulation of fluids, and the functioning of the heart can be understood by the laws of mechanics. However, another perspective is necessary, which does not annul that of the mechanic but deepens his medical knowledge by permitting the understanding of the formation of fluids as well as the formation of chyle. Only chemistry allows the understanding that the human body is not a machine like others because it is provided with a sort of perpetual motion: the process of nutrition, the functioning of the brain, which is as a distillation by alambic,[31] and the regeneration of tissues or the secretion of glands must lead back to the operations of fermentation or of distillation. Geoffroy can then, in the fifth and last part of his thesis, deduce from this that the doctor must be at the same time a chemist and a mechanic.

From 1703, Geoffroy refused to choose between the two camps of chemical medi-

---

[26] Fontenelle, "Eloge de M. Geoffroy" (cit. n. 7), 96.
[27] See David Cantor, ed., *Reinventing Hippocrates* (Aldershot, 2002).
[28] Brockliss, "Consultation" (cit. n. 25), 96.
[29] Geoffroy, *An medicus, Philosophus Mechanico-Chymicus?* (1703), 4.
[30] Ibid. "Chymistry has double operation, Solution, Concretion."
[31] Ibid., 7.

cine and iatromechanism. His concern for reconciliation is not so much a character trait[32] as an epistemological position: far from all reductionism, Geoffroy regarded medicine as the field where mechanics and chemistry complement each other rather than conflict with one another. Moreover, his vision of chemistry as the study of fundamental forces that are at the origin of the movements of matter gives this discipline a place that can seem all the more legitimate in that it does not encroach on the place of other sciences. In these conditions, chemistry has no need to justify its presence within the new natural philosophical knowledge, nor is it subordinate to doctrines that, like Cartesianism, reduce these operations and their objects to mechanical ones.[33] On the contrary, it can pride itself on its antiquity, without either being forced to break with a past for which it should be embarrassed or needing to fit itself into new doctrines that do not respect its specificity. At the same time, we see here that Geoffroy was a figure more attached to an empirical approach to knowledge than to the invocation of the great scientific doctrines of his time.

QUARRELS WITH THE CARTESIANS

This view guided Geoffroy throughout his career as a chemist, as seen in his quarrel with his colleague Louis Lémery from 1704 to 1708 on the artificial fabrication of metals.[34] It is the theory of the principle Sulfur developed by Wilhelm Homberg, which Lawrence Principe has shown to be rooted in the alchemical tradition,[35] that constitutes the point of departure of this quarrel. Reexamining the analysis of the composition of sulfur developed by Wilhelm Homberg in 1703,[36] Geoffroy, in a memoir presented to the Academy on 12 November1704,[37] proposed to "imitate nature," that is, to produce sulfur with substances that he thought to have been isolated by the chemical analysis of this product (a salty acid, a bituminous substance imagined to contain the Sulfur Principle, and an earthy alkali).[38] He added that the same process must allow the artificial production of metals and, in particular, of iron from their components, presumed to be bitumens, vitriolic salt, and an earth. He

---

[32] Fontenelle, in his "Eloge de M. Geoffroy" (cit. n. 7), describes him as a man with a nature "mild, circumspect, moderate, and perhaps even a bit timid" (96).

[33] In my work *Descartes et la chimie* (Paris, 2011), I showed why and how Descartes had wanted to reduce chemistry to mechanism, and the reasons for the failure of such an undertaking. See also Joly, "Could a Practicing Chemical Philosopher Be a Cartesian?" in *Cartesian Empiricisms*, ed. Mihnea Dobre and Tammy Nyden (Dordrecht, forthcoming).

[34] At this time, metals and sulfur were still considered as mixed bodies. I elaborated the different moments in this dispute and analyzed the stakes in "Quarrels between E. F. Geoffroy and Louis Lémery at the Académie Royale des Sciences in the Early Eighteenth Century: Mechanism and Alchemy," in *Chymists and Chymistry: Studies in the History of Alchemy and Early Modern Chemistry*, ed. Lawrence Principe (Sagamore Beach, Mass., 2007), 203–14; Joly, "Le mécanisme et la chimie dans la nouvelle Académie royale des sciences: les débats entre Louis Lémery et Etienne-François Geoffroy," in *Chimie et mécanisme au tournant XVIIe–XVIIIe siècle, Methodos*, no. 8 (2008), http://methodos.revues.org/1403 (accessed 25 May 2014).

[35] See the works cited in n. 6.

[36] Homberg, "Essai de l'analyse du soufre commun," *MARS* 1703 (1720), 31–40.

[37] Geoffroy, "Manière de recomposer le souffre commun par la réunion de ses principes, et d'en composer de nouveau par le mélange de semblables substances, avec quelques conjectures sur la composition des métaux," *MARS* 1704 (1745), 278–86.

[38] Fontenelle presented the method followed by Geoffroy in this way: "One is never so sure of having decomposed (or broken down) a mixt into its true components as when one is able to reconstitute it out of those same components. This reconstitution is not always possible, but when it is not, this failure does not necessarily prove conclusively that the analysis of the mixt is incorrect, but when [the reconstitution] succeeds, it does prove the analysis correct" (*HARS* 1704 [1745], 37–8).

promised numerous experiments to prove this hypothesis, and he indeed presented a new memoir entitled "Problème de chimie. Trouver des cendres qui ne contiennent aucunes parcelles de fer"[39] at the end of 1705. This memoir highlighted a new aspect of the problem: the use of a magnetized blade showed that all plant ash contained some fragments of iron. It was Homberg who first, in a memoir on 8 May 1706 entitled "Observations sur le fer au verre ardent,"[40] presented the hypothesis that Geoffroy would consequently accept again and treat as his own: "the products resulting from incinerating all types of vegetable matter contain iron."

Louis Lémery had rejected this conclusion several weeks earlier in his memoir on 14 April 1706 ("Diverses expériences et observations chimiques et physiques sur le fer et sur l'aimant"),[41] where he maintained that the properties of iron came not only from chemistry but also from physics, alone capable of understanding these magnetic properties of iron by a theory of pores of the material in which the corpuscles of the magnetic material circulated.[42] He outlined the thesis that he would develop in his later memoirs: by combining with the oily parts of the iron, the acids from plants form vitriols that plug up the pores of the metal and prevent the circulation of magnetic material; the iron was "hidden," since it no longer reacted to the presence of a magnet. The violent fire of the calcination destroyed the vitriol by driving off the oily part, thus rendering the iron detectable by the magnet again. Thus there was no need, for Lémery, to imagine the artificial production of iron, because it was already present in the plant before its calcination.

Six months later, on 13 November 1706, Louis Lémery would use this thesis to explicitly attack Geoffroy's position in the memoir: "Que les plantes contiennent réellement du fer, et que le métal entre nécessairement dans leur composition naturelle."[43] The experiments of the "végétation chimique," that is, the crystallization that was produced when one threw oil of tartar on iron filings dissolved in spirit of niter, shows that the iron can rise despite the action of gravity;[44] this rendered Geoffroy's hypothesis useless. Lémery came back to it several weeks later, in a memoir on 8 January 1707 entitled "Expériences nouvelles sur les huiles, et sur quelques autres matières où l'on ne s'était point encore avisé de chercher du fer."[45] Affirming that he worked both with experience and reason, he concludes: "the materials that M. Geoffroy used, and that he blended together to produce his artificial iron, are all rightly suspected of containing genuine iron."[46]

It is important to note here that Lémery's criticisms of Geoffroy extended beyond the sole question of the production of artificial metals. They indeed include epistemological and institutional dimensions because, by contesting the specific chemical character of the operations of chemistry, they questioned the existence of chemis-

[39] MARS 1705 (1730), 362–3.
[40] MARS 1706 (1731), 158–65.
[41] MARS 1706 (1731), 119–35.
[42] Without getting into the details, this theory returns to the works of Descartes in the fourth part of Principia philosophiae, articles 133–83, AT VIII 275–311 for the 1644 Latin edition, AT IX-2 271–305 for the 1646 French edition.
[43] MARS 1706 (1731), 411–8.
[44] Meanwhile (memoir from 10 July 1706), Louis Lémery's father, Nicolas, had in effect played the two protagonists off each other, denying at the time that combustion could be an instrument in the synthesis of metals and that iron could go back to plant fibers, which, according to him, would have contradicted the laws of gravity ("Du miel et de son analyse chimique," MARS 1706 [1731], 272–82).
[45] MARS 1707 (1730), 5–11.
[46] Ibid., 6.

try as a scientific discipline distinct from others. Indeed, for Lémery, chemical substances that contributed to the formation of a new mixed body juxtaposed themselves to constitute new assemblages without losing any part of their identity or of their properties; they were at most temporarily camouflaged. Thus, vitriol is not a new body, but a simple juxtaposition of iron and acid that takes place in such a way that the iron seems hidden in the acidic solution. The unions and separations of the simple bodies are in the end only mechanical operations. Thus he can speak, in a memoir on 12 November 1707 devoted to the production of vitriolic inks,[47] about the "mécanique" by which nutgall traps the acid spirit of vitriol in its pores and releases iron that gives a black color to its ink, according to the same "mécanique" that makes verjuice, an acid juice, erase the spots of ink on linen by absorbing the iron. Lémery uses the same term again three separate times in his last attack against Geoffroy, on 5 December 1708,[48] when he alludes to the "mécanique" by which "vitriol flows in plants,"[49] and then that by which this "alleged recomposition of iron"[50] is made.

In the responses to his adversary, Geoffroy supported his position by using the conceptual frame furnished by a chemistry of principles inherited from the alchemical tradition that works as well for minerals as for plants: this allows him to keep the specificity of chemistry as a science that distinguishes itself from physics. Thus in a memoir from 21 May 1707 entitled "Eclaircissements sur la production artificielle du fer, & sur la composition des autres métaux,"[51] he writes:

> I am going to show that the principles of plants and those of minerals are essentially the same and that minerals can be broken down quickly and without much effort by separating their principles and recomposed by substituting principles taken from plants for the originals that have been removed.[52]

He then shows that "salts of plants do not differ at all from salts of minerals"[53] and that, in the composition of iron, an inflammable principle of plant origin replaces the inflammable principle of mineral origin that the plant contained before its calcination. He stresses the proof establishing the reversibility of operations. The calcination of iron indeed produces a calx that can be transformed into iron by calcination with an inflammable material, like flaxseed oil: "in this operation, the iron earth regains the sulfur principle that it had lost."[54] The formula of artificial fabrication of iron is thus only a variant of the more general operation by which calxes regain "the metallic form that they had lost" when the addition of the sulfur principle to an earthy material allows the reconstitution of metal.[55] One sees the similarity with Stahl's theory of phlogiston. Moreover, Geoffroy recognized unequivocally in 1720 during the debates that developed about his *table des rapports entre les différentes substances* that

---

[47] Lémery, "Eclaircissement sur la composition des différentes espèces de vitriols naturels & explication physique & sensible de la manière dont se forment les Ancres [*sic*] vitrioliques," *MARS* 1707 (1730), 538–49.

[48] Lémery, "Nouvel éclaircissement sur la prétendue production artificielle du fer, publiée par Becher, & soutenue par M. Geoffroy," *MARS* 1708 (1730), 376–402.

[49] Ibid., 394.

[50] Ibid., 400.

[51] *MARS* 1707 (1730), 176–88.

[52] Ibid., 180.

[53] Ibid., 181.

[54] Ibid., 184.

[55] Ibid., 187.

the Sulfur Principle, which he named "Principe huileux," corresponds to what Stahl called "Principe phlogistique."[56]

In the last memoir that he dedicated to this question on 2 May 1709, "Expériences sur les métaux faites avec le verre ardent du palais Royal,"[57] Geoffroy presents the analyses made by the burning glass on iron, copper, tin, and lead: in all these cases, we see that these metals are composed of a metallic sulfur that is not different from the oil that we find in animals and plants, along with a base that is unique to each one. This recourse to the constituent principles of all bodies renders the evocation of the porous nature of metals useless and, consequently, avoids a mechanistic type of interpretation. At the same time, the reversibility of the transformation of a metal into its calx and of its calx into metal proves the fact that the "artificial" production of iron is only the implementation of a perfectly natural process.

We now come across a curious situation. In effect, Homberg and Geoffroy used a modern device, in this case, the large burning glass fabricated by Tschirnhaus that the Duke of Orléans had had installed in the Palais Royal in Paris at the request of Homberg,[58] while Nicolas and Louis Lémery used traditional processes of chemistry.[59] But the instrumental modernity of Homberg and Geoffroy was used for demonstrations that depended on a theoretical base consisting of concepts and representations inherited from the alchemical tradition, while the Lémerys claimed to be modern by referring to Cartesian concepts. It is therefore not surprising that Louis Lémery had wanted, at the end of the quarrel, to discredit his adversary by recalling his sources. Thus, in the memoir of 5 December 1708 that I have already mentioned, Lémery indicates that Geoffroy was inspired by the writings of Becher, "Doctor-Chemist, known as such from his many writings addressed to the public, wishes to revive the courage of those who work on *metallification*, and defend alchemy from public insults."[60] In Lémery's eyes, Geoffroy is then a man of the past whose science is out-of-date, alchemy being the name of the outdated chemistry of previous centuries.

Since his first memoir in 1704, Geoffroy had not hidden the fact that he drew his hypothesis concerning the production of metals from what he called "le procédé de Becker." Johann-Joachim Becher, in a 1671 supplement to the *Physica subterranea*, developed his theory of the natural generation of metals, based on three types of earth rather than the three Paracelsian principles.[61] Geoffroy explicitly inscribed himself in

---

[56] Geoffroy, "Eclaircissements sur la table insérée dans les mémoires de 1718 concernant les rapports observés entre les différentes substances," *MARS* 1720 (1722), 29. Geoffroy was in contact with Stahl at the time through the intermediary Caspar Neumann, apothecary to the King of Prussia, whom Stahl had persuaded to come to Paris to get in contact with the French alchemists. See Partington, *History of Chemistry* (cit. n. 2), 2:702–6.

[57] *MARS* 1709 (1733), 162–76.

[58] On this subject, see John Powers, "Measuring Fire: Herman Boerhaave and the Introduction of Thermometry into Chemistry," in this volume.

[59] This is what Fontenelle notes, *HARS* (1706), 32–3: "M. Lemery (the son) and M. Homberg have both studied iron, one by traditional chemistry, the other by the new chemistry, where the only furnace is the burning glass of the Palais Royal."

[60] *MARS* (1708), 377. Fontenelle, summarizing the memoir of Lémery in l'*Histoire de l'Académie*, could conclude in an ironic manner: "It is not yet time to conceive of the pleasing prospect of the artificial production of metals" (*HARS* [1708], 65).

[61] *Experimentum chymicum novum quo artificialis et instantanea metallorum generatio et transmutatio ad oculum demonstratur*. Supplement published separately in 1671 in Frankfurt am Main and then integrated in the second edition of *Physica subterranea* (Frankfurt am Main, 1681).

the tradition of transmutational alchemy. This is what Fontenelle comments on, not without irony:

> If the composition of this metal were once worked out with complete certainty, this would apparently be a step toward [understanding] those of other metals. Chemistry has nothing grander or more difficult to offer than to know [their composition] down to their very principles and perhaps then this famous object of so much useless study [transmutation of metals] would cease to be fanciful.[62]

And this is what Geoffroy subsequently observed, in his memoir on 21 May 1707:

> I hope . . . to show . . . that not only can iron be produced but all the other metals; they can be built up or broken down by uniting or separating the principles of which they are composed.[63]

We then see that, contrary to what certain preconceived schemas of the history of French science would have us believe about the beginning of the eighteenth century, it is not in the name of Newtonianism, but rather from the alchemical tradition that Geoffroy opposed the mechanistic conceptions of some of his colleagues who passed for Cartesians. More precisely, it is to Becher and Stahl that Geoffroy refers, and not to Newton.

A TABLE WITH CONTROVERSIAL ORIGINS

In 1718 Geoffroy presented a memoir titled "Table des différents rapports observés en chymie entre les différentes substances" to the Académie Royale des Sciences. He writes:

> Whenever two substances having some disposition to join with one another are found united together, if a third appears having more rapport with one of these two, it will unite with it by causing the other to let go.[64]

There have been many debates on the origins and the significance of this publication.[65] Let us simply remark here that Geoffroy generalized observations that chemists had reported for a long time and that we find abundantly in the "Cours de chymie" of the seventeenth century.[66] In his *Cours de chymie* of 1675, Nicolas Lémery questions why spirit of salt precipitates what aqua fortis had dissolved; according to him, it is that the points of the acid that form the spirit of salt are stronger and heavier

---

[62] *HARS* 1704 (1745), 39.

[63] *MARS* 1708 (1730), 176–7.

[64] *MARS* 1718 (1741), 202–12, on 203. On the history of the theory of chemical affinities, see Michelle Goupil, *Du flou au clair? Histoire de l'affinité chimique* (Paris, 1991), 145; Mi Gyung Kim, *Affinity, That Elusive Dream: A Genealogy of the Chemical Revolution* (Cambridge, Mass., 2003). Kim's work includes an abundant bibliography on the question.

[65] See, e.g., the debate between Ursula Klein and Frederic Holmes: Ursula Klein, "The Chemical Workshop Tradition and the Experimental Practice: Discontinuities within Continuities," *Sci. Context* 9 (1996): 251–87; Frederic L. Holmes, "The Communal Context for Etienne-François Geoffroy's 'Table des rapports,'" *Sci. Context* 9 (1996): 289–311.

[66] See also on this subject William R. Newman, "Elective Affinity before Geoffroy: Daniel Sennert's Atomistic Explanation of Vinous and Acetous Fermentation," in *Matter and Form in Early Modern Science and Philosophy*, ed. Gideon Manning (Leiden, 2012), 99–124.

than those of the aqua fortis, which weakens the union of the aqua fortis with a metal, provoking their precipitation. But, not wanting to be wedded to this reduction of the chemical operations to a mechanical process, Lémery immediately adds:

> Some have written that one need not impute this precipitation either to the weight or to the force [acting there] or to any shaking or concussion that the spirit of salt gives to the aqua fortis or to dissolved substances. But instead [one should impute the precipitation] to the connection of the acid of that spirit [of salt] to the volatile and sulphureted alkali of the aqua fortis or to spirit of niter which, constrained by [connection], this latter gives up the metal which it had dissolved.[67]

In short, the chemist is continuously confronted, in his laboratory activities, with operations in which the introduction of an acid or an alkali in a solution containing a metal provokes the precipitation of this metal at the end of an agitation, of heating, of an effervescence that makes one think of a battle.

Geoffroy had collected everything that he had been able to learn from reading old formulas and from his practices and from those of his colleagues in the Académie Royale des Sciences. He held true to observed facts and guarded against adopting the explanatory hypotheses of any chemist or alchemist. The originality of his work lies in the idea of putting the results of his experiments in tabular form, and taking what he calls the "rapports," that is to say the capacity of a substance to constantly associate itself with another in preference to a third, as criteria for ordering them. It is possible he had drawn the idea of reassembling the experimental results in a table from Francis Bacon's *Novum organum*. Bacon indeed wrote:

> But *Natural* and *Experimental History* is so various and scattered that it may bewilder and distract the intellect unless it be set down and presented in suitable order. So we must fashion *Tables*, and *Structured Sets of Instances*, marshalled in such a way that the intellect can get to work on them.[68]

What remains is the specific concept of "rapport," often thought of as an equivalent of Newtonian attractions in the eighteenth century, all the more readily in light of Geoffroy's connections with England. Fontenelle recognized the importance of Geoffroy's work when he maintained that "this table became somewhat prophetic, because when one mixes the substances together, one can predict the effect and result of that mixture"; but he also remarked that "it is here that the affinities and attractions would become appropriate, if they were something."[69] Thirteen years later in his *éloge* for Geoffroy, he wrote again that "these affinities make some of us concerned because we fear that they are only disguised attractions."[70] Subsequently, the New-

---

[67] Nicolas Lémery, *Cours de Chymie*, 7th ed. (Paris, 1690), 356.

[68] Francis Bacon, *Novum organum*, London, 1620, book II, aphorism 10, in Graham Rees and Maria Wakely, *The "Instauratio magna" Part II: "Novum organum" and Associated Texts* (Oxford, 2004), 215. We do not know whether Geoffroy had read Bacon, but it would be very surprising if his friends in the Royal Society had not spoken to him about it.

[69] *HARS* 1718 (1719), 30–7.

[70] Fontenelle, "Eloge de M. Geoffroy" (cit. n. 7), 93. On Fontenelle's subtle and complex attitude toward chemistry, see François Pépin, "Fontenelle, l'Académie et le devenir scientifique de la chimie," *Methodos*, no. 12 (2012), http://methodos.revues.org/2898 (accessed 25 May 2014); Luc Peterschmitt, "Fontenelle et la chimie: la recherche d'une 'loi fondamentale' pour la chimie," *Methodos*, no. 12 (2012), http://methodos.revues.org/2873 (accessed 25 May 2014).

tonian character of Geoffroy's table was affirmed by numerous authors such as Buffon, Bergman, and Maupertuis.[71] It is not likely that Geoffroy would have accepted this "newtonization" of the relationships between chemical substances. Indeed, we are struck by the care with which he held himself back from using any other term than "rapport" and kept Newtonian theories that he knew so well at a distance,[72] which has led many historians of science to question the Newtonianism of Geoffroy.[73]

Far from referencing the theses of Newton, in the first paragraphs of his memoir, Geoffroy insists on the empirical character of the arrangement of the table. So, he writes,

> One *observes* in chemistry certain rapports between different substances which enable them to unite easily among themselves. These rapports have their degrees [of strength] and their *laws*. One *observes* their differential degrees when, where several substances having some *disposition* to unite together are intermixed, *we perceive* that one of these substances always unites constantly with another particular one in preference to all the others.
>
> *I have observed that*, among those substances that have this *disposition* to unite together, when two are found united, some [of the others], *in approaching and mixing*, are joined with one of these and make it let go of the other substance whereas others neither join with either [of the united pair] or separate them at all.[74]

The "laws" and "dispositions" are thus only the result of constant and concordant observations for Geoffroy, the fruit of a generalizing Baconian induction allowing him to pass from "I have observed that . . . " to "whenever . . . ."[75] For Geoffroy, law is neither metaphysical deduction, like the laws of impact of bodies in the second part of the *Principes de la philosophie* by Descartes, nor axiomatic prerequisites such as

[71] Maupertuis, *Vénus physique* (Paris, 1745), 102–3; Buffon, "De la nature. Seconde vue," *Histoire naturelle*, vol. 13 (Paris, 1765), 12–3; Torbern Bergman, *De attractionibus electivis* (Uppsala, 1775), French translation, *Traité des affinités chymiques* (Paris, 1788), 1–5. Maupertuis maintained in *Vénus physique*, after having cited Geoffroy's memoir: "I cannot help noting here that these forces and relationships are nothing other than what other, bolder thinkers have called attraction. This ancient term, revived in our time, at first scared off the physicians who believed that they could explain all natural phenomena without it. The astronomers were the first to feel the need for a new principle for the movement of heavenly bodies, and they believed they had discovered it in these same movements. Chemistry has since recognized this necessity: and the chemists are the most prominent today in accepting attraction and expanding it further than the astronomers."

[72] We have an example of this distancing in a letter to Sloane from 10 April 1715, about a communication from John Freind refuting the objections of the adversaries of the force of attraction. Geoffroy writes: "I have seen articles in the Philosophical Transactions addressing attraction, magnetism, or the electricity of bodies, and the centripetal and centrifugal forces. You have spoken to me about an account of several experiments relating to it before the Royal Society. Is it a simple index of experiments, or is it an inventory? Either way it would give me pleasure if you would send it to me. These experiments appear very curious to us, but it is difficult to adopt the term attraction that for us suggests occult qualities. The term magnetism gives a more sensible idea of the way to approximate or gauge these bodies. The truth is that it carries with it a prejudice that one can avoid by using the word attraction, which simply signals an effect for which the cause, according to Mr. Newton, is still unknown" (British Museum, Sloane MSS 4044, fol. 51). Cited by Cohen, "Isaac Newton" (cit. n. 10), 100.

[73] E.g., William Smeaton, "E. F. Geoffroy Was Not a Newtonian Chemist," *Ambix* 18 (1971): 212–4; Ursula Klein, "E. F. Geoffroy's Table of Different 'Rapports' Observed between Different Chemical Substances: A Reinterpretation," *Ambix* 40 (1995): 79–100.

[74] *MARS* 1718 (1741), 202–3 (emphasis mine).

[75] Let's recall the quote given above: "*Whenever* [in French, "toutes les fois que"] two substances having some disposition to join with one another are found united together, if a third appears having more rapport with one of these two, it will unite with it by causing the other to let go" (emphasis mine) *MARS* 1718 (1741), 203.

the three laws of motion that open Newton's *Principia mathematica*; rather, it is only the expression of the result of an empirical process. To construct a table, he says, "I have collected more experiences and observations from other Chemists than from my own work."[76]

Geoffroy was aware of the inevitable limits brought to his work by such a method. He explains this at the end of his memoir of 1718. First he comments that his table does not pretend to be exhaustive: "Although this Table contains quite a few Substances whose rapports we have compared, I have no doubt at all that one could still add many more whose rapports could be determined by dint of experiment."[77] He then observes that he does not justify the order in which the different substances are organized in each of the columns. The precise data that he promises to give later will not be deduced from a theory, but only from new experimental observations:

> If it be thought fit, I shall subsequently provide all the experiments on which the rapports of different substances contained in this Table are based and which have enabled me to determine how to arrange them in the order in which one sees them disposed.[78]

Finally, he remarks that the predictions that the table promises are approximate, owing to the infinite diversity of material substances:

> It should be noted that, in several of these experiments, the separation of substances is not always perfectly exact and precise. This results from several unavoidable causes, such as the viscosity of the liquid, its movement, the shape of the precipitating [substance] and the precipitate and other similar things which do not allow a rapid drop or an exact separation of all the particles; this is nevertheless so inconsiderable that it does not prevent us from regarding the rule as constant.[79]

It is precisely these "other similar things . . . so inconsiderable" that would lead chemists at the beginning of the nineteenth century, notably Berthollet in his *Essai de statique chimique* of 1803, to consider these exterior circumstances that Geoffroy judged as negligible to play an essential role in chemical reactions.

CONCLUSION

Over the last few decades, many studies of the history of chemistry have shown that alchemy comprised a form of rational knowledge in the seventeenth century, taking its place among other disciplines of the time, to the point that the distinction between chemistry and alchemy familiar to us today makes hardly any sense in that period.[80] Even Descartes, in a text rarely cited,[81] recognized that the project of transmutation of metals was a rational activity, and his strong reservations toward alchemists were

---

[76] *MARS* 1718 (1741), 203. I developed this analysis of the empiricism of Geoffroy's positions in "Was Early Eighteenth Century's Chemistry an Empirical Science?" in *What Does It Mean to Be an 18th Century Empiricist?* ed. Anne-Lise Rey and Siegfried Bodenmann (Dordrecht, forthcoming).

[77] *MARS* 1718 (1741), 212.

[78] Ibid.

[79] Ibid.

[80] See the works cited in n. 6.

[81] *Notae in programma*, AT VIII 353. In this critique of the positions of his former disciple Regius, Descartes distinguishes things brought up by faith or theological debate from those "that are only subjected to study by human reason, such as the quadrature of the circle, the philosopher's stone, and other similar things."

actually directed toward the chemical doctrines of his contemporaries, which seemed to him incompatible with his metaphysical conception of matter.[82] It is in this context that it is necessary to evaluate the debates focusing on chemistry at the Académie Royale des Sciences at the beginning of the eighteenth century. The alchemy that some criticized was not irrational knowledge, hermetic, or esoteric; such interpretations of alchemy—essentially anachronistic—would come much later. Alchemy, for a man like Louis Lémery, is only the name of a chemistry that he considers out-of-date, essentially because it does not allow itself to be reduced to the demands of Cartesian mechanics.[83] But for Geoffroy, on the contrary, it constitutes a past that would be absurd to deny, for it is so rich in experiments and hypotheses about the constitution of material bodies that provide fodder for a modern chemist.

Such an approach renders traditional debates between partisans of a continuist reading of the history of sciences and those who consider that modern science is made up of ruptures with its past obsolete. There is, in fact, just as much continuity as rupture in the works of chemists with whom a faithfulness to the past can be considered the engine of their innovation. Lémery and his Cartesian friends claimed to be estranged from the chemical understanding of the previous century. But it is necessary to comment on the weak heuristic power of their new doctrines; chemistry gained nothing from reasoning in terms of points and pores. Geoffroy, who did not reject the principle-based theories of material that chemists with a keen interest in alchemy like Becher or Homberg put forth, produced innovation by inventing his *Table des rapports*.

But we can also remark that the fecundity of Geoffroy's research lies in the fact that it ultimately held itself apart from all systems, whether Cartesian mechanism or Newtonian theory of attraction. Fundamentally, Geoffroy was an empiricist, not in the pejorative sense that the term still had at the time among learned physicians, but in the sense we can give to this term to designate the epistemological position of those who, at all times, give priority to results obtained by experimental research. The circumstances of the education that he had received in his youth had, without a doubt, favored such a state of mind, but this empiricism seems to me above all to be the result of the practice of a chemist who knows that the material whose secrets he would like to penetrate will always present itself to him in diverse forms, moving and shimmering, which no doctrine can pin down. This was, it seems to me, the view of the alchemists, the actual inventors of the laboratory,[84] who were themselves indefatigable experimenters; in the end it was by remaining faithful to them that Geoffroy became an innovator.

---

[82] Joly, *Descartes et la chimie* (cit. n. 33).

[83] Joly, "Practicing Chemical Philosopher" (cit. n. 33).

[84] Antoine Furetière, in his *Dictionnaire Universel* (The Hague, 1690), s.v. "Laboratoire," defines the laboratory as a "terme de Chymie." It is, he says, "the place where chymists conduct their operations, with their furnaces, their drugs, their vessels."

CHEMISTRY IN THE 18TH CENTURY

Communications of
Chemical Knowledge:

Georg Ernst Stahl and the Chemists at the French Academy of Sciences in the First Half of the Eighteenth Century

by Ku-ming (Kevin) Chang*

ABSTRACT

Histories of eighteenth-century chemistry often assert that the works of the German chemist Georg Ernst Stahl (1659–1734), especially his ideas about phlogiston, were largely unknown to French chemists until the 1740s. A careful analysis of Stahl's writings and the publications of the Royal Academy of Sciences in Paris shows that academy chemists were well informed about, and even integrated, Stahl's chemical theories, experiments, and methods beginning in the 1710s, and that Stahl kept abreast of the work by his colleagues at the Paris Academy. It also reveals the frequency and significance of the communication between French and German chemical communities in the first half of the eighteenth century.

The historiography of eighteenth-century chemistry has long been dominated by the so-called Chemical Revolution. This historiography portrays Antoine-Laurent Lavoisier (1743–94), the leading figure of the revolution, as a scientist who distinguished himself from his chemical forebears in his rationality, experimental insight, and lucidity of language. He overcame the open contestations of the chemists who, following German chemist Georg Ernst Stahl (1659–1734), erroneously argued that substances owed their inflammability to a mysterious material in their composition called phlogiston.

Compared with the considerable scholarship devoted to Lavoisier's establishment of modern chemistry, only minimal efforts have been made to know when and how Stahl's work was received in Enlightenment France. Yet the reception of Stahl's work is a necessary part of the complete historiography of the Chemical Revolution. Hélène Metzger suggested that this reception started with Jean Baptiste Sénac (1693–1770), who published a textbook based on the principles of Newton and Stahl in 1723.[1]

* Institute of History and Philology, Academia Sinica, Taipei City, 11529, Taiwan; kchang@sinica.edu.tw.

I wish to thank the editors of the present volume for their thorough reading of, and helpful advice on, two different previous versions of this essay.

[1] Hélène Metzger, *Newton, Stahl, Boerhaave et la doctrine chimique* (Paris, 1930), 94–5.

However, Sénac said nothing about Stahl's phlogiston theory and left no lasting impact on the French chemical community. Later research credits Guillaume-François Rouelle (1703–70) with discovering Stahl in the 1740s, only after decades of the French chemical community's ignorance of the German chemist.[2] Though he published little during his career, through his popular lectures at the Jardin du Roy in the 1740s and 1750s, Rouelle enthusiastically introduced Stahl's chemistry to the luminaries in Enlightenment Paris, such as Denis Diderot (1713–84) and Jean-Jacques Rousseau (1712–78), and began a school of French Stahlianism whose members included Pierre-Joseph Macquer (1718–84) and Gabriel François Venel (1723–75). Macquer in particular published influential chemical textbooks and dictionaries that made Stahlian chemistry the ruling doctrine before Lavoisier's Revolution.

This portrait suggests that Stahl was largely unknown in Enlightenment France until the mid-eighteenth century. This lack of familiarity with Stahl outside Germany, the historiography suggests, can be attributed to several causes. Stahl's style was archaic and entangled, and his character was mean-spirited, both factors contributing to his alienation from the chemical community abroad.[3] He wrote in crude German and difficult Latin, inaccessible to the intellectual mainstream of the high Enlightenment.[4] It might seem natural that his work was ignored in Enlightenment France until the magic moment when Rouelle discovered the appeal of Stahl's chemistry.

Stahl's career lent some truth to the impression of his isolation, especially in comparison with his German contemporaries. Thirteen years Stahl's senior, Gottfried Wilhelm Leibniz (1646–1716) was polite and cosmopolitan. Unlike Stahl, he had studied in Paris in his youth, kept up a correspondence with the intellectual communities abroad, and was elected to the scientific academies in London and Paris. Friedrich Hoffmann (1660–1742), Stahl's colleague at the University of Halle and later his fierce rival, made a grand tour through Europe before settling in his teaching position at Halle. In contrast, Stahl made no such peregrination before or after receiving his doctorate at Jena (1684). He first taught as an unsalaried lecturer at his alma mater, then served for a few years as a court physician in Weimar, assumed a professorship at Halle in 1694, and finally worked in Berlin as the Personal Physician to the King of Prussia Friedrich Wilhelm (r. 1713–40) from 1715 until his death in 1734, all without leaving what is today's Germany. While Leibniz wrote fluently in French, the lingua franca of the Republic of Letters, Stahl published his works in pedantic Latin and began to write in the vernacular language, German, only late in his career.[5] In comparison with Leibniz, who represented intellectual cosmopolitanism in Germany at its best, or in contradistinction to Lavoisier, the archetypal Enlightenment scientist, Stahl was provincial and removed from the growing Enlightenment forces. All these factors might appear to justify the French scientists' lack of interest in his work.

---

[2] Rhoda Rappaport, "Rouelle and Stahl—the Phlogistic Revolution in France," *Chymia* 7 (1961): 73–102; Martin Fichman, "French Stahlism and Chemical Studies of Air, 1750–1770," *Ambix* 18 (1971): 94–122; Jon Bledge Eklund, "Chemical Analysis and the Phlogiston Theory, 1732–1772: Prelude to Revolution" (PhD diss., Yale Univ., 1971).

[3] J. R. Partington, *A History of Chemistry* (London, 1961), 2:654–5.

[4] This is summarized in Bernadette Bensaude-Vincent and Isabelle Stengers, *A History of Chemistry*, trans. Deborah van Dam (Cambridge, Mass., 1996), 61.

[5] For Stahl's chemico-medical publications in German, see Ku-ming (Kevin) Chang, "Georg Ernst Stahl's Alchemical Publications: Anachronism, Reading Market, and a Scientific Lineage Redefined," in *New Narratives in Eighteenth-Century Chemistry: Contributions from the First Francis Bacon Workshop*, ed. Lawrence M. Principe (Dordrecht, 2007), 23–43.

The recent literature on eighteenth-century chemistry has begun to fill in some of the lacunae of Stahl's reception in the first half of the 1700s. A number of historians have turned their attention to the chemical work in the Royal Academy of Sciences in Paris (hereafter the French Academy) during this half century. These scholars have focused mainly on the development of the idea of relative affinities from Etienne-François Geoffroy (1672–1731) to Lavoisier.[6] The motive of this scholarship is, by and large, still to account for the success of the Chemical Revolution, though not to approach it through the overthrow of the phlogistic chemistry of combustion, but through the development of the understanding of chemical composition that began with Geoffroy's affinity table and continued with the new chemical nomenclature that Lavoisier introduced. As these scholars have studied Geoffroy and his institution, they have discovered some connections between Stahl and the chemists at the French Academy. Ursula Klein, for instance, has located some references to Stahl in the publications of the French Academy.[7] Mi Gyung Kim has further pointed out the references of Geoffroy and his colleague Gilles-François Boulduc (1675–1742) to Stahl in the 1720s.[8] Hence, current scholarship has filled in some points of contact between Stahl and his French colleagues in the late 1710s and early 1720s. Yet this work leaves the impression that there was little or no knowledge of Stahl in the French Academy from about the mid-1720s to the 1740s, when Rouelle rediscovered Stahl.

A thorough study of the mutual relations between Stahl and the chemical elite at the French Academy gives a very different picture, however. In fact, the references to Stahl by the chemists at the French Academy were remarkably continuous, frequent, and comprehensive. Eleven Academy chemists cited Stahl in the *Histoire* and *Mémoires* of the Academy in twenty-one of the thirty-eight years between 1713 and 1750. Between 1730 and 1750, phlogiston frequently turned up in their discussions in the Academy publications. This frequency and intensity certainly belies the previous impression that Rouelle discovered the isolated Stahl only after decades of ignorance. It also sheds light on a high degree of scientific communication between the German and French chemical communities that has not been noted before.

This essay reviews the communication between Stahl and the chemists at the French Academy. The first section examines Stahl's references to his French colleagues. The rest of the essay investigates the reverse direction of the communication. I will document these materials by means of tables compiling the references to Stahl and those to phlogiston, respectively. They are separate, as the two types of references are not identical to each other. Then I will examine a number of early cases in which Stahl's works were discussed in the Academy. Finally, the essay will

---

[6] Ursula Klein, *Verbindung und Affinität: Die Grundlegung der neuzeitlichen Chemie an der Wende vom 17. zum 18. Jahrhundert* (Boston, 1994); Mi Gyung Kim, *Affinity, That Elusive Dream: A Genealogy of the Chemical Revolution* (Cambridge, Mass., 2003). Frederic L. Holmes, on the other hand, pays attention to the experimental tradition in the chemical work of the French Academy in the first few decades of the eighteenth century; Lawrence M. Principe studies the alchemical theory and practice at the academy. See Holmes, *Eighteenth-Century Chemistry as an Investigative Enterprise* (Berkeley, Calif., 1989); Holmes, "Chemistry in the Académie Royale des Sciences," *Hist. Stud. Phys. Biol. Sci.* 34 (2003): 41–68; Principe, *Wilhelm Homberg and the Transmutation of Chymistry* (manuscript).

[7] E.g., Ursula Klein, "E. F. Geoffroy's Table of Different 'Rapports' Observed between Different Chemical Substances—a Reinterpretation," *Ambix* 42 (1995): 86.

[8] Kim, *Affinity* (cit. n. 6), 146–53.

analyze some passages in which Stahl's phlogistic accounts of chemical phenomena were portrayed favorably and survey the Stahlian chemical methods and experimental knowledge that his French colleagues admired. What emerges from all this is a remarkable degree of communication between Stahl and the French Academy, and a continuous and comprehensive knowledge of Stahl's ideas, techniques, and publications in the French Academy. This will serve to revise the previous historiography and to supplement recent efforts to trace the trajectory of the reception of Stahl's chemistry in France in the eighteenth century.

STAHL ON THE FRENCH ACADEMICIANS HOMBERG AND GEOFFROY, 1710s AND 1720s

Stahl first referred to chemists at the French Academy in a student's dissertation that he supervised in 1712, *Dissertatio . . . qua solutio martis in puro alcali et anatomia sulphuris communis, sistuntur*. A subject of the dissertation was the analysis of the composition, or what Stahl called *anatomia*, of common sulfur. The author of the dissertation cited a work by Homberg on the anatomy of sulfur, in which the academician described a dark gum as the residual substance of the analysis of sulfur.[9] After referring to Homberg, the author cited a work by Geoffroy in the *Histoire* of the French Academy, in which the latter tried to recompose common sulfur.[10] The references stopped short of discussing the content of these two works.

The short references nonetheless indicate Stahl's awareness of the chemical work that Homberg and Geoffroy had been carrying on in Paris. The writings in question were two memoirs published by the French Academy, namely, Homberg's "Essais de l'analyse du soufre commun (1703)," and Geoffroy's "Manière de recomposer le soufre commun par la réuniun de ses principes . . . (1704)." Therefore, Stahl referred to these works only a few years after their publication. Though not dwelling at length on these works, Stahl quoted the remark of the editor of the *Histoire* on Geoffroy's memoir, in French: "Si la decouverte que M. Geoffroy a faite, se verifie dans la suite, elle sera . . . importante" (If the discovery that Mr. Geoffroy has made is verified as true later, it will be . . . important).[11] Although the works of Homberg and Geoffroy might not have come to his attention immediately after their publication, once they did Stahl hurried to defend his intellectual priority by citing another dissertation that he had supervised, *Metallurgiae pyrotechnicae et docimasiae metallicae fundamenta . . .* (1700), asserting that he had studied the same subject.[12] The relatively short time period between the publication of Stahl's *Metallurgiae pyrotechniae* and the works by Homberg and Geoffroy seems to have created a sense of urgency and competition, for Stahl at least.

The closeness in time and direct references also show that Stahl and his French colleagues shared an interest that lasted for a significant period of their careers, namely, the composition of sulfur and its relation to the sulfurous principle. Stahl pursued this interest further a few years later in his *Zufällige Gedancken und nützliche Bedencken über den Streit, von dem so genannten Sulphure, und zwar sowol dem gemeinen, verbrennlichen, oder flüchtigen, als unverbrennlichen, oder fixen* (1718), often known

---

[9] Georg Ernst Stahl, *Dissertatio medico-chymica inauguralis, qua solutio martis in puro alcali et anatomia sulphuris communis, sistuntur . . .* (Halle, 1712), 23–4.

[10] Ibid.

[11] Ibid., 24.

[12] Ibid.

as *Treatise on Sulfur*. This treatise, Stahl explained, was written to present in German his experiments and findings on sulfur and the sulfurous principle in works that he had previously published in Latin, dating back to his first monograph *Zymotechnia fundamentalis* (1697) and including his dissertation of 1712.[13] He retold much of what he had said in the 1712 dissertation, including his reference to Homberg, if not to Geoffroy. In fact, Stahl enlarged an experiment presented by the former, quoting it at length.

The particular experiment in question was one in which Homberg tried to analyze sulfur by solution and then distillation. Distillation was a method very commonly used by the chemists of the French academy to analyze the composition of herbal and other materials.[14] Homberg first dissolved four ounces of flowers of sulfur in a pound of turpentine (or fennel) oil, which, after being heated gently "in digestion" for eight days, produced a dark red liquid. When cooled, there emerged sulfur crystallized in the shape of needles. He added another pound of the same oil, dissolved the crystals, and distilled the solution over a small fire for twelve to fifteen days. About two-thirds of the original quantity of turpentine oil was collected in the recipient, and in addition four ounces of whitish heavy fluid. He then intensified the heat and collected a thick and strongly colored oil. What remained behind in the retort was a black gummy residue. To investigate the nature of the residue, Homberg heated it with a magnifying lens and observed little change.[15]

Stahl gave his approval to several aspects of Homberg's experiment and interpretation. For instance, he spoke favorably of the French academician's quantitative approach. He also agreed with Homberg's assumption that the colorless oil in the recipient was turpentine oil and that the heavy whitish liquid was oil of vitriol, that is, vitriolic acid. Like Homberg, he considered vitriolic acid to be a constituent of sulfur that was driven out by the analysis. He also accepted the suggestion that the true chemical principles—the sulfurous, saline, and mercurial—could never be obtained perfectly pure, nor could they be rendered sensible.[16]

Stahl also wrote as if he agreed with Homberg on the quality of what he considered the oily principle, if not on the composition of sulfur. This principle gave volatility and inflammability to sulfur and other resinous substances. Homberg called this principle the "sulfur of common sulfur" [*Soufre du Soufre commun*]. Stahl, also a believer in the oily principle or phlogiston, accepted this with no problem. The problem was that Homberg proposed that sulfur was composed of three constituent substances: the oily principle, vitriolic acid, and an earthy material. Homberg, Stahl suggested, could not explain how a substance as volatile as sulfur would contain in its composition the heavy gummy earth that was so "fixed" or resistant to fire. If volatility all derived from the oily principle, how much of the oily principle was necessary to make its compound with vitriolic acid and gummy earth volatile? For Stahl, Homberg's quantitative account did not support his speculation.[17]

Stahl's interpretation was that the earthy part was not a direct constituent part

[13] I will cite the page numbers in the French translation of Stahl's *Zufällige Gedancken*, i.e., *Traité du soufre, remarques sur la dispute qui s'est élevée entre les chymistes, au sujet du soufre, tant commun, combustible ou volatil, que fixe, &c.* (Paris, 1766), 54–5.

[14] Holmes, "Chemistry"; Holmes, *Eighteenth-Century Chemistry*, 61–83 (both cit. n. 6).

[15] Stahl, *Traité du soufre* (cit. n. 13), 302–4.

[16] Ibid., 305–6.

[17] Ibid.

of sulfur but the product of vitriolic acid deprived of its water. Johannes Kunckel (1630–1703) had reported the creation of shining black substances made by distilling spirit of niter (nitric acid) with turpentine oil, and then adding oil of vitriol (concentrated vitriolic acid) to the residue.[18] Such substances could also be made, Stahl noted, by directly mixing turpentine oil with the vitriolic acid. According to Stahl, in Homberg's experiment sulfur was decomposed into vitriolic acid and phlogiston, and the vitriolic acid so produced was further stripped of its constituent water, leaving behind a fixed earth with the appearance of black gum.[19]

Stahl made these assertions in his *Treatise on Sulfur*, published after Homberg's death in 1715, thus precluding any response. The composition of sulfur and especially its relation to sulfuric acid would be further developed in Stahl's *Ausführliche Betrachtung und zulänglicher Beweiss von den Saltzen* (1723), often known as *Treatise on Salts*.

Thereafter Stahl continued to critique Homberg's interpretation of the composition of sulfur. In his *Treatise on Salts*, he returned to Homberg's experiment on sulfur and again disputed the late academician's proposition that there was a "purely earthy" substance in sulfur. As in the *Treatise on Sulfur*, it was vitriolic acid that produced the earthy dark gum described by Homberg.[20] Revisiting his interpretation of the composition of sulfur, Stahl asserted that the corpuscle of sulfur was a combination of one particle of vitriolic acid and one of oil. Thinking in corpuscularian terms, Stahl asserted that the particle of vitriolic acid was "a very subtle combination of a very soft earthy particle and an aqueous particle." The particle of turpentine oil was made up in turn of aqueous and inflammable particles [*Stäubchen*].[21]

In the *Treatise on Salts*, Stahl also discussed Homberg's work on borax. In 1702, Homberg had presented a memoir on his pioneering experiments on drying, dissolving, and turning borax into glass.[22] Evidently aware of this work, Stahl considered borax a species of the saline principle, together with vitriol, marine salt, and niter, which were traditionally recognized as three species of salt.[23]

Although Stahl did not cite Geoffroy, there was a resonance between Stahl's discussions and Geoffroy's affinity table. Chapter 22 of the *Treatise on Salts* was devoted to the relative "forces" [*Kräfte*][24] of acids in dissolving different kinds of bodies. Vitriolic acid, for example, dissolved fixed alkali (known as sodium carbonate, Na_2CO_3, or potassium carbonate, K_2CO_3, today) with an intensity greater than occurred in any other substances. Although fixed alkali was not compared with phlogiston in this chapter, his mention of the experiment on vitriolated tartar would serve to remind readers, at least those who were familiar with his work, of his theory that phlogiston displaced fixed alkali from its union with vitriolic acid. After fixed alkali,

[18] Ibid., 307.

[19] Ibid., 310.

[20] Georg Ernst Stahl, *Traité des sels, dans lequel on demontre qu'ils sont composés d'une terre subtile, intimement combinée avec de l'eau* (Paris, 1771), 318.

[21] "Das vitriolische Saltz-Wesen sey an sich, eine solche subtileste Vermischung, aus einem zahrtesten Erdischen, und Wasser-Stäubchen"; Georg Ernst Stahl, *Ausführliche Betrachtung und zulänglicher Beweiss von den Saltzen* (Halle, 1723), 325. See also Stahl, *Traité des sels* (cit. n. 20), 319–20.

[22] Wilhelm Homberg, "Essais de chymie: Article premier, des principes de la Chymie en général," *Mémoires de mathématique et de physique de l'Académie royale des sciences* (hereafter cited as *MARS*) 1702 (1702), 44–50.

[23] Stahl, *Traité des sels* (cit. n. 20), 22–3.

[24] Stahl, *Betrachtung von den Saltzen* (cit. n. 21), 214.

vitriolic acid formed combinations, in decreasing order of strength, with volatile alkali, earthy substances (such as chalk), zinc, cadmium, iron, copper, and silver.[25] Stahl also compared the relative strength of the acids in their reactions with fixed alkali.[26] This order, he suggested, essentially agreed with the columns for vitriolic acid and for fixed alkali on Geoffroy's affinity table. Neither Geoffroy nor Stahl was the first to note displacement reactions.[27] It would thus make no sense to speculate who learned from whom. The fact that Stahl spoke of the relative forces or strengths and placed them in order, however, makes it seem likely that he was aware of his French colleague's affinity table.

In contradistinction to Geoffroy, Stahl pondered the reason for the relative forces of chemical unions, for example, between acids and metals. Geoffroy is well-known for having refrained from speculating on the cause of different degrees of affinity. In contrast, Stahl made two suggestions based on his view that the acid was a compound of an earthy particle and an aqueous particle. First, the acid attacked the metals on their earthy particles, and, second, in consequence it acted more strongly on the metals that contained more earthy principles.[28]

In addition to the *Treatise on Salts*, Stahl's critique of Homberg's position on the composition of sulfur was repeated in another work published in 1723, namely, *Billig Bedencken, Erinnerung und Erläuterung über D. J. Bechers Natur-Kündigung der Metallen* (1723), a book that also severely criticized Becher's chrysopoetic ideas. In that work, once again, Stahl rejected Homberg's black residue as a part of sulfur.[29] This seems to conclude Stahl's references to the chemists of the French Academy.

CHEMISTS OF THE FRENCH ACADEMY ON STAHL, 1710s TO 1750: AN OVERVIEW

The French Academy chemists first paid attention to Stahl in the 1710s. The first who did (in 1713) was Geoffroy, the author of the affinity table as seen above. He was followed by Louis Lémery, Nicolas's son.

As seen from table 1, eleven members of the French Academy discussed Stahl in four years from 1711 to 1720, five years from 1721 to 1730, five years from 1731 to 1740, and eight years from 1741 to 1750. These members were Etienne-François Geoffroy (1672–1731), Louis Lémery (1677–1743), Gilles-François Boulduc (1675–1741), Louis-Claude Bourdelin (1696–1777), Claude-Joseph Geoffroy (1685–1752, Etienne-François's younger brother), Henri-Louis Duhamel du Monceau (1700–1782), Jean Grosse (?–1744), Jean Hellot (1685–1766), Paul-Jacques Malouin (1701–88), Guillaume-François Rouelle (1703–70), and Pierre-Joseph Macquer (1718–84). This list includes almost all the chemists active in the French Academy from its reorganization in 1699 until the mid-eighteenth century, except Wilhelm Homberg (1652–1715), Nicolas Lémery (1645–1715), Simon Boulduc (1652–1729), and Charles François du Fay (1698–1739). These chemists can be

---

[25] Stahl, *Traité des sels* (cit. n. 20), 214–6.
[26] Ibid., 222–3.
[27] William Newman documents some predecessors to the notion of relative affinity in his "Elective Affinity before Geoffroy: Daniel Sennert's Atomistic Explanation of Vinous and Acetous Fermentation," in *Matter and Form in Early Modern Science and Philosophy*, ed. Gideon Manning (Boston, 2012), 99–124. I thank Newman for kindly bringing his new work to my attention.
[28] Stahl, *Traité des sels* (cit. n. 20), 305, 343.
[29] Ernst Georg Stahl, *Billig Bedencken, Erinnerung und Erläuterung über D. J. Bechers Natur-Kündigung der Metallen* (Frankfurt am Main, 1723), 289, 381.

Table 1. *References to Stahl in the* Histoire *and* Mémoires *of the French Academy*

Year	Author	Title	Stahl's works cited
1713	E.-F. Geoffroy	Observations sur le vitriol et sur le fer	Journal of Halle[a]
1717	Louis Lémery	Second mémoire sur le nitre	Not specific [perhaps work on niter]
1719	*Histoire*	Sur le chacril	Genereal reference to a work on pleuresies or peripneumonics
1720	E.-F. Geoffroy	Eclaircissements	*Specimen Beccherianum;*[b] question passed on personally by Neumann, *Zymotechnia;* Halle Observation (note a)
1724	*Histoire*	Sur un sel cathartique d'Espagne	Question passed on to Geoffroy; *Opusculum*
1724	G.-F. Boulduc	Sur la qualité & les propriétés d'un Sel découvert en Espagne . . .	Not specific
1726	*Histoire*	[Referring to Boulduc]	Not specific (reference to a certain work on ignited sulfur and the salt of tartar)
1726	G.-F. Boulduc	Mémoire d'analyse en general des nouvelles eaux minerales de Passy	*Dissertation de acidulis & thermais;*[c] *Specimen Beccherianum; Treatise on Salts*
1728	G.-F. Bourdelin	Mémoire sur la formation des sels lixiviels	*Fundamenta chemiae*
1729	*Histoire*	Examen du vinaigre concentré par la gelée	Not specific
1729	C.-J. Geoffroy	Examen du vinaigre concentré par la gelée	Not specific
1730	G.-F. Bourdelin	Mémoire sur le sel lixiviel du Gayac	Not specific
1733	Grosse	Recherche sur le plomb	*Specimen Beccherianum*
1734	*Histoire*	Sur le sel de soufre	Not specific
1735	Hellot	Analise chimique du zinc, Parts 1 & 2	Dissertation on the salts of metals[d]
1736	*Histoire*	Sur l'antimoine et sur un nouveau phosphore détonnant	Not specific
1736	Duhamel	Sur la base du sel Marin	*Specimen Beccherianum*
1736	C.-J. Geoffroy	Quatrième mémoire sur l'antimoine. Nouveau Phosphore détonnant fait avec ce Minéral	*Opusculum*
1739	Hellot	Sur la liquer etherée de M. Froebenius	Not specific

1741	C.-J. Geoffroy	Moyens de congeler l'esprit de vin	Not specific
1742	C.-J. Geoffroy	Moyens de volatiliser l'huile de vitriol . . .	Not specific
1743	Malouin	Sur le zinc	Dissertation on the salt of metals, *Specimen Beccherianum*; Dissertation *vitulus aureus igne combustus*[e]
1744	Rouelle	Mémoire sur les sels neutres	Not specific
1745	Rouelle	Sur le sel marin: premiere partie: De la crystallisation du sel marin	*Treatise on Sulfur*
1745	Malouin	Sur le sel de la chaux	Not specific
1746	*Histoire*	Sur l'arsenic	Not specific
1746	Macquer	Recherches sur l'arsenic	Not specific
1747	*Histoire*	Sur la chaux et sur le plastre	Not specific
1747	Rouelle	Sur l'inflammation, de l'huile de Térebenthine par l'acide nitrux pur	Not specific
1747	Macquer	Observations sur la chaux et sur le plastre	Not specific
1749	*Histoire*	[Summary of Macquer's *Elémens de chymie théorique*]	Not specific
1749	Macquer	Sur une nouvelle espèce de Teinture bleue	Not specific

[a] This refers to *Observationes selectae ad rem litterariam spectantes* (cit. n. 42).

[b] Georg Ernst Stahl, *Specimen Beccherianum, sistens, fundamenta, documenta, experimenta . . .* (Leipzig, 1703).

[c] This is most likely the dissertation supervised by Stahl's student Michael Alberti, who succeeded Stahl on the faculty of medicine at Halle: *Epistola gratulatoria qua thermarum et acidularum idolum medicum destruit . . .* (Halle, 1713).

[d] Georg Ernst Stahl, *Dissertatio medico-chymica inauguralis de salibus metallicis* (Halle, 1708).

[e] It is in fact not a doctoral thesis, but an article included in Stahl, *Opusculum*.

seen as belonging to three generations. Etienne-François Geoffroy was one of the three founding-generation chemists of the reorganized Academy (along with Nicolas Lémery, Homberg, and Simon Boulduc), although he was considerably younger than the others. Louis Lémery (Nicolas's son), Gilles-François Boulduc (Simon's son), and Claude Joseph Geoffroy (thirteen years younger than his elder brother Etienne-François) succeeded their fathers or, in Geoffroy's case, brother, at their deaths to become pensioned members of the academy in the 1710s and 1720s. They thus can be seen as the second generation, along with Bourdelin, Grosse, Hellot, and Duhamel, who joined the academy in the 1720s and 1730s. The third generation included Malouin, who, though only a few years younger than Duhamel, was admitted to the academy significantly later (1742), Rouelle (admitted in 1744), and Macquer (1745), the youngest of this group. Together Rouelle and Macquer popularized Stahlian chemistry in France starting in the 1740s.

These chemists cited a large number of Stahl's works. Among them were monographs like *Zymotechnia fundamentalis* (1697), *Specimen Beccherianum* (1703), the so-called *Treatise on Sulfur* (1718), *Fundamenta chemiae* (1723), and the so-called *Treatise on Salts* (1723). They also included articles that were first published in hard-to-find periodicals but were later republished in a collection of Stahl's early works, *Opusculum chymico-physico-medicum*, or *Opusculum* (1715). There were also individual dissertations that Stahl supervised, such as the dissertation on the salts of metals. These works include a great majority of Stahl's publications in chemistry, thus suggesting that the academy chemists had, at least collectively, an almost comprehensive knowledge of Stahl's publications. Moreover, they often cited Stahl's works only a few years after their publication. Therefore their knowledge of Stahl's works was quite up-to-date. In addition, there were personal communications between him and his French colleagues in the late 1710s.

The appearances of phlogiston in the academy publications are particularly interesting. Surprisingly it appeared very often, and sometimes in articles in which Stahl's name was not specifically cited. Such appearances are listed in table 2. Phlogiston first appeared in the academy publications in 1720, and then again in 1724. From 1731 to 1740, there were four years in which phlogiston was mentioned, and from 1741 to 1750, five years. Altogether seven authors referred to phlogiston, and they all cited Stahl in their works, though not all in the articles in which they mentioned phlogiston. Indeed, from 1732 onward these chemists often referred to phlogiston without making explicit references to Stahl. Table 3 shows the number of years in which Stahl or phlogiston was referred to in academy publications in specific decades.

It is thus obvious that the knowledge of Stahl and his work continued uninterrupted in the French Academy from the 1710s to 1750. Admittedly some chemists, especially C.-J. Geoffroy, referred to Stahl and phlogiston more frequently than others. Yet since their memoirs were, as required, presented in the joint meetings of the French Academy, the chemists there should have all heard the references to the name and works of Stahl uttered frequently and explicitly. The frequency provides clear proof that the French elite chemists' knowledge of Stahl's work continued from at the latest 1713 through the first half of the eighteenth century.

This is not to suggest that all chemists at the French Academy received Stahl favorably. Some obviously did, but others did not. A few examples are examined below.

Table 2. *Appearances of Phlogiston in Publications of the French Academy*

Year	Author	Title
1720	E.-F. Geoffroy	Eclaircissements
1724	Boulduc	Sur la qualité & les propriétés d'un Sel découvert en Espagne . . .
1732	Duhamel & Grosse	Des différentes manières de render le tartre soluble (without mentioning Stahl)
1734	Duhamel & Grosse	Recherche chemique sur la composition d'une liqueur très volatile (without Stahl)
1734	C.-J. Geoffroy	Mémoire sur l'éméticité de l'antimoine, sur le tartre émétique (without Stahl)
1736	Duhamel	Quelques expériences sur la liqueur colorante que fournit la Pourpre (without Stahl)
1737	Hellot	Le Phosphore de Kunckel et Analyse de l'urine (without Stahl)
1742	Duhamel	Deux procédés nouveaux pour obtenir sans le secours du Feu une Liqueur éthérée fort (without Stahl)
1744	C.-J. Geoffroy	Observations sur la terre de l'alun (without Stahl)
1745	*Histoire*	Sur le sel de la chaux (without Stahl)
1745	*Histoire*	Sur une preparation de verre d'antimoine, spécifique pour la dysenterie (without Stahl)
1745	*Histoire*	Observations chymiques (without Stahl)
1745	C.-J. Geoffroy	Examen d'une preparation de verre d'Antimoine
1745	Duhamel	Observations botanico-météologiques (without Stahl)
1748	Macquer	Second Mémoire sur l'arsenic (without Stahl)
1749	*Histoire*	Sur une nouvelle espèce de teinture bleue
1749	*Histoire*	[Summary of Macquer's *Élémens de chymie théorique*]

Table 3. *Number of Years in Which Stahl or Phlogiston Was Referred to in Publications of the French Academy*

	1711–20	1721–30	1731–40	1741–50
Number of years	3	5	7	9

FIRST REFERENCES TO STAHL IN THE FRENCH ACADEMY: 1710s

Etienne-François Geoffroy and Louis Lémery were the first two academy chemists to refer to Stahl. In a 1713 report of his work on vitriol and iron, Geoffroy cited a process that Stahl published in a journal in Halle, that is, *Observationes selectae ad rem litterariam spectantes* (1700–1705), which Stahl coedited.[30] The process was on the preparation of what the latter called volatile sulfurous acid spirit or volatile spirit of sulfur.[31] The acidic gas (thus "spirit") that Stahl described was characterized by its

[30] Etienne-François Geoffroy, "Observations sur le vitriol et sur le fer," *MARS* 1713 (1713), 168–86.
[31] Ibid., 172.

pungent smell and is known as sulfur dioxide today. It was used by Geoffroy to produce what he called a styptic liquid of sulfur, a brown-red, unctuous liquid that could
be made in several ways.[32]

Louis Lémery referred to Stahl's method of producing saltpeter in a memoir on
niter. Without citing a specific publication, Lémery approvingly discussed an artificial method, proposed by Stahl, of producing saltpeter with a wall made of thatch
and mud. For Lémery, this method "agreed perfectly" [*s'accorde parfaitement*] with
what happened by means of natural methods, such as the efflorescence of saltpeter
from decaying vegetable matter covered with excrement taken from a stable.[33]

The 1719 *Histoire* of the Academy reported the discussion "Sur le Chacril" of
Boulduc and Guy-Crescent Fagon (1638–1718), honorary member of the Academy
and head physician to Louis XIV until the king's death. Chacril was the French name
for the bark of the *Kinakina* tree found in Peru. The author of the *Histoire* recounted
that Boulduc and Fagon, who had died a year before, compared it to *Quinquina*, the
Peruvian bark that contained quinine. He then indicated the uses of chacril that had
been introduced by physicians, among whom he included *l'Illustre M. Stahl, Medicin
du Roi de Prusse*.[34] Though the author did not specify his source, it is reasonable to
assume that it was Boulduc and Fagon.

The next reference was made by Geoffroy again in his "Eclaircissements: Sur la
Table inserée dans les Mémoires de 1718 concernant les Rapports observés entre
differentes Substances" (1720). As its title suggests, this was a clarification of his
Table concernant les Rapports (1718), now known as the affinity table. Geoffroy
had hinted at his knowledge of Stahl's formulation of phlogiston, or the principle of
inflammability in his affinity table. It was, however, not until his "Eclaircissement"
that Geoffroy explicitly acknowledged that he had integrated Stahl's discussion of
phlogiston into his table, and indeed his thinking about sulfur and salt.

Geoffroy's affinity table consisted of sixteen columns, the fourth of which described the relative combinative intensity of different substances with vitriolic acid.
That acid stood on the top of the column. Immediately below it stood "Principe
huileux ou Soufre Principe," oily principle or Sulfur Principle. Following the oily
principle down the column were, in order, fixed alkali salt, volatile alkali salt, "absorbent earths," iron, copper, and silver. This indicated that the oily principle forms
the strongest union with vitriolic acid. It could displace, for example, fixed alkali salt
from the latter's combination with vitriolic acid, and indeed any other substances
in the column from their combinations with that acid. Silver, at the bottom of the
column, therefore formed the weakest union with vitriolic acid and could be displaced by any other substances in the column from its compound with that acid.[35]

In the text that explained the table, Geoffroy spoke of the sulfurous principle only
once. On the extraordinary corrosive power of *aqua regia* (the mixture of saline and
nitric acids), he wrote, "This results, according to the views of some, from a subtle
sulfurous principle, which is contained in the spirit of niter and which is communicated in this way to the acid of sea salt." Without naming whose views they were,

[32] Geoffroy provided three ways. The last and the most productive was to use the volatile sulfurous
spirit. Ibid., 171–2.

[33] Louis Lémery, "Second mémoire sur le nitre," *MARS* 1717 (1717), 129.

[34] "Sur le chacril," *Histoire de l'Académie royale des sciences* (hereafter cited as *HARS*) (1719), 55.

[35] Etienne-François Geoffroy, "Table of the Different Relations Observed in Chemistry between
Different Substances 27 August 1718," *Sci. Context* 9 (1996): 320.

Geoffroy ended this thread by saying, "This is not the place to examine this issue in depth."[36] The source of this view would only be revealed two years later in Geoffroy's "Eclaircissements."

In the "Eclaircissements," Geoffroy first referred to Stahl in the discussions of the relative affinities, or *rapports*, of acids with different substances. Geoffroy's table showed that acids had greater rapport with alkali, either fixed or volatile, than with metals. Therefore metals in general did not displace alkali in their combination with an acid. To support his position with an alkali that was less well known, he cited an experiment from Stahl's *Specimen Beccherianum* (1703), where the "volatile urinous salt," an alkali, precipitated the metal substance that had been compounded with the acid of salt.[37]

Geoffroy then uncovered the identity of the source of the principle of inflammability, and also his mutual contact with this source. The source was Stahl, and the contact or medium was Caspar Neumann (1683–1737), who had come through Paris in 1717 or 1718. Having previously held different private or state positions in Prussia and England, Neumann had met Stahl, then the Personal Physician to the King of Prussia in Berlin.[38] Neumann, Geoffroy recounted, showed him a letter from Stahl presenting a puzzle that served as a challenge. Taking up this challenge, Geoffroy spent a significant portion of his "Eclaircissements" solving the puzzle. He acknowledged that Stahl's puzzle was the specific reason for his inclusion of the oily principle in his affinity table.[39]

As Stahl's puzzle has been examined by Kim,[40] the discussion here can be short. The question reads: "When one has saturated and crystallized vitriolic acid with salt of tartar, find a means of separating this acid from the fixed salt in a moment of time and in the palm of a hand."[41] As salt of tartar (potassium carbonate, K_2CO_3) was a well-known fixed alkali, this question about its combination with vitriolic acid and its separation immediately related it to Geoffroy's affinity table, especially the column for vitriolic acid. If something, as Stahl's puzzle suggested, could displace salt of tartar, then this substance would have to stay above fixed alkali salt in his table.

At the heart of this puzzle was an experiment on "vitriolated tartar" (K_2SO_4) that Stahl had proudly presented in three places: first in his *Zymotechnia fundamentalis*, then in an article that he published in the same year, and in an article in the Halle journal mentioned above.[42] Vitriolated tartar was the product of salt of tartar satu-

[36] Ibid., 318.

[37] Etienne-François Geoffroy, "Eclaircissements: Sur la Table inserée dans les Mémoires de 1718 concernant les Rapports observés entre differentes Substances," *MARS* 1720 (1722), 26.

[38] Neumann was later appointed Court Apothecary in the Prussian Court in Berlin and served together with Stahl on the kingdom's Higher Medical Board. See Karl Hufbauer, *The Formation of the German Chemical Community (1720–1795)* (Berkeley and Los Angeles, 1982), 173–4.

[39] Geoffroy, "Eclaircissements" (cit. n. 37), 28.

[40] Kim, *Affinity* (cit. n. 6), 146–51.

[41] Geoffroy, "Eclaircissements" (cit. n. 37), 28.

[42] Georg Ernst Stahl, *Zymotechnia fundamentalis, seu fermentationis theoria generalis . . .* (Halle, 1697); "Experimentum novum verum sulphur arte producendi (1697)," "De arcani duplicati et tartari vitriolati genealogia (1701)." The second title was published in a monthly serial by Stahl, *Observationum chymico-physico-medicarum curiosarum: mensibus singulis bono cum Deo continuandarum* (1697–8), while the third was published in the third volume (1701) of a short-lived journal that Stahl coedited with his colleagues at the University of Halle; Christian Thomasius and Johann Franz Buddeus, *Observationes selectae ad rem litterariam spectantes* (Halle, 1700–1705). All three were included in Stahl, *Opusculum chymico-physico-medicum* (Halle, 1715) (hereafter cited as *Opusculum*). I will cite the pagination in the *Opusculum*.

rated by vitriolic acid, a substance that appeared in Stahl's question for Geoffroy.[43] Stahl judged that this salt was among the most difficult things to decompose. Proud that he had overcome this difficulty, he indicated that he placed vitriolated tartar and ground charcoal together in a white-hot crucible, with salt of tartar added as a flux for the hardly fusible salt. The vitriolic acid in the compound, Stahl reasoned, seized phlogiston from charcoal, resulting in common sulfur and fixed alkali.[44] This can be represented as:

$$\text{vitriolated tartar} + \text{charcoal} \xrightarrow[\Delta]{\text{salt of tartar (as flux)}} \text{sulfur} + \text{fixed alkali.}$$

Stahl then proceeded to reproduce vitriolated tartar by combining fixed alkali and sulfur. Two or three parts of the alkaline salt and one part of pulverized sulfur were fused in a crucible: they produced *hepar sulphuris* (liver of sulfur), red in color and very bad in smell.[45] Heated for some time, this "liver of sulfur" in turn produced vitriolated tartar, in the form of a white bitter octahedral crystal. This can be represented as:

$$\text{fixed alkali} + \text{sulfur} \xrightarrow{} \text{liver of sulfur} \xrightarrow{\Delta} \text{vitriolated tartar.}$$

Thus for Stahl the combination of sulfur and fixed alkali indeed led to the production of vitriolated tartar.[46] He was able not only to decompose vitriolated tartar but also to confirm its composition by both analysis and synthesis.

Geoffroy's solution of the puzzle reflected his close knowledge of Stahl's discussions in *Zymotechnia fundamentalis*, which he explicitly cited.[47] Like Stahl, he melted vitriolated tartar in a crucible with the help of a little salt of tartar and added inflammable matter such as charcoal or oily or resinous fatty matter. That produced *hepar sulfuris*. He placed this liver of sulfur, dissolved in water, on the palm of his hand and poured on some drops of distilled vinegar, which created a precipitation of sulfur and separated salt of tartar. Geoffroy also provided a second solution, for which he cited an article by Stahl.[48] He began with burning sulfur and a strong lye (*lessive*) of alkali, soaked into some linen. It also produced vitriolated tartar, except that the particles of vitriolic acid in this operation were extremely rarefied by the

---

[43] Stahl arrived at vitriolated tartar by a chain of displacement reactions. First, he dissolved silver in some spirit of niter. Silver was replaced if copper was added. Copper was further replaced by iron, iron by zinc, zinc by volatile alkali, and finally volatile alkali by fixed alkali. The combination of fixed alkali and acid of niter were joined by acid of vitriol and placed on a fire. Acid of vitriol attacked acid of niter, Stahl figured, so that the latter was dispelled. The final product was vitriolated tartar. Stahl, *Zymotechnia fundamentalis* (cit. n. 42), in his *Opusculum* (cit. n. 42), 143. Stahl was aware of Robert Boyle's synthesis of vitriolated tartar from oil of vitriol and turpentine oil, although he added that Boyle's method was easier said than done. Ibid., 142.

[44] Ibid., 143.

[45] *Hepar sulfuris* is a mixture not very well defined, consisting of potassium sulfide (K_2S), potassium polysulfide (K_2S_x), potassium thiosulfate, $K_2S_2O_3$, and probably potassium bisulfide, HKS. See Eklund, "Chemical Analysis" (cit. n. 2), 29.

[46] Stahl, *Opusculum* (cit. n. 42), 143–4.

[47] Geoffroy, "Eclaircissements" (cit. n. 37), 29.

[48] This article was published in the so-called Halle journal as "De copiosa, facili, et concentrata collectione spiritus acidi summe volatilis sulphureo-vitriolici, et theoretico-practica apodeixei generationis eiusdem (1700)," in *Opusculum* (cit. n. 42), 246–58.

"element of fire," or, as M. Stahl claims, by the "inflammable principle," compared with the concentrated acid in the ordinary preparation of vitriolated tartar.[49] To separate this acid from its vitriolated tartar, he added spirit of vitriol, spirit of niter, or spirit of salt, as the strong acids had greater affinity with the alkali salt than with the rarefied or volatile vitriolic acid.

For Geoffroy, Stahl had demonstrated two very important points. First, sulfur consisted of at least vitriolic acid and the sulfurous principle or phlogiston. Agreeing with Stahl, Geoffroy thought that the oily principle insinuated itself into the component particles of vitriolated tartar and united with vitriolic aid. Second, the experiment served as a clear indication that the sulfurous principle, or phlogiston, had a stronger affinity with vitriolic acid than fixed alkali, thus deserving to stay above fixed alkali on the affinity column for vitriolic acid.

STAHL IN THE FRENCH ACADEMY, 1720–50

The interest in Stahl's chemical work took root and grew in the academy after 1720. By 1735, a year after Stahl's death, almost all the academy chemists of the second generation had discussed Stahl in their memoirs. Some were favorable, such as G.-F. Boulduc (admitted as *élève* or pupil in 1699, promoted to assistant chemist in 1716, and ordinary chemist-associate in 1727), C.-J. Geoffroy (admitted as botanist pupil in 1707, botanist-associate in 1711, chemist-associate in 1715, and pensioned chemist in 1723), and Jean Hellot (admitted as assistant chemist in 1735).[50] Others were critical or had reservations about certain explanations of chemical processes or phenomena provided by Stahl. These included Louis-Claude Bourdelin (admitted to the academy as assistant chemist in 1725, chemist-associate since 1731), Duhamel (admitted as assistant chemist in 1728 and promoted as botanist-associate in 1738), and Jean Grosse (admitted as assistant chemist in 1731).[51] Favorable or critical, they all showed their familiarity with Stahl's phlogistic account of chemical compositions and reactions and even adopted it when criticizing Stahl's specific explanations.

Boulduc, who served as royal apothecary as well as chemist-assistant at the academy at the time, played a role that renewed the academy's studies on sulfur and continued the interest in Stahl's work. In a memoir of 1724, in which he studied a certain salt discovered in Spain, Boulduc mentioned the question on the production of vitriolated tartar that Stahl presented to E.-F. Geoffroy through Neumann.[52] He also described the teaching of "the famous Stahl" in the *Opusculum* on the replacement

---

[49] Geoffroy, "Eclaircissements" (cit. n. 37), 33; Kim, *Affinity* (cit. n. 6), 149.

[50] These figures' positions show that the boundary between the academy's fields of apothecary and chemistry was thin and fluid. For biographical information on these figures, see Paul Dorveaux, "Apothicaires membres de l'Académie Royale des Sciences: IV. Gilles-François Boulduc, V. Etienne-François Geoffroy," *Rev. Hist. Pharm.* 19 (1931): 113–26; Dorveaux, "Les grands pharmaciens: Apothicaires membres de l'Académie royale des Sciences, I: Claude Bourdelin," *Bull. Soc. Hist. Pharm.* 17 (1929): 289–98; Dorveaux, "Apothicaires membres de l'Académie royale des Sciences. VII: Claude-Joseph Geoffroy," *Rev. Hist. Pharm.* 20 (1932): 113–26.

[51] "Éloge de M. de Bourdelin," *HARS* (1777), 118–26; Claude Vie, "Duhamel du Monceau, naturaliste, physicien et chimiste," *Rev. Hist. Sci.* 38 (1985): 55–71; Paul Dorveaux, "Jean Grosse, médecin Allemand, et l'invention de l'éther sulfurique," *Bull. Soc. Hist. Pharm.* 17 (1929): 182–7.

[52] Gilles-François Boulduc, "Sur la qualité & les propriétés d'un Sel découvert en Espagne, qu'une Source produit naturellement; & sur la conformité & identité qu'il a avec un Sel artificiel que Glauber, qui en est l'auteur, appelle Sel Admirable," *MARS* (1724), 124.

of a weaker acid by a stronger acid in a combination with a base.[53] On the cause of the ease with which vitriolic acid displaced the other acids bound with metallic substances, he referred his reader to Stahl and E.-F. Geoffroy's affinity table. He also praised Stahl for providing a good theory on why vitriolic acid could take away the principle of inflammability (which he identified as Stahl's phlogiston or Homberg's sulfurous principle) in a substance to form sulfur.[54] Boulduc related Stahl's work on the analysis of salts, such as vitriolated tartar and Glauber's salt (sodium sulfate), to his own analysis of the Spanish salt.

Boulduc extended the personal communication of the French Academy with Stahl. The anonymous account in the *Histoire* of 1724 reporting Boulduc's memoir suggests that "M. Stahl the son" taught Boulduc techniques to make vitriolated tartar by transporting vitriolic acid from one base to another through substitution or exchange reactions.[55] It is not absolutely clear whether this "Stahl the son" was the royal physician or his son, Georg Ernst Stahl Jr. (1713–72).[56] At least it is evident that Stahl and his French colleague(s) communicated further after the indirect contact between him and Geoffroy in the late 1710s, either by correspondence or in person.

The personal communication was supplemented in due course by publications. Stahl's French colleagues were apparently vigilantly looking out for his publications. In 1724, Boulduc cited Stahl's *Opusculum* (1715), and two years later he cited a work that had been published only in 1723.

In his 1726 memoir, Boulduc analyzed the water of Passy, a neighborhood in Paris, which was known to have healing properties. In his analysis of the acidity of the water, he quoted Friedrich Hoffmann (1660–1742), Stahl's former colleague at Halle, Nicolas Lémery, and Stahl to support his proposition that there was a saline, acid principle in the waters he studied.[57] In addition, Boulduc turned to Stahl for authority for the salt he found in the thermal waters. Calling this a sulfurous salt, he identified it as the salt that Stahl had produced by mixing the vapor of burnt sulfur and the salt of tartar (thus a sulfite). This salt, Boulduc points out, was described in Stahl's *Specimen Beccherianum* (1703) and the *Treatise on Salts* (1723). Like Stahl, Boulduc considered this salt a "sel composé [compound salt]," a species of salts to which, he thought, Glauber's salt (sodium sulfate) also belonged.[58]

Then, in the context of analyzing the selenite in the water, Boulduc cited Stahl's experiment on vitriolated tartar as an exemplar. Stahl had demonstrated that this neutral salt, or middle salt [*sel moyen*], could be decomposed to its constituent principles, with the appearance of *hepar sulfuris* as its medium step. And in reverse, the constituents could be combined to produce the salt. Boulduc said approvingly that he himself was able to repeat this experiment.[59] For him, Stahl's work became an exemplary case in the chemistry of salts.

[53] Ibid., 120.

[54] Ibid., 130.

[55] "Sur un sel cathartique d'espagne," *HARS* (1724), 57.

[56] The report states, "M. Stahl le fils a appris à M. Boulduc celui de transporter cet Acide sur une autre base." This statement is puzzling because Stahl's best-known son, Georg Ernst Stahl Jr., who would later succeed to his father's title as *Hofrat* (Court Councilor), was only eleven years old. It is not known whether Stahl had an older son.

[57] Gilles-François Boulduc, "Essai d'analyse en general des nouvelles eaux minerales de Passy, avec des raisons succinctes . . .," *MARS* (1726), 312.

[58] Ibid., 315.

[59] Ibid., 324–5.

Although in this memoir he attributed to Stahl a dissertation that he probably never supervised, Boulduc shared suppositions that were proposed in his German colleague's *Treatise on Salts*.[60] In the end it was no mistake to claim that Stahl supposed that vitriolic acid existed in most acids, and that vitriolic acid could be, or be made to be, very volatile, referring to Stahl's identification of volatile vitriolic acid (sulfur dioxide).[61]

Two years later Bourdelin published a memoir on lixivial salts, that is, salts extracted from wood ashes.[62] He cited Stahl's work, *Fundamenta chemiae* (1723), for the formation and composition of the alkali salts that resulted from the combustion of plants. Stahl had identified these salts as new compounds produced by fire. After quoting a passage in which Stahl detailed the process of treating the salts, Bourdelin cited his German colleague's conclusion that the combination of the greasy part of the plants and the essential salt produced alkali salt.[63] Bourdelin brought up his own experimental experience that the distillation of plants gave off considerable oil. That served as the proof for the composition of oily or greasy substances in plants. When burned, plants were converted to ashes that tasted salty and dry, which Bourdelin saw as evidence of the lack of grease or oil. He then postulated that fire removed the greasy portion of plants. The alkali salt therefore could not be formed by the junction of the essential salt of the plant and oil.[64]

In order to clarify his disagreement with Stahl over the conception of fixed salt, Bourdelin considered the nature of two related salts, niter and fixed niter. Fixed niter (K_2CO_3) was the fixed alkali left in the crucible after the detonation of niter (KNO_3) mixed with ground charcoal.[65] What Bourdelin called "Stahl's system" interpreted fixed niter to be a union of the sulfurous principle and niter. For the academy chemist, fixed niter was a product of niter decomposed by the sulfurous principle. It was, therefore, a decomposed salt, not a *sel composé*, a compound salt, as Stahl suggested.[66] To support his objection to Stahl's system, Bourdelin ignited a mixture of seven ounces of charcoal power and sixteen ounces of niter. The ignition produced only six ounces of fixed niter. If Stahl had been right, the weight of the product would have been greater than niter, the starting material.[67] An experiment on the "regeneration of niter" further confirmed that niter was a compound of fixed niter and nitric acid, as niter was produced by pouring acid of niter over fixed niter.[68] Instead of seeing fixed niter as the sulfurous principle joined with niter, Bourdelin conceived of the principle as taking away the constituent nitric acid in niter. To pour nitric acid over

---

[60] Hoffmann instead wrote or supervised several works on similar subjects: Friedrich Hoffmann, *Dissertatio solennis medica de acidularum et thermarum ratione ingredientium et virium convenientia* (Halle, 1712); Hoffmann, *Dissertatio inauguralis medica, Observationes et cautelas circa acidularum et thermarum usum et abusum* (Halle, 1717). Stahl's former student and protégé Michael Alberti, who succeeded to Stahl's professorial position at Halle, also supervised a dissertation and published a letter on this subject: Alberti, *Dissertatio medica inauguralis, de haemorrhoidariorum prudenti therapia per acidulas et thermas . . .* (Halle, 1719); Alberti, *Epistola gratulatoria qua thermarum et acidularum idolum medicum destruit . . .* (Halle, 1713).

[61] Stahl, *Traité des sels* (cit. n. 20), 16, 47, 222.

[62] Louis-Claude Bourdelin, "Mémoire sur la formation des sels lixiviels," *MARS* (1728), 384–400.

[63] Ibid., 388.

[64] Ibid., 389.

[65] William R. Newman and Lawrence M. Principe, *Alchemy Tried in the Fire: Starkey, Boyle, and the Fate of Helmontian Chymistry* (Chicago, 2002), 242–4.

[66] Bourdelin, "La formation des sels lixiviels" (cit. n. 62), 394–5.

[67] Ibid., 396–7.

[68] Ibid., 395–8.

fixed niter was to let the pointed particles of nitric acid find their way back to their previous pores, matrices, or lodges in the particles of fixed niter.[69] He would continue his objection to Stahl in another memoir published two years later.

Although Bourdelin was critical of Stahl in his interpretation of fixed alkali, he referred to his work as a coherent system of chemical teachings. He cited only his German colleague's *Fundamenta chemiae*, which had appeared just a few years earlier (1723). This work was a compilation of lecture notes from 1684–5 in which the term "phlogiston" had not yet appeared. The properties of the sulfurous or oily principle in either Stahl's *Fundamenta* or Bourdelin's *Mémoire*, however, did not deviate from those of phlogiston or the sulfurous principle commonly accepted at the time. Although Bourdelin did not use the term "phlogiston" in his discussion, his colleagues would use it frequently starting in the 1730s.

One of those who spoke freely of phlogiston was E.-F. Geoffroy's son Claude-Joseph. He first referred to Stahl in a memoir of 1729 on the German chemist's method of concentrating vinegar with frost [*gelée*], which was available only in the cold weather. As the cold froze the water in the vinegar as ice and left the "spirit of vinegar" (acetic acid) in liquid form, it separated the "spirit" from water. This early modern form of refrigeration thus achieved an easy and effective concentration, or what Stahl called "dephlegmation" (i.e., dehydration) of aqueous solutions.[70] In this discussion Stahl served as the authority for a chemical technique.

In a 1734 memoir, the younger Geoffroy discussed the emetic quality of the glass of antimony in terms of phlogiston, though without referring to Stahl. Glass of antimony was known to be emetic, meaning that it caused vomiting. According to Stahl, its emetic quality is resident in a composite [*composé*] or coarse sulfur composed of very little vitriolic acid and phlogiston, a sulfur that is united to a vitrifiable earth. If this earth has more interstices to hold the sulfur, it will not be emetic. This is the case with crude antimony. In the reverse situation, if it has few interstices to hold the sulfur, as with glass of antimony, it will be very emetic.[71]

This phlogistic account continued in Geoffroy's writing in the 1740s. In 1744, he spoke of the phlogiston contained in vitriol.[72] In 1745 he saw the conversion of the glass of antimony to regulus of antimony as the result of phlogiston added to the glass.[73]

Such phlogistic accounts of chemical phenomena were repeated in the works of other chemist academicians. Duhamel and Grosse copublished an article in 1732 on ways to make tartar soluble. They described the formation of an alkali salt as a result of combining the phlogiston of tartar and the earthy sediment of *chaux* (lime).[74] They denied,

---

[69] Ibid., 395, 398.

[70] Claude Joseph Geoffroy, "Examen du vinaigre concentré par la gelée," *MARS* (1729), 68–78. I thank William Newman for suggesting this plausible interpretation of *gelée*. The Latin original that Stahl used is *gelu*; Georg Ernst Stahl, "Mensis October, commendans concentrationem sive dephlegmationem vini, aliorumque fermentatorum, et salinorum liquorum, salvis universis eorum viribus," in *Opusculum* (cit. no. 42), 417–8, Latin original on 417. Stahl applied this method to the concentration of beer first, and to vinegar thereafter.

[71] Claude-Joseph Geoffroy, "Mémoire sur l'émécitié de l'antimoine, sur le tartre émétique, et sur le kermes minéral," *MARS* (1734), 420–1.

[72] Claude-Joseph Geoffroy, "Observations sur la terre de l'alun; Manière de le convertir en Vitriol, ce qui fait une exception à la Table des Rapports en Chymie," *MARS* (1744), 71.

[73] Claude-Joseph Geoffroy, "Examen d'une préparation de verre d'Antimoine, spécifique pour la dysenterie," *MARS* (1745), 163.

[74] Henri-Louis Duhamel du Monceau and Jean Grosse, "Des différentes maniéres de rendre le Tartre soluble," *MARS* (1732), 328. Duhamel and the French translation of Stahl's work consistently used the term *chaux* (usually meaning lime) to refer to what we call metallic calx.

however, that the dissolution of tartar was due to the phlogiston in lime or fire parti-cles.[75] Duhamel, in some memoirs published by himself alone, also applied phlogiston in his chemical reasoning. For instance, in 1736, he worked to verify whether the red color of calcinable bodies came from the phlogiston that was concentrated in the pores of compounded material, although his conclusion was that it might be due to the red-colored ray of sunlight.[76] Hellot in his memoir on phosphorus and urine spoke of ammo-nia as a compound of a volatile urinous salt, a marine acid, a phlogistic oil, and a subtle earth.[77] A summary of Malouin's work on the salt of lime in the *Histoire* of 1745 took for granted the composition of sulfur as a compound of vitriolic acid and phlogiston.[78]

Phlogistic reasoning was not limited to the French Academy. The academicians cited publications or communications of chemists based in London and the prov-inces, who also spoke of phlogiston in the 1730s and 1740s. Duhamel and Grosse in their work on diethyl ether quoted at length the account by August Sigmund Fro-benius (dates unknown) published in the *Philosophical Transactions of the Royal Society of London* in 1730. For Frobenius, a German who worked in London at the time, ether was a compound of a volatile urinous salt, the phlogiston of a plant, and an extremely subtilized acid.[79] Ether was also studied by Pierre Toussaint Navier (1712–79), a physician based in the French town of Châlons-sur-Marne and a cor-respondent of the French Academy. In 1742, Duhamel quoted Navier's passages de-scribing phlogiston in fire and iron and recounting a chemical procedure to keep a va-por rich in phlogiston from being lost.[80] A general anonymous account in the *Histoire* of 1745 even suggested that it is a "well-received opinion among the chemists" [*une opinion assez reçue chez les Chymistes*] that most minerals of iron are not attracted to the magnet before having been calcined with phlogiston-rich material.[81] This no doubt meant that this opinion was not limited to the chemists at the French Academy but was shared by the wider chemical world.

Stahl constantly served as the academicians' authority for chemical analyses. Vit-riolated tartar is a good example. In addition to E.-F. Geoffroy and Boulduc, Ma-louin also relied on Stahl's account of vitriolated tartar for his analyses in 1745.[82] As seen above, vitriolated tartar was closely tied to *hepar sulfuris* or liver of sulfur in Stahl's work. Consequently, liver of tartar, especially the analyses and uses that Stahl introduced, attracted the attention of the academy chemists. Grosse singled out Stahl's treatment of gold with heated liver of sulfur.[83] Malouin cited Stahl's formula for making the liver and then tried to produce it from lime.[84] In Stahl's preparation,

[75] Ibid., 333.

[76] Henri-Louis Duhamel du Monceau, "Quelques expériences sur la Liqueur colorante que four-nit la Pourpre, espece de Coquille qu'on trouve abondamment sur les Côtes de Provence," *MARS* (1736), 55.

[77] Jean Hellot, "Le phosphore de Kunckel et analyse de l'urine," *MARS* (1745), 375.

[78] "Sur le sel de la chaux," *HARS* (1745), 41.

[79] Henri-Louis Duhamel du Monceau and Jean Grosse, "Recherche chimique sur la composition d'une liqueur tres-volatile, connuë sous le nom d'Éther," *MARS* (1734), 42.

[80] Henri-Louis Duhamel du Monceau, "Deux procédés nouveaux pour obtenir sans le secours du Feu une Liqueur éthérée fort approchante de celle à laquelle M. Froboenius Chymiste Allemand, a donné le nom d'Éther," *MARS* (1742), 382, 383, 387.

[81] "Observations chymiques," *HARS* (1745), 47.

[82] Paul-Jacques Malouin, "Sur le sel de la chaux," *MARS* (1745), 102.

[83] Jean Grosse, "Recherche chimique sur la composition d'une liqueur tres-volatile, connuë sous le nom d'Éther," *MARS* (1733), 315.

[84] Paul-Jacques Malouin, "Sur le zinc. Second Mémoire," *MARS* (1743), 81; Claude-Joseph Geof-froy, "Moyens de congeler l'esprit de vin, et de donner aux Huiles grasses quelques-uns des caractéres d'une Huile essentielle," *MARS* (1741), 19.

vitriolated tartar was also associated with the addition or subtraction of phlogiston in alkali salt or tartar. Indeed, Stahl was cited in such contexts by Duhamel in 1732, C.-J. Geoffroy in 1741, and a *Histoire* summary in 1749.[85] Moreover, for Rouelle, Stahl was the ultimate authority on the "etiology" or source of the inflammation of niter as well as that of the inflammation of oils.[86]

In addition, Stahl served as an important point of comparison or reference in chemical methods and experimental knowledge. As already seen, academy chemists often consulted Stahl's work on the reduction of antimony. In 1736 C.-J. Geoffroy proposed a new method to draw the regulus from antimony more effectively than Stahl's well-known method.[87] Hellot referred in 1735 to Stahl's technique of making the so-called flowers of zinc (i.e., zinc oxide) when zinc was united to gold.[88] Duhamel presented in 1736 his examination on the base of marine salt or common salt by consulting the German chemist's methods of using the acids of vitriol and of niter.[89] In 1746 Macquer cited Stahl's preparation of a volatile spirit of niter in the midst of his own finding that arsenic gave rise to the release of niter from its acid better than vitriolic acid.[90]

Rouelle cited Stahl almost as soon as he was admitted to the Academy in 1744. His first two memoirs, published in 1744 and 1745, studied the circumstances of the dissolution and crystallization of salts, respectively, citing Stahl's *Treatise on Salts*. The first memoir addressed neutral salts in general and the second common salt in particular.[91] His 1747 memoir explored the inflammation of turpentine oil, which, as seen above, had been used by Homberg and Stahl in combination with sulfur and vitriolic acid. Rouelle instead inflamed this oil with nitric acid.[92]

Although Macquer referred to Stahl and phlogiston in almost all his memoirs, starting with the first (1746), none of them could compare with his chemical textbooks and dictionaries for influence and popularity. The first edition of his textbook, *Élémens de chymie théorique* (1749), was based squarely on the phlogiston theory. The *Histoire* of the Academy published a summary of this book in the same year: phlogiston is the matter of fire. Only in this form can fire enter the composition of the body. When phlogiston combines with vitriolic acid, it produces sulfur. Like vitriolic acid, nitric acid also binds with great affinity with phlogiston and indeed inflames

[85] Duhamel du Monceau and Grosse, "Des différentes maniéres" (cit. n. 74), 328; Geoffroy, "Moyens de congeler l'esprit de vin" (cit. n. 84), 19; "Sur une nouvelle espèce de teinture bleue," *HARS* (1749), 113.

[86] Guillaume-François Rouelle, "Sur l'inflammation, de l'huile de Térebenthine par l'acide nitrux pur, suivant le procédé de Borrichius; et sur l'inflammation de plusieurs huiles essentielles, & par expression avec le même acide, & conjointement avec l'acid vitriolique," *MARS* (1747), 39, 54.

[87] Claude-Joseph Geoffroy, "Quatrième mémoire sur l'antimoine. Nouveau phosphore détonnant fait avec ce Minéral," *MARS* (1736), 68. Geoffroy also compared his own method with that of Kunckel.

[88] Jean Hellot, "Analise chimique du Zinc," *MARS* (1735), 236.

[89] Henri-Louis Duhamel du Monceau, "Sur la base du sel marin," *MARS* (1736), 230. Stahl had shown that sulfur was the base of sulfuric acid (i.e., vitriolic acid), which prompted chemists to look for the base of common salt, especially when he used vitriolic acid and nitric acid to decompose the salt.

[90] Pierre-Joseph Macquer, "Recherches sur l'arsenic," *MARS* (1746), 224.

[91] Guillaume-François Rouelle, "Mémoire sur les sels neutres, dans lequel on propose une division méthodique de ces Sels, qui facilite les moyens pour parvenir à la théorie de leur crystallisation," *MARS* (1744), 353–64; Rouelle, "Sur le sel marine. Première Partie. De la Crystallisation du Sel marin," *MARS* (1745), 57–79.

[92] Rouelle, "Sur l'inflammation" (cit. n. 86).

and detonates with a loud noise.[93] Likewise, the acid of salt is also a compound of phlogiston and a singular sulfur. Metals—for example, copper, iron, tin, lead, mercury, antimony, and bismuth—are composed of phlogiston and vitrifiable earths. They all owed important chemical qualities to phlogiston, according to Macquer. Having lost their phlogiston by calcination, metals become *chaux* or calces. Imperfect metals like antimony can even turn into "glasses" when exposed to more extreme heat. The calces and glasses of metals can be reduced to the metallic state by being fed with phlogiston-rich material.[94] Phlogiston occurs most richly in oils, which are phlogiston united with water by means of an acid, and in charcoal, which is phlogiston united to the most fixed earth.[95]

Macquer's textbook thus shifted the discussions of phlogiston from the analysis of salts to calcinations and combustions. It highlighted the elements of Stahlian chemistry that became well-known in the second half of the eighteenth century. This textbook thus can be seen as marking the "official" beginning of French Stahlianism, although, as I have shown in this essay, a long incubation period in the French Academy had preceded it.

CONCLUSION

The knowledge of Stahl's chemical work thus was continuous, frequent, and rather up-to-date at the Royal Academy of Sciences in Paris from the 1710s up to the so-called Chemical Revolution. Even before Rouelle and Macquer began advocating Stahlian chemistry, their colleagues at the academy had been familiar with the German chemist's work and had regularly discussed it in public. Although some researchers disputed specific explanations that Stahl provided, all were familiar with his chemical reasoning in terms of phlogiston and his chemical methods or techniques. Admittedly, the academy chemists did not refer to Stahl in all of their publications. But the sum of their references kept the knowledge of Stahl's chemistry alive at their institution. These academicians, including Rouelle and Macquer, had to look no further than their own academy for Stahl's publications and his chemical teachings. Likewise, Macquer in fact had no need to learn of Stahl's chemistry through Rouelle, for their senior colleagues had prepared the way for French Stahlianism.

The relationship between Stahl and his colleagues at the French Academy, though bilateral, was not symmetrical. Throughout his career, Stahl seems to have cited only two memoirs by Homberg and Geoffroy from 1703 and 1704. Among the first-generation chemists after the reorganization of the French Academy in 1699, only Geoffroy paid attention to Stahl. Nonetheless, by 1735, a year after Stahl's death, all chemists of the second generation, except du Fay, had cited Stahl, either favorably or critically. Thereafter, the academy's interest in the German chemist's work continued to grow.

The academy chemists cited a considerable number of Stahl's works. Among those works were Stahl's earliest monograph (*Zymotechnia*) and some of his last publications. They also included articles published in a journal of mostly local circulation in the early 1700s, individual dissertations that Stahl supervised, a collection of

---

[93] "[A summary of Macquer's *Élémens de chymie théorique*]," *HARS* (1749), 116–8.
[94] Ibid., 120–30.
[95] Ibid., 130–2.

his early works in the *Opusculum* (1715), and a chemical textbook based on lecture notes. Indeed, the reading list of the chemist academicians, when put together, comprised the great majority and the latest of Stahl's important chemical works.

In addition, the academicians also communicated with Stahl personally. E.-F. Geoffroy communicated with Stahl through Neumann, a German colleague. Boulduc learned chemical operations through Stahl himself or his son, either in person or by correspondence. The chemists in Berlin and Paris thus had personal, if not direct, connections.

Having started without much cosmopolitan background, Stahl enjoyed fame at the prestigious and lavishly endowed French Academy and beyond. E.-F. Geoffroy referred to him as "the illustrious M. Stahl" as early as 1720, and Boulduc referred to him as the "famous Stahl" in 1724. His stature only grew in later years. Stahl probably owed some of this recognition to his appointment as the Personal Physician to a king who elevated Brandenburg-Prussia to the rank of a powerful kingdom. Yet as the references in the academy publications show, the French chemists accepted his chemical reasoning and revered his chemical methods and practical operations. Moreover, as the references indicate, Stahl's phlogiston theory was accepted not only in the French Academy but also in the Royal Society in London, and in the French provinces.

A review of this relationship has shown that Stahl and his French colleagues shared common interests. In the first two decades of the eighteenth century, Homberg, Geoffroy, and Stahl all embarked on the investigation of chemical principles, especially sulfur, the namesake of the sulfur principle. They all agreed that, instead of a simple substance, sulfur was composed of at least vitriolic acid and the oily principle, for which Stahl coined the term "phlogiston." Stahl shared with Boulduc an interest in acids, and with Bourdelin a similar interest in salts and alkalis. The other chemist academicians accepted his phlogistic thinking and his chemical methods and techniques.

The survey above gives us a richer and deeper picture of the reception of Stahl's chemistry in France in the first half of the eighteenth century. There is no longer any missing link between Stahl's chemistry and the French chemical community that embraced phlogiston in the second half of the eighteenth century. Although other connections may not and need not be ruled out, one link was continuous in the French Academy. The knowledge of Stahl's chemical teachings was very much alive at that institution. Stahl's teachings appealed to his French colleagues in part for their phlogistic accounts of chemical phenomena. The academy chemists spoke of phlogiston as a material principle and applied it in their explanations of the medicinal effect of glass of antimony, the formation of alkali salts, the composition of ammonia, and the production of the pigment Prussian blue, for example. The appeal of Stahl's teachings also included exemplary chemical analyses that practical chemists admired and followed, such as the analysis of vitriolated niter, the uses of liver of sulfur, the decomposition of salts, and the combustion of niter and oils. The appeal, moreover, comprised chemical methods and techniques on which the academy chemists improved or with which they compared their own, such as Stahl's method of preparing "antimony" (antimony sulfide) from its regulus, the production of flowers of zinc, the analysis of the base of common salt, and the preparation of the volatile spirit of niter (nitrogen dioxide).

The survey provided in this essay serves to show that the German and French chemical communities that Stahl and the French Academy represented were well informed of each other in the first half of the eighteenth century. Far from being an isolated figure, Stahl was famous, and his phlogiston already known, across important parts of Europe like Paris and London as early as the 1730s. That is not to suggest that Rouelle and Macquer contributed nothing to the popularization of Stahlian chemistry in the second half of the eighteenth century. It does, however, suggest that a continuous familiarity with Stahl's chemistry at the French Academy from Stahl's lifetime to the Chemical Revolution can no longer be ignored.

Measuring Fire:

Herman Boerhaave and the Introduction of Thermometry into Chemistry

by John C. Powers*

ABSTRACT

This essay examines Herman Boerhaave's work with the instrument maker, Daniel Gabriel Fahrenheit, on integrating the thermometer into the practice of eighteenth-century chemistry. Boerhaave utilized the thermometer to generate empirical evidence for the existence and actions of his instrument, "fire," by incorporating the instrument into pedagogical demonstrations, chemical research on heat, and, finally, the performing of operations. I examine how the use of the thermometer altered the chemists' traditional approach to heat, based on skilled sense perception and experiential judgment, and suggest that the threat to traditional practice posed by the instrument explains some of the resistance to it among some chemists in the mid-eighteenth century.

INTRODUCTION

The century before Antoine-Laurent Lavoisier was far from being the period of stagnation that early (and some later) historians of the Chemical Revolution have characterized.[1] Lawrence Principe, for example, has recently argued that the period from 1675 to 1725 constituted an earlier chemical revolution "that nobody noticed," marked by the increase in the status of chemical practitioners, a host of new aims and applications for chemistry, and new theoretical innovations based on "experimental and practical results."[2] In making this claim, Principe builds on the foundational work of Frederic Holmes, who has convincingly shown that Lavoisier's experimental work derived from earlier innovations in salt and plant chemistry, new methods and instruments, and developments in the chemical industry.[3] While historians have

---

* Department of History; Science, Technology, and Society Program; 811 S. Cathedral Place, Virginia Commonwealth University, Box 842001, Richmond, VA 23284; jcpowers@vcu.edu.

[1] On the depiction of pre-1750 chemistry as a period of "stagnation," see, for example, Robert Siegfried, *From Elements to Atoms: A History of Chemical Composition* (Philadelphia, 2002); William Brock, *The Norton History of Chemistry* (New York, 1992). Note, the stagnation hypothesis dates back to the beginning of the history of science as a profession; see Herbert Butterfield, *The Origins of Modern Science, 1300–1800* (London, 1950), chap. 11.

[2] Lawrence Principe, "A Revolution Nobody Noticed? Changes in Early Eighteenth-Century Chemistry," in *New Narratives in Eighteenth-Century Chemistry: Contributions from the First Francis Bacon Workshop*, ed. Lawrence M. Principe (Dordrecht, 2007), 1–22.

[3] Frederic Lawrence Holmes, *Eighteenth-Century Chemistry as an Investigative Enterprise* (Berkeley, Calif., 1989). For other studies based on this perspective, see Principe, *New Narratives* (cit. n. 2); Mi Gyung Kim, *Affinity, That Noble Dream: A Genealogy of the Chemical Revolution* (Cambridge,

---

traditionally focused on the theoretical changes in eighteenth-century chemistry, Ursula Klein has argued that chemists were "hybrid" practitioners who manipulated substances to make things as well as to understand the properties of those things and their constituent substances.[4] This suggests that eighteenth-century chemistry was the blend of at least two chemical traditions: the craft chemistry of apothecaries, herbalists, distillers, and other chemical artisans and the natural philosophy of experimentalists and academics like Robert Boyle, Georg Stahl, and Herman Boerhaave. Thus, this chemistry was a form of experimental philosophy, but one dedicated to both the investigation of nature and material improvement. Chemist, physician, and public lecturer Peter Shaw (1694–1763) called this new chemistry "Philosophical Chemistry," which, he explained, would "by means of appropriate Experiments, scientifically explained, lead to the discovery of *Physical Axioms*, and *Rules of Practice*, for producing useful effects" in order to "improve the State of natural knowledge, and the Arts thereon depending."[5] One proponent of this new chemistry, Herman Boerhaave, summarized this attitude by stating, "When the Chemist explains to you the nature of glass, he at the same time teaches you a sure way of making it."[6]

What made eighteenth-century chemistry so lively was that the traditional ways of making things were under constant scrutiny, and many of its practitioners were willing to go beyond traditional chemical methods, concepts, and approaches. One way to examine the changes in chemistry during this time is to look at the plethora of new instruments and apparatuses that came into use. The period from Boyle to Lavoisier saw the introduction into chemistry of air pumps, thermometers, burning lenses, and the pneumatic trough, just to name the most prominent new instruments.[7] What is striking about this group was that not one of these instruments was used in the traditional recipes and operations of the artisan-chemists but, rather, came to the notice of chemical practitioners via the work of experimental philosophers and instrument makers. Their inclusion in chemical experimentation and debate was, in almost all cases, brought about by the desire to elucidate questions regarding the properties, effects, and behavior of matter, rather than practical application.[8] Although some new apparatuses were variations or improvements on extant chemical technology, most

Mass., 2003); Ursula Klein and Wolfgang Lefèvre, *Materials in Eighteenth-Century Science: A Historical Ontology* (Cambridge, Mass., 2007).

[4] Ursula Klein, "The Laboratory Challenge: Some Revisions of the Standard View of Early Modern Experimentation," *Isis* 99 (2008): 769–82; Klein, "Blending Technical Innovation and Learned Knowledge: The Making of Ethers," in *Materials and Expertise in Early Modern Europe: Between Market and Laboratory*, ed. Ursula Klein and E. C. Spary (Chicago, 2010), 125–57. See also Lissa Roberts, Simon Schaffer, and Peter Dear, eds., *The Mindful Hand: Inquiry and Invention from the Late Renaissance to Early Industrialization* (Amsterdam, 2007).

[5] Peter Shaw, *Chemical Lectures, Publickly Read at London in the Years 1731 and 1732, and since at Scarborough, in 1733* (London, 1734), 1.

[6] Herman Boerhaave, *Elements of Chemistry*, trans. Timothy Dallowe (London, 1735), 1:51.

[7] See John C. Powers, *Inventing Chemistry: Herman Boerhaave and the Reform of the Chemical Arts* (Chicago, 2012), 79–80, 124–7, 133–9; Jan Golinski, "'Fit Instruments': Thermometers in Eighteenth-Century Chemistry," in *Instruments and Experimentation in the History of Chemistry*, ed. Frederic L. Holmes and Trevor H. Levere (Cambridge, Mass., 2000), 185–210; William Smeaton, "Some Large Burning Lenses and Their Use by Eighteenth-Century French and British Chemists," *Ann. Sci.* 4 (1987): 265–76; Henry Guerlac, "The Continental Reputation of Stephen Hales," *Arch. Int. Hist. Sci.* 4 (1951): 393–404; John Parascandola and Aaron J. Ihde, "History of the Pneumatic Trough," *Isis* 60 (1969): 351–61.

[8] Deborah Gene Warner has termed these kinds of instruments "philosophical instruments"; Warner, "What Is a Scientific Instrument, When Did It Become One, and Why?" *Brit. J. Hist. Sci.* 23 (1990): 83–93.

of these new instruments were designed to increase the precision of measurement or collection of chemical substances within an experimental context.[9]

The new instrumental practices, however, instigated a more fundamental change: a challenge to the traditional skills and status of artisan-chemists. A large part of a chemist's skill centers on how to manipulate instruments and apparatuses to know and make things. Pamela Smith argues that in using these skills artisan-chemists deploy a "vernacular epistemology," which relies on using one's body, specifically one's senses, to obtain knowledge about the world, in this case, chemical species and operations. A chemist's status as an artisan depends on his experience and judgment in reading and interpreting the signs, both in nature and in his operations.[10] This is exactly the kind of authority, however, that many of the new chemical instruments undermined. As Lissa Roberts has shown, the chemical artist in the early eighteenth century usually grounded his practice, both in the laboratory and in chemistry courses, in the recognition and interpretation of the empirical, qualitative characteristics of matter and chemical operations. By the end of the century, however, he was required to have skills in manipulating instruments, which mediated and subordinated his sensory expertise.[11] Jan Golinski, in an insightful study of the thermometer in eighteenth-century chemistry, follows this theme by arguing that as the use of the thermometer became more common, the idea of what it measured became more theoretically abstract, and the sensory expertise of the chemist became less relevant.[12] The new instruments, then, led to divisions among chemical practitioners, not just over theoretical innovation but also, and perhaps more fundamentally, over challenges to extant modes of knowledge-making, pedagogy, and expertise.

This essay focuses on the introduction of the thermometer into chemical practice by one philosophical chemist, Herman Boerhaave (1669–1738). This is a telling case, because it illustrates the ways in which a new instrument changed pedagogical, experimental, and operational practices in chemistry and, at the same time, reveals the subtle interactions between artisanal and academic chemistry and between chemical and nonchemical approaches to the art, which define philosophical chemistry. In his own practices Boerhaave switched from a reliance on the senses to the mediation of thermometers to understand and control heat, which he justified by integrating the new instrument into a network of theoretical, pedagogical, experimental, and operational uses. The Leiden professor of medicine and chemistry first used a thermometer in his chemistry lectures in 1718 after beginning a relationship with the instrument maker, Daniel Gabriel Fahrenheit (1686–1735). The two men corresponded on the principles and problems of thermometer construction and graduation and on the kind of measurements the instrument was capable of making. While Boerhaave did not accept all of Fahrenheit's proposed uses for his thermometers, this interaction convinced him that a reliable instrument could be used to demonstrate the presence and action of "fire." Within Boerhaave's system of chemistry, fire was a subtle, imponder-

[9] Cf. Holmes and Levere, *Instruments and Experimentation* (cit. n. 7), esp. chaps. 4–9.

[10] Pamela H. Smith, *The Body of the Artisan: Art and Expertise in the Scientific Revolution* (Chicago, 2004), 142–9. See also Ursula Klein, "Apothecary-Chemists in Eighteenth-Century Germany," in Principe, *New Narratives* (cit. n. 2), 97–137.

[11] Lissa Roberts, "The Death of the Sensuous Chemist: The 'New' Chemistry and the Transformation of Sensuous Technology," *Stud. Hist. Phil. Sci.* 26 (1995): 503–29.

[12] Golinski, "'Fit Instruments'" (cit. n. 7). On the difficulty of establishing what a thermometer is actually measuring, see Hasok Chang, *Inventing Temperature: Measurement and Scientific Progress* (New York, 2004).

able fluid, whose interactions with normal matter accounted for all phenomena relating to heat, including actual fire, but also a host of other effects. Previously, he had presented the effects of fire in his course via simple, empirical examples, which signaled the presence of fire in action. After 1718, however, he presented the Fahrenheit thermometer as the proper instrument to detect and indicate the relative strength of fire produced in both natural phenomena and chemical operations. Through the popularity of his courses at Leiden, and through his even more popular textbook, *Elementa Chemiae* (1732), chemical thermometry gained a foothold in eighteenth-century chemistry, although not without some resistance.

CHEMISTRY AND FIRE

In January 1702, the Leiden University curators granted Boerhaave permission to offer courses in chemistry. He offered his first lecture course in chemistry later that month, and in the fall term he offered this same course, followed in early 1703 by a course in chemical operations. In both of these courses, he attempted to present chemistry according to a regular, academic method, incorporate recent developments in chemical theory and experimentation, and describe chemical entities, such as fire, in the philosophical language of corpuscles.[13] Nevertheless, the chemistry that he presented was firmly grounded in the operations and practices of the artisan-chemist, which one could find in contemporary textbooks and courses of chemistry.[14] Boerhaave himself obtained practical training in chemistry while he was a student at Leiden. In 1692 or 1693, he worked in the shop of a local apothecary, David Stam (1633–1711), who taught the young medical student how to conduct basic chemical operations, which would have included the techniques to manipulate and control the fires, furnaces, and baths used to conduct these operations.[15] This approach to knowledge would not have been alien to Boerhaave, as medical training in Leiden was also aimed at "educating the senses" to recognize healthy and diseased bodies through courses in anatomy, botany, and clinical medicine.[16] In addition, following the work of the illustrious professor Franciscus Sylvius (1614–72), Leiden boasted of a tradition in chemical medicine, which focused on the analysis of bodily fluids and other substances in terms of their tastes, smells, and colors. Sylvius, following common

[13] See Powers, *Inventing Chemistry* (cit. n. 7), 63–72.

[14] On chemistry textbooks, see Antonio Clericuzio, "Teaching Chemistry and Chemical Textbooks in France: From Beguin to Lemery," *Sci. & Educ.* 16 (2006): 335–55; Ursula Klein, "Nature and Art in Seventeenth-Century French Textbooks," in *Reading the Book of Nature: The Other Side of the Scientific Revolution*, ed. Allen G. Debus and Michael Walton (Kirksville, Mo., 1998), 239–50; J. R. R. Christie and J. V. Golinski, "The Spreading of the Word: New Directions in the Historiography of Chemistry, 1600–1800," *Hist. Sci.* 20 (1982): 235–66; Lynn Thorndike, *A History of Magic and Experimental Science* (New York, 1958), 8:104–69; Hélène Metzger, *Les doctrines chimiques en France au début du XVIIe à la fin du XVIIIe siècle* (Paris, 1923).

[15] Powers, *Inventing Chemistry* (cit. n. 7), 57–9. On Stam, see G. A. Lindeboom, "David en Nicholaas Stam, apothekers te Leiden," *Pharm. Weekblad* 108 (1973): 153–60.

[16] On "educating the senses" via medical training, see Susan C. Lawrence, "Educating the Senses: Students, Teachers, and Medical Rhetoric in Eighteenth-Century London," in *Medicine and the Five Senses*, ed. W. F. Bynum and Roy Porter (New York, 1993), 154–78. On empiricism in medical education at Leiden, see G. A. Lindeboom, "Medical Education in the Netherlands, 1575–1750," in *The History of Medical Education*, ed. C. D. O'Malley (Berkeley and Los Angeles, 1970), 201–16; J. Heninger, "Some Botanical Activities of Herman Boerhaave, Professor of Botany and Director of the Botanic Garden at Leiden," *Janus* 58 (1971): 1–78; Andrew Cunningham, *The Anatomist Anatomis'd: An Experimental Discipline in Enlightenment Europe* (London, 2010).

practices of chemical analysis, correlated these empirical properties both to chemical properties and to the corpuscular arrangements and motions of substances.[17] It is, then, not surprising that Boerhaave deployed the same empirical methods and language to describe chemical entities in his first chemistry course.

The topic of "fire" proved challenging for Boerhaave. Fire was traditionally seen as the central instrument and agent of change in chemistry, and as such, the extant chemistry textbooks of the day presented various means of understanding fire and its effects, from the use of chemical principles, like sulfur, to empirical methods to gauge the heat of chemical operations. Many of Boerhaave's views on fire, however, had been shaped by academic sources on the subject prior to his first chemistry course. Historians over the years have speculated regarding the origins of Boerhaave's "fire," some arguing that this concept derived from Cartesian notions of "first matter," while others have asserted the similarities between fire and Newton's ether.[18] Examining his experiences as a student and instructor at Leiden, I have shown that his philosophical ideas on fire derived from two sources. The primary source was the lectures on natural philosophy from his undergraduate mentor, Wolfred Senguerd (1646–1724). In his textbook, *Philosophia Naturalis* (1680), Senguerd described fire as a ubiquitous fluid composed of subtle particles present in all space and capable of insinuating themselves into the pores of normal matter. The motion of this fluid caused the phenomena of heat and flame.[19] The second source was the *Dissertationes Chemico-Physicae* (1685) of the Leipzig medical professor, Johannes Bohn (1640–1718). Bohn's book introduced Boerhaave to the six chemical instruments, which included the five natural instruments—fire, air, water, earth, chemical menstrua—and one category of artificial instruments—furnaces, glassware, and other apparatuses. These instruments were the tools that chemists manipulated during chemical operations to effect changes in matter. Bohn had composed a dissertation for each instrument, and Boerhaave cribbed heavily from these dissertations when he composed his first chemistry lecture course in 1702. In effect, the chemical instruments provided a pedagogical framework through which he could organize chemical phenomena he wished to discuss in his course—in the case of fire, these included heat, light, flame, and combustion.[20]

In Boerhaave's first lecture course, his discussion of fire extended far beyond simple or "vulgar" fire that the chemists used in their laboratories, but it was never-

[17] Evan Ragland, "Chymistry and Taste in the Seventeenth Century: Franciscus Dele Boë Sylvius as a Chymical Physician between Galenism and Cartesianism," *Ambix* 59 (2012): 1–21. On the wide use of chemical methods outside of the university, see Harold J. Cook, *Matters of Exchange: Commerce, Medicine, and Science in the Dutch Golden Age* (New Haven, Conn., 2007), 267–76, 293–303.

[18] Cf. Hélène Metzger, *Newton, Stahl, Boerhaave et les doctrines chimiques* (Paris, 1930); Rosaleen Love, "Some Sources of Herman Boerhaave's Concept of Fire," *Ambix* 19 (1972): 157–74; Love, "Herman Boerhaave and the Instrument-Element Concept of Fire," *Ann. Sci.* 31 (1974): 547–69; Rina Knoeff, "The Making of a Calvinist Chemist: Herman Boerhaave, God, Fire and Truth," *Ambix* 48 (2001): 102–11. For Boerhaave as a "Newtonian" chemist, see Arnold Thackray, *Atoms and Powers: An Essay on Newtonian Matter Theory and the Development of Chemistry* (Cambridge, 1970); Robert E. Schofield, *Mechanism and Materialism: British Natural Philosophy in an Age of Reason* (Princeton, N.J., 1970).

[19] Wolfredus Senguerdius, *Philosophia Naturalis, quatuor partibus primarius corporum species, affectiones, differentias, productiones, mutationes, & interitus exhibens*, 2nd ed. (Leiden, 1680), 340–2.

[20] Johannes Bohn, *Dissertationes Chymico-Physicae, Chemiae Finem, Instrumenta & Operationes Frequentiores Explicantes, cum Indice Rerum & Verborum* (Leipzig, 1685). On the instrument theory, see Powers, *Inventing Chemistry* (cit. n. 7), 72–83; Powers, "Chemistry without Principles: Herman Boerhaave on Instruments and Elements," in Principe, *New Narratives* (cit. n. 2), 45–61.

theless shaped by his use of empirical, sensory-based examples to support his philosophical claims. He attempted to explain a wide range of phenomena such as friction, the action of burning lenses, and the burning of phosphorus by proposing the existence of a universally distributed, subtle fluid, whose motions and interactions with normal, ponderable matter accounted for the various observed effects. This subtle fluid possessed the property of repelling ponderable matter and was the ultimate cause of all phenomena involving heat, including flame, although Boerhaave's fire was not flame itself.[21] In the course, he presented a taxonomy of effects, mostly cribbed from Bohn, but some, such as the intense heat generated by burning lenses and mirrors, reflected Boerhaave's interest in recent developments in chemistry.[22] Since this was a lecture course, he did not perform chemical operations but instead described relevant phenomena in terms of how they are perceived by the senses. So, for example, he described the heat of friction by pointing out the warmth that one felt as one rubbed one's hands together. He then explained this sensation in terms of fire particles: the motion of the hands agitated the fire particles occupying the space between the hands, and their motion in turn caused the sensation of heat. Proceeding to more complex examples from chemistry, he described the ebullition that one sees during the reaction of a strong acid with an alkali as a signifier of the action of fire. In this case, the swift motion of acid and alkali particles displaced and agitated the fire and air particles in the solution, which generated air bubbles and heat.[23]

Boerhaave's description of the effects of fire followed the artisan-chemist's approach of reading the empirical signs to understand phenomena and control processes. Put into practice, this allowed the chemist to manipulate his apparatus properly, control the fire and level of heat, and as a result achieve the desired effects of his operation. In these early lectures, Boerhaave differentiated the level of heat a chemist employed in his operations into four general grades. The first degree of heat was that of a gentle water bath equivalent to the warmth of dung [fimo] or a hen sitting on an egg [fovens instar gallinae circubantis]. The second degree was that of a summer day or warm-water bath; the third degree extended up to the point of boiling water or a sand bath; and the fourth degree included the melting point of metals and point of combustion.[24] As Boerhaave explained, each of these degrees of fire was connected to specific types of heating technology (i.e., baths, stills, assaying furnaces) and employed for producing specific types of changes in substances during chemical operations.[25] For Boerhaave, the association of magnitudes of heat with chemical processes was a pedagogical classification designed to help the student compare the effects of various chemical operations. In practice, reading the sign with one's senses, the warmth of a vessel or ebullition of a liquid inside it, was an integral

[21] Cf. "Collegium Chemicum" in Fundamental Library, Military Medicine Academy, St. Petersburg, Russia (hereafter VMA), Fund XIII, MS no. 3, fol. 28r. Note that Boerhaave never gave a detailed description of "fire" as a subtle fluid in one place. His best, yet incomplete, description may be found in Boerhaave, Elements (cit. n. 6), 1:78–9. For the best reconstruction of Boerhaave's "fire" by historians, see Love, "Herman Boerhaave"; Metzger, Newton, Stahl, Boerhaave, 209–28 (both cit. n. 18).

[22] Boerhaave's interest in burning lenses and mirrors stemmed from work with these instruments at the Académie Royale des Sciences. Cf. Powers, Inventing Chemistry (cit. n. 7), 79; Boerhaave, Elements (cit. n. 6), 1:134–51. On the French program, see Lawrence Principe, "Wilhelm Homberg et la chimie de la lumière," Methodos 8 (2008), http://methodos.revues.org/1223 (accessed 8 August 2012).

[23] "Collegium Chemicum" (cit. n. 21), fol. 28v.

[24] Ibid., fol. 31v.

[25] Ibid., fol. 33r.

part of performing the operation correctly. This traditional approach to heat, found in seventeenth-century textbooks of chemistry, illustrated the extent to which chemists, including Boerhaave, relied on empirical judgment rather than rigorous measurement to control their operations.[26]

THE FAHRENHEIT THERMOMETER

In early 1717, Daniel Gabriel Fahrenheit arrived in Amsterdam to establish himself as a technician and instrument maker. He was a self-taught, peripatetic entrepreneur, who spent the early part of his life traveling around northern and eastern Europe visiting the shops and laboratories of mechanical artisans and experimentalists. As a practitioner of mixed or practical mathematics, he made his living by selling instruments, such as barometers, thermometers, and hydrometers, consulting on projects as a technical expert, and giving public, subscription lectures on optics, hydrostatics, and chemistry.[27] In Amsterdam, Fahrenheit revived a course of research that he had begun several years earlier in Berlin. He constructed a thermometer that employed mercury as the thermometric fluid instead of his usual spirit of wine. He then sent two of these new thermometers to the newly appointed chair of chemistry at Leiden University, Herman Boerhaave. This act prompted a meeting with Boerhaave and, ultimately, a twelve-year correspondence in which the two discussed the problems of thermometry along with a myriad of other scientific topics.[28] Boerhaave became Fahrenheit's most important patron, promoting Fahrenheit and his work to his scientific contacts and utilizing Fahrenheit thermometers in his post-1718 chemistry lectures and textbook, *Elementa Chemiae* (1732). Since Fahrenheit himself published only five short papers in the *Philosophical Transactions*, three of which concerned thermometry, much of the information relating to his work derives from material that Boerhaave opted to include in the *Elementa*. Later, eighteenth-century scholars and technicians searched the *Elementa* for facts regarding thermometer construction, the Fahrenheit temperature scale, and Fahrenheit's own experiments on heat and cold.

Most thermometers until the late seventeenth century were idiosyncratic devices that were not graduated according to set scales. According to their makers, early thermometers displayed the expansion and contraction of fluids—liquids or air—under

---

[26] For other chemists' classification of heat in this way, see, e.g., Nicolas Lémery, *Cours de Chymie, contenant la maniere de faire les Operations qui sont en usage dans la Medicine, par une Methode facile*, 10th ed. (Paris, 1713), 57–62.

[27] On Fahrenheit's life, see Pieter Van Der Star, ed., *Fahrenheit's Letters to Leibniz and Boerhaave* (Amsterdam, 1983), 1–17 (hereafter cited as *FL*); Horst Kant, *G. D. Fahrenheit, R.-A. F. de Réaumur, A. Celsius*, Biographien hervorragender Naturwissenschaftler, Techniker und Mediziner 73 (Leipzig, 1984), 26–49; Ernst Cohen and W. A. T. Cohen-de Meester, "Daniel Gabriel Fahrenheit," *Verhandlung der Koninklijke akademie van Wetenschappen*, 16 (1936): 1–37, 16 (1937): 682–9; J. B. Gough, "Fahrenheit, Daniel Gabriel," in *Dictionary of Scientific Biography*, vol. IV (New York, 1971), 516–8. On mixed mathematics and mathematical instruments, see James A. Bennett, "The Challenge of Practical Mathematics," in *Science, Culture and Popular Belief in Renaissance Europe*, ed. Stephen Pumfrey, Paolo L. Rossi, and Maurice Slawinski (Manchester, 1991), 176–90; Bennett, "Practical Geometry and Operative Knowledge," *Configurations* 6 (1998): 195–222; Bennett, "Early Modern Mathematical Instruments," *Isis* 102 (2011): 697–705; Stephen Johnson, "Mathematical Practitioners and Instruments in Elizabethan England," *Ann. Sci.* 48 (1991): 319–44. On the instrument trade and makers in the Netherlands, see Peter R. de Clercq, *At the Sign of the Oriental Lamp: The Musschenbroek Workshop in Leiden, 1660–1750* (Rotterdam, 1997).

[28] See *FL* (cit. n. 27). On Fahrenheit sending Boerhaave the two thermometers, see Fahrenheit to Boerhaave, 20 March 1729, in *FL*, 144–7.

various conditions of heat and cold, but there were no standard calibration practices or common points of reference that would allow observations with one instrument to be compared to observations with another. When instrument makers and philosophers first began to attempt to establish set scales of heat, they found that replicating a scale from one instrument to another proved to be an enormous technical challenge. Often the same maker could not get his instruments to read in concordance with each other. An eighteenth-century study revealed, for example, that Francis Hauksbee's so-called Royal Society thermometers often diverged up to 10°F from one another under similar conditions.[29] In theory, a stable scale was anchored to reliably reproducible phenomena for which the "degree of heat" remained constant and, thus, could act as "fixed points" for the scale. Practitioners typically employed two such fixed points and constructed their thermometers by assigning numerical values to these points and simply dividing the tubes of their thermometers along even intervals between these points. The main technical problem with this method of graduation was determining which fixed points were, in fact, stable. For example, the French academicians employed the ambient conditions in the cellars of the Royal Observatory in Paris as a fixed point until the eighteenth century, when this method was deemed unreliable. Some practitioners thought that ice maintained a constant degree of heat until Philippe de la Hire (1640–1718) showed that, while an ice and water bath remained constant, solid ice may get colder after it freezes. Similarly, Guillaume Amontons (1633–1705) argued that water boiled at a degree of heat that varied according to the atmospheric pressure.[30]

Fahrenheit established himself as an instrument maker by claiming to have solved the problem of thermometer replication. In 1708, he visited the Copenhagen workshop of Ole Rømer (1644–1710), from whom he learned the construction and graduation of thermometers and who provided him with the first version of his famous temperature scale.[31] In 1714 Fahrenheit called on Christian Wolff in Halle, presenting him with two spirit thermometers of his own construction that read to within one-sixteenth of a degree of each other despite the fact that they were of different sizes. Wolff was so impressed with this that he published a description of the instruments in the *Acta Eruditorum*. Fahrenheit did not reveal his method of construction, but he boasted that his method was so "constant" that he could construct a thermometer anywhere in the world that would read in concordance with any other thermometer he made.[32] When Fahrenheit called on Boerhaave in 1717, he claimed the same stability

[29] See W. E. Knowles Middleton, *A History of the Thermometer and Its Use in Meteorology* (Baltimore, 1966), 59; Louise Diehl Patterson, "The Royal Society's Standard Thermometer, 1663–1709," *Isis* 44 (1953): 51–64. On the problem of establishing set scales in the seventeenth century, see Middleton, *History*, 41–64; Chang, *Inventing Temperature* (cit. n. 12), 8–56.

[30] Middleton, *History* (cit. n. 29), 50–7; Guillaume Amontons, "Discours sur quelques propriétés de l'Air, & le moyen d'en connoitre la temperature dans tous les climats de la Terre," *Mémoires de l'Académie Royale des Sciences* (hereafter *MARS*) 1702 (1737): 216–43; Amontons, "Le Thermometre réduit à une mesure fixe & certaine, & le moyen d'y rapporter les observations faites avec les anciens Thermometres," *MARS* 1703 (1739): 64–72; Philippe de la Hire, "Experiences sur le Thermometre," *MARS* 1711 (1715): 188–95.

[31] On Rømer and Fahrenheit, see N. Ernst Dorsey, "Fahrenheit and Roemer," *J. Washington Acad. Sci.* 11 (1946): 361–72; Middleton, *History* (cit. n. 29), 66–79.

[32] [Christian Wolff], "Relatio de novo barometrorum & thermometrorum concordantium genere," *Acta Eruditorum* (1714): 380–1; cf. Middleton, *History* (cit. n. 29), 74. Note that Fahrenheit was still utilizing a version of Rømer's scale at this time. Each degree on Fahrenheit's initial twenty-four-degree scale was equivalent to 4° in his later, revised scale, which is still in use today. Thus, a one-sixteenth of a degree disparity corresponds to approximately one-fourth of a degree Fahrenheit.

of readings among his mercury thermometers and, significantly, between his spirit and mercury thermometers. With this technical feat, Fahrenheit hoped to convince Boerhaave of his prowess as an instrument maker and enroll him as a patron. In fact, to promote this claim to other potential patrons, he later constructed a demonstration instrument, which consisted of mounted spirit- and mercury-filled thermometer tubes that read in concordance on a shared scale.[33] As a result of this first meeting, Boerhaave commissioned Fahrenheit to construct for him a spirit and a mercury thermometer, which read in concordance under the same conditions.[34]

While Boerhaave waited for the thermometers, Fahrenheit cultivated their relationship through correspondence in which he recounted how he developed his temperature scale and method of graduation. He explained that the first version of his method of thermometer construction derived from that of Ole Rømer. As he reported, Rømer employed two fixed points: a water and ice bath and another water bath at "blood heat" (i.e., body temperature). To calibrate his instruments, Rømer simply placed an ungraduated thermometer bulb and tube, which was already partially filled with spirit of wine, in each bath successively. After the liquid level had stabilized in each bath, he marked the level on the tube. The lower point (water and ice) was marked 7 1/2° and the upper point (blood heat) 22 1/2°. Fahrenheit also explained that the 0° level on Rømer's scale was generated by dissolving sal ammoniac or sea salt in ice water, which cooled the solution until it was saturated, but that this procedure was too laborious to be used in practice. When he began to construct his own thermometers, Fahrenheit multiplied Rømer's scale by four to eliminate the awkward fractions, and by 1717 he adjusted the scale so that "blood heat" was 96° instead of 90°, probably because he found it physically easier to divide thermometer tubes into ninety-six units.[35]

As Fahrenheit pursued his own work in thermometry, he ultimately became convinced that there was only one reliable fixed point. While attempting to construct his own thermometers, he realized that human body temperature (i.e., blood heat) was not a stable fixed point but varied according to the age and health of the individual.[36] In pursuit of finding suitable alternatives, he taught himself to read French, having learned Latin as a boy, and he read books on thermometry and barometry, such as those of Robert Boyle that appeared in Latin, as well as papers by Amontons and de la Hire on these subjects that he found in the *Mémoires* of the *Académie royale des sciences*.[37] Fahrenheit, in fact, began experimenting with mercury as a thermometric fluid in 1713 after he had read a memoir by Amontons detailing how the mercury in barometers expanded with respect to temperature.[38] As a result of his study of the lit-

[33] Fahrenheit presented this instrument when he visited the Royal Society in 1724; Royal Society of London, *Journal Book of Society Meetings*, vol. 12, March 5, 1723/4, 443.

[34] Fahrenheit to Boerhaave, 30 March 1729, in *FL* (cit. n. 27), 147; Boerhaave, *Elements* (cit. n. 6), 1:87–8. Note that Boerhaave revealed Fahrenheit's method of graduation for his spirit and mercury thermometers in his lectures on fire in early 1719; cf. VMA, XIII, MS no. 7, fol. 5r.

[35] Fahrenheit to Boerhaave, 17 April 1729, in *FL* (cit. n. 27), 170–1. On the origins of Fahrenheit's scale, see Dorsey, "Fahrenheit and Roemer" (cit. n. 31); I. B. Cohen, "Roemer and Fahrenheit," *Isis* 39 (1948): 56–8; Cyril Stanley Smith, "A Speculation on the Origin of Fahrenheit's Temperature Scale," *Isis* 56 (1965): 66–9; Middleton, *History* (cit. n. 29), 66–79.

[36] Fahrenheit to Boerhaave, 17 April 1729, in *FL* (cit. n. 27), 162–3.

[37] Ibid., 172–3.

[38] G. Amontons, "Que tous les Barometres, tant doubles que simples qu'on construits jusquoici, agrissent seulement par le plus ou le moins de poind da l'air, mais encore par son plus ou moins de chaleur . . . ," *MARS* 1704 (1747): 234–45. On Fahrenheit's idea to use mercury, see Fahrenheit to Boerhaave, 27 March 1719, in *FL* (cit. n. 27), 140–3.

erature, combined with his own experimentation, Fahrenheit concluded that the only reliable fixed point was an ice-water bath brought to thermal equilibrium.[39]

Because he only had confidence in one fixed point, Fahrenheit devised an alternative method of thermometer construction and graduation. His new method involved defining his scale according to the measured volumetric expansion of the thermometric fluid as it was heated. When he constructed a new thermometer, he precisely and painstakingly measured the volume of the empty thermometer tube and bulb and the volume of mercury or spirit with which he filled it. He then placed the new thermometer in an ice-water bath and marked the fluid level in the tube as 32°. Since he only had one reliable fixed point, he graduated the tube based on the fraction of the fluid's original volume that expanded over 96° on his scale. This method required an accurate measurement of the fluid's thermal expansion, which had to be determined ahead of time. In theory he determined this value by measuring the volume of a sample of the liquid at 0° and again at 96°, but in practice he measured the volume at two more convenient points and extrapolated linearly for a 96° spread. By 1729 Fahrenheit reported that the mercury he used expanded by $1/116\frac{9}{16}$ of its original volume as it was heated 96° on his scale. By representing the expansion of a fluid as the fractional increase in its volume over a set range of degrees of heat, he could apply this measurement directly to thermometer construction by converting the fraction of expansion into parts of mercury. As he explained to Boerhaave, "Multiplying [the expansion of mercury] $116\frac{9}{16}$ by 96 [degrees], I now get 11,190, and this is the number of parts of quicksilver that the cylinder must contain in order to increase by 96 parts [over] 96 degrees of heat."[40] Thus, at 32° the thermometer will contain 11,222 (i.e., 11,190 + 32) parts of mercury, with each additional volumetric part that the mercury expanded corresponding to one degree on Fahrenheit's scale. By relating "parts" of mercury to the carefully measured volume of a real bulb and tube, Fahrenheit believed he could reproduce his scale in any instrument regardless of size.

In the letters that followed his initial contact, Fahrenheit revealed his motives for approaching Boerhaave. He wished to convince the new chair of chemistry that his mercury thermometer could be a useful tool for chemical analysis. The new thermometer significantly extended the range of thermometric measurement. Spirit of wine thermometers were useful to about 174° on Fahrenheit's scale, at which point the spirit began to boil in the thermometer tube. Since quicksilver boiled at about 600°, Fahrenheit's new mercury thermometers made possible a wide range of new measurements. The most important of these were measurements of the boiling points for common chemical reagents, such as water, "highly purified spirit of wine," oil of vitriol, and spirit of niter, each of which boiled at a higher degree of heat than could be measured with a spirit thermometer.[41] Fahrenheit contended that each pure chemical species had a stable boiling point, specific gravity, and thermal expansion, which, if known, could be used to determine the composition and purity of unknown samples. He devised new techniques and instruments for measuring these properties and often included in his letters tables presenting his latest measurements of specific

[39] Fahrenheit to Boerhaave, 17 April 1729, in *FL* (cit. n. 27), 162–5; de la Hire, "Experiences" (cit. n. 30).

[40] Fahrenheit to Boerhaave, 20 March 1729, in *FL*, 138–9. See also *FL*, 24–5 (both cit. n. 27).

[41] Fahrenheit to Boerhaave, 12 December 1718, in *FL*, 84–7; Fahrenheit to Boerhaave, 23 January 1719, in *FL*, 96–9 (both cit. n. 27).

gravity and thermal expansion for the same fluids on which he reported the boiling point.[42] In a letter from December 1718, Fahrenheit described how he utilized his "hydrostatic" (as he called them) measurements to determine that a sample of spirit of niter he had purchased was contaminated. He reported that the sample boiled 9° higher than previous samples of spirit of niter that he had tested, and that the measured specific weight and volumetric expansion of the sample were also higher. He noted that each characteristic he measured fell between that of spirit of niter and oil of vitriol. "I conclude[d]," wrote Fahrenheit, "that the same spirits of nitre were not pure, but perhaps adulterated with oil or spirits of vitriol."[43]

Despite Fahrenheit's enthusiasm for his analytic methods, Boerhaave remained skeptical. While Fahrenheit could construct thermometers utilizing the same thermometric fluid, which read in concordance, he initially had a difficult time achieving this feat between spirit and mercury thermometers. In late 1718, Boerhaave informed Fahrenheit that the spirit and the mercury thermometers that he had purchased did not agree in their readings. As Fahrenheit investigated this problem, he reported that at times the two instruments diverged by as much as 6°.[44] He devoted the next several years to investigating and correcting this problem. In 1729, Boerhaave asked Fahrenheit to report on his progress, since the Dutch professor wished to discuss thermometry in his chemistry textbook, which he was then composing. Fahrenheit reported a litany of possible sources for the discrepancy between the thermometers, including that fact that he had not measured the expansion of mercury with sufficient accuracy in 1718 and that the two thermometers were constructed with two different types of glass, each of which expanded at different rates when heated.[45] Even after he corrected the problem of glass expansion, however, Fahrenheit discovered that the two types of thermometers would not agree in their readings at all points along his scale. In a subsequent letter, he reported that the mercury thermometer rose faster between 36° and 64° and the spirit thermometer faster above 96°. Even after he attempted to correct these discrepancies by adding various proportions of water to the alcohol in his spirit thermometer, he could not make the two types of thermometer agree at temperatures higher than 112°.[46]

In spite of this setback, Boerhaave became a strong advocate for Fahrenheit and his thermometers. In effect, Fahrenheit's patronage strategy was successful. In addition to purchasing instruments, Boerhaave provided Fahrenheit with news and letters of introduction, notably to members of the Royal Society when he traveled to London

[42] Cf. Fahrenheit to Boerhaave, 12 December 1718, in *FL*, 84–7; Fahrenheit to Boerhaave, 23 January 1719, in *FL*, 96–9 (both cit. n. 27). For a new aerometer devised by Fahrenheit to measure specific gravities above that of water, see Fahrenheit, "Areometri novi descripto & usus," *Phil. Trans. Roy. Soc. Lond.* 33 (1724), 140–1; *FL* (cit. n. 27), 37–8.

[43] Fahrenheit to Boerhaave, 12 December 1718, in *FL* (cit. n. 27), 89. Fahrenheit also described his analytical methods in his public courses given in Amsterdam, cf. Cohen and Cohen-de Meester, "Daniel Gabriel Fahrenheit" (cit. n. 27), 14–7; see also the notes from one of Fahrenheit's courses found at the University of Leiden Library: Rijksuniversiteit Leiden, MS BPL 772, pp. 67–183.

[44] Fahrenheit to Boerhaave, 30 March 1729, in *FL* (cit. n. 27), 147.

[45] Ibid., 146–57. For Fahrenheit's initial speculations about the cause of the discrepancy, see Fahrenheit to Boerhaave, 23 January 1719, in *FL*, 94–5 (both cit. n. 27).

[46] Fahrenheit to Boerhaave, 17 April 1729, in *FL* (cit. n. 27), 162–3. Note that the reason for this discrepancy was probably the fact that the expansion of fluids when heated is not a linear process as Fahrenheit assumed. On this problem, see Hasok Chang, "Spirit, Air and Quicksilver: The Search for the 'Real' Scale of Temperature," *Hist. Stud. Phys. Sci.* 31 (2000): 249–84; Chang, *Inventing Temperature* (cit. n. 12), chap. 2.

in 1724.[47] But most importantly, Boerhaave utilized and promoted Fahrenheit thermometers in the chemistry courses at Leiden starting in the fall of 1718 and, ultimately, in the *Elementa Chemiae*. But rather than promote the thermometer as a tool
for chemical analysis, Boerhaave used them first as pedagogical tools and then as instruments to help control certain chemical operations. Neither of these uses required
the level of precision to which Fahrenheit strove. The thermometer, as a chemical
instrument, only had to provide a more reliable gauge of heat than the unaided senses
of the chemist.

MAKING THE THERMOMETER A CHEMICAL INSTRUMENT

When Boerhaave was appointed to the Chair of Chemistry at Leiden in March 1718,
he was obligated to offer one course on chemistry that was available to all matriculated students at the university without lecture fees. In October, he gave his first lecture on the chemical instruments, intending to present an exhaustive account of each
instrument—fire, air, water, earth, and chemical menstrua—in turn over a period of
several years.[48] The "instruments course" differed from his earlier chemistry courses
in that, where he could, he presented important claims about the properties of each
instrument through demonstration experiments. The aim of these experiments was
to reinforce the empirical foundations of chemistry by showing that the principles
that guided chemical practice derived from natural phenomena and not speculative
theories. This was a new pedagogical method for chemistry, which Boerhaave modeled on the tradition of experimental physics demonstrations conducted by Wolfred
Senguerd and Burchard de Volder (1643–1709) in Leiden's *Theatrum Physicum*.
These demonstrations, which often focused on instruments, such as air pumps, barometers, and lenses, were designed to illustrate and reinforce the theoretical principles of physics introduced in lectures.[49] Boerhaave, however, flipped the pedagogical aim of his demonstrations by showing how an ordered set of experiments could
be used to generate theoretical claims.[50] Boerhaave saw Fahrenheit's thermometer as
a valuable tool in this process because he thought that it could demonstrate empirically the existence and properties of instrumental fire. In the first few lectures of the
instruments course, he constructed a theory of the thermometer as the tool for exhibiting the presence, motion, and intensity of fire. He then utilized the thermometer
throughout the rest of the course when he needed to demonstrate the presence and
action of fire. In effect, by using the thermometer to show the presence and action of
fire repeatedly in various contexts, Boerhaave collected abundant empirical evidence
to support his claims about the subtle fluid.

Boerhaave initiated his course on the chemical instruments with a series of dem-

[47] See, e.g., Boerhaave to William Sherard, 10 March 1724, in G. A. Lindeboom, ed., *Boerhaave's
Correspondence* (Leiden, 1962), 1:130.
[48] The "instruments course" lasted until 1728; cf. VMA, MS no. 7; Powers, *Inventing Chemistry*
(cit. n. 7), chap. 5.
[49] On experimental physics at Leiden, see Gerhard Wiesenfeldt, *Leerer Raum in Minervas Haus:
Experimentelle Naturlehre an der Universität Leiden, 1675–1715* (Amsterdam, 2002); C. de Pater, "Experimental Physics," in *Leiden University of the Seventeenth Century*, eds. Th. H. Lunsingh
Sheurleer and G. H. M. Posthumus Meyjes (Leiden, 1975), 309–27; Edward G. Ruestow, *Physics
at Seventeenth- and Eighteenth-Century Leiden: Philosophy and the New Science at the University*
(The Hague, 1973), 96–112. On the use of instruments in demonstrations, see lecture notes from De
Volder's course: British Library, Sloane Collection, MS 1292, fols. 78r–141v.
[50] See Powers, *Inventing Chemistry* (cit. n. 7), 121–9.

onstration lectures designed to connect his theory of fire with Fahrenheit's understanding of the thermometer. In effect, he constructed a philosophical argument, which held that the given expansion of a fluid through heat was caused by a proportional amount of fire. He began this argument with a clarion call for the use of strict empirical methods in chemistry. Only phenomena that were "universally demonstrated"—rendered observable to the senses and intelligible to the mind—would be accepted as facts.[51] These criteria presented an interesting problem in the case of fire. Although Boerhaave still maintained that fire was the primary instrument of the chemist, his conception of fire was that of a subtle, weightless fluid composed of particles beyond the direct apprehension of the senses. He framed this problem in his lectures and later in his textbook, *Elementa Chemiae* (1732), as the need to find a "sensible property," a "sign" or "mark" of fire, that was present in all phenomena involving fire and by which one could demonstrate the presence of fire experimentally. As he asserted in the *Elementa*, "It is absolutely necessary that this mark [of fire] should be perfectly obvious to our senses, and that we should all agree about it; otherwise the word Fire . . . would signify nothing at all."[52]

In pursuit of this goal, he presented a list of empirical qualities that were traditionally associated with fire and were, thus, candidates for being such a sign: heat, light, the rarefaction of fluids and solids, combustion, and fusion. He rejected light, combustion, and fusion as possibilities on the grounds that they were not present in all phenomena involving fire. Heat (*calor*) posed an interesting problem because Boerhaave agreed that heat was, in fact, a direct, sensible result of the action of fire. He argued, however, that our direct sensation of heat was unreliable. To substantiate this claim, he presented the example of underground caves that maintain their temperature throughout the year. When one enters such a cave after having been in the hot summer sun, the air inside the cave feels very cool. On the contrary, when one enters the same cave after having been in the cold, winter air, the air feels comfortably warm.[53] To correct this error in perception, Boerhaave proposed the expansion or rarefaction of bodies as a more reliable measure of heat. He explained in his *Elementa*:

> It seems, therefore, as if the easy expansion of . . . fluids might serve as a certain mark of the presence, increase, or decrease of Fire: For this no way depends upon our senses, which we find in these inquiries to be so uncertain a guide, [and] will not easily lead us into mistakes. . . . This then alone is what we shall make use of as we proceed in the investigation of [heat]; and we shall take it for granted, that in every *phænomenon* where we see this rarefaction excited, there is there a proportional degree of Fire [that is] the cause of it.[54]

The expansion of fluids would not fool the senses like the direct sensation of heat and was observable, he claimed, in all phenomena involving fire.[55]

Once Boerhaave had asserted that expansion was the sign of fire, he presented a series of demonstration experiments to show this empirically. In these experiments, he constructed situations in which he could exhibit the expansion and contraction of bodies due to fire in an unambiguous manner. In his first experiment, Boerhaave mea-

[51] Cf. VMA, Fund XIII, MS no. 7, fols. 1v–2r.
[52] Boerhaave, *Elements* (cit. n. 6), 1:80.
[53] VMA, MS no. 7 (cit. n. 51), fol. 2r; Boerhaave, *Elements* (cit. n. 6), 1:82.
[54] Boerhaave, *Elements* (cit. n. 6), 1:85.
[55] VMA, MS no. 7 (cit. n. 51), fol. 3r–v.

sured the length of an iron rod and showed that the rod passed snugly through an iron ring. He then heated the rod in a flame until it was red-hot, measured the increase in length of the rod, and demonstrated how the rod had also expanded in diameter, showing that it would no longer fit through the iron ring. He explained that all solid bodies would expand when heated in this manner and suggested that each incremental increase in the degree of heat added would likewise increase the expansion of the rod. He then allowed the rod to cool, exhibiting how the rod contracted to its former length and diameter. In his interpretation of this experiment for his students, Boerhaave suggested that these phenomena demonstrated the presence and action of fire. As he explained, fire imparted motion to the particles comprising solid bodies, causing the body as a whole to dilate. As a body cooled, the motion lessened, causing the body to contract. Ultimately, he asserted that the expansion of a body "always increases in proportion, as a greater quantity of Fire is admitted into the expanding body."[56]

Having established that the expansion of bodies is proportional to the amount of fire they contain, Boerhaave next presented ways to measure the proportion of fire in fluids. To show the expansion of air, he employed a J-shaped "Dutch" or "Drebbel"-type air thermometer. Breathing on the bulb and holding it in his hands, he was able to show the "proportional" expansion of the air.[57] For liquids, he displayed one of Fahrenheit's spirit of wine thermometers. He explained to his auditors, using descriptions and figures that he obtained from Fahrenheit himself, the relationship between the expansion of the spirit and the scale according to which the thermometer was graduated. He stated that the bulb contained 1,933 parts and the tube ninety-six parts, one for each degree as marked on the tube. The spirit expanded by about one-twentieth of its volume (i.e., 96/1933) when warmed from the "greatest cold" (0°) to the "greatest heat" (96°). He also noted that the spirit boiled at 174°, at which point it had expanded by one-eleventh of its volume.[58] In addition, he displayed thermometer tubes filled with oil of turpentine and "pure rainwater," noting the expansion of each from room temperature (about 52°) to the point at which the water boiled. Finally, Boerhaave displayed one of Fahrenheit's mercury thermometers. As with the spirit thermometer, he reported that the bulb contained 11,124 parts, and like the oil of turpentine and rainwater, he reported that the expansion of the mercury was 1/52 between 0° and 212°.[59]

The purpose of this extended series of demonstrations was to establish the Fahrenheit thermometer as the main tool with which one would study the effects of fire in chemistry. From that point in his lectures and textbook, Boerhaave deployed Fahrenheit's thermometers and expressed degrees of heat in Fahrenheit's scale without making extended references to the instrument that he used or arguments regarding what it supposedly measured. Within the context of his course, the thermometer became what Bruno Latour had termed a "black box."[60] Boerhaave removed questions about

[56] Ibid., fol. 4r; Boerhaave, *Elements* (cit. n. 6), 1:86–93, on 87.

[57] VMA, MS no. 7 (cit. n. 51), fol. 4v. Note that Boerhaave did not describe the thermometer in the lecture notes, but when he addressed this demonstration in the *Elementa Chemiae*, he employed a Drebbel-type instrument. Cf. Boerhaave, *Elements* (cit. n. 6), 1:94–5. On this type of thermometer, see Middleton, *History* (cit. n. 29), 19–21; F. Sherwood Taylor, "The Origin of the Thermometer," *Ann. Sci.* 5 (1942): 129–56.

[58] VMA, MS no. 7 (cit. n. 51), fol. 4v.

[59] Ibid., fol. 5r. See also, Boerhaave, *Elements* (cit. n. 6), 1:106.

[60] Bruno Latour, *Science in Action: How to Follow Scientists and Engineers through Society* (Cambridge, Mass., 1987), 68.

the operation of the thermometer itself from the phenomena that it measured, thus allowing the instrument to be used unproblematically in further demonstrations to indicate the presence and relative strength of fire. With each new application of the thermometer, Boerhaave implicitly generated more proof for the existence of his subtle fire, hardening the concept of fire into a substance that could seemingly be measured and manipulated by chemists, just as they could manipulate any other substance.[61]

The larger pedagogical aim of Boerhaave's lectures on fire was to outline what an experimentally based chemistry might look like and to convey a method through which this chemistry might be achieved. At the outset of the instruments course, he presented an outline of an experimental philosophy for chemistry, which he attributed to Francis Bacon. He argued that the axioms of any natural philosophy were generated from sensory observations, which must be collected and organized into natural or experimental histories. For chemistry, these histories were to be constructed by observing the effects when various species of bodies were resolved into their "simplest elements" or "agitated by various grades of motion." By proceeding in this manner, chemistry revealed the latent properties of bodies, and thus "the vast utility and absolute necessity of chemistry is understood."[62] He reiterated a version of this scheme at the conclusion of his lectures on fire. The programmatic goal of his experimental approach was to generate a "complete and certain history of Heat." He suggested to his students that all "simple bodies" from the three kingdoms of nature should be examined, first by mixing them with other bodies of their "class," and then with bodies of other classes.[63] The final product of this endeavor would be a systematic account of each species of body, listing both its innate sensible properties and those properties that were generated by being brought into contact with other bodies.

Within this experimental program, the Fahrenheit thermometer was the instrument to reveal the presence and relative intensity of fire actuated by chemical interactions. Boerhaave utilized the thermometer to construct several sets of demonstration experiments designed to illustrate the action of "fire" in chemical combinations and reactions and, most importantly, how these experiments could be conducted methodically to generate a "history of Heat." In the first series, he took a quantity of distilled rainwater and measured its temperature with a thermometer. He then mixed in an equal quantity of spirit of wine at the same "degree of heat." A large thermometer placed in the mixing vessel exhibited to his auditors the increase in "degrees of heat" generated by the solution. Subsequent demonstrations expanded on this experiment as Boerhaave varied the concentration of the spirit of wine and purity of the water, which he adulterated with small amounts of spirit. These experiments were not intended to provide rigorous measurements. Boerhaave pointed out that the size of the thermometer needed for the demonstration affected the accuracy of measurement, since more fire was absorbed to expand the fluid in a larger instrument. Rather, they were meant

---

[61] See the similar arguments regarding the existence of phlogiston in Mi Gyung Kim, "The 'Instrumental' Reality of Phlogiston," *HYLE* 14 (2008): 27–51; Victor D. Boantza and Offer Gal, "The 'Absolute Existence' of Phlogiston: The Losing Party's Point of View," *Brit. J. Hist. Sci.* 44 (2011): 317–42.

[62] VMA, MS no. 7 (cit. n. 51), 1v–2r; Boerhaave, *Elements* (cit. n. 6), 1:50–1; Powers, *Inventing Chemistry* (cit. n. 7), 119–21. See also Ursula Klein, "Experimental History and Herman Boerhaave's Chemistry of Plants," *Stud. Hist. Phil. Biol. Biomed. Sci.* 34 (2003): 533–67.

[63] Boerhaave, *Elements* (cit. n. 6), 1:222.

to demonstrate merely that heat or, rather, fire was evolved through the interaction of different chemical species and that some interactions generated more fire than others. The overall aim of these experiments was to construct general, qualitative principles regarding the generation of fire. At the conclusion of the water and spirit of wine experiments, for example, Boerhaave summarized the results by stating, "the purer the Water is, the more heat it will generate in the Alcohol, and *vice versa*."[64]

Boerhaave, of course, could not treat all chemical interactions with this level of detail. His aim was to provide a model for conducting experiments on heat that his students could then follow themselves. After performing the demonstrations on water and spirit of wine, he continued with the experimental history of heat, first by mixing other vegetable substances (vinegar, oil of tartar, oil of turpentine) with water and then with each other. He then proceeded to animal substances (urine, salt of urine) and, finally, to fossils (various salts, acids, and metals).[65] During these demonstrations, he would occasionally pose questions regarding the interpretation of the experiment at hand in terms of the action of the material and fiery particles. For example, he admitted that he did not know the cause that attracted fire to the interactions of chemical species. He asked, "Is it the reciprocal attractive power of these elements, by which, when they rush towards each other, excite little fires? Or does an alternate attraction and repulsion . . . produce a very swift attrition between them, and by this means produce heat?"[66] In all these cases, he did not present a definitive answer but merely left the questions for his students to ponder. In effect, Boerhaave presented to his students a research program, which included a methodology for conducting experiments and general framework for interpreting experimental results complete with guiding questions to be answered. At the conclusion of his mixing experiments, he encouraged his students to pursue this program and complete this "certain history of heat." He advised them "to make use of those beautiful thermometers of Fahrenheit" when pursuing this course, because "they are exceeding[ly] sensible of Heat and Cold."[67]

Perhaps more striking than his experimental program for the thermometer was Boerhaave's regular use of the instrument as a pedagogical tool in other demonstration experiments. He readily deployed thermometers whenever he needed to demonstrate the presence or intensity of fire in support of a theoretical point he was constructing. In one demonstration during his lectures on fire, he showed how a red-hot iron radiated heat by moving a thermometer close to the iron, without touching the metal, and pointing out how the fluid level in the thermometer rose.[68] Boerhaave utilized the thermometer in lectures on the other instruments as well. In his lectures on "air" (ca. 1720 or 1721), he wished to demonstrate the principle that lowering the air pressure also lowered the boiling point of liquids. To show this, he placed in the receiver of his air pump three flasks of water, each at different temperatures as indicated by a thermometer. As he evacuated the pump, the flask whose thermometer read 150° boiled first, followed by the flask at 91°, and finally by the flask at 44°.[69] In

[64] Ibid., 215.
[65] See ibid., 217–21.
[66] Ibid., 216.
[67] Ibid., 222.
[68] Ibid., 156.
[69] Ibid., 297.

both of these cases, the thermometer was not required to make Boerhaave's theoretical point; his students could have used their hands to feel the heat or differences
in degrees of heat. The thermometers performed the pedagogical function of convey-
ing the presence and magnitude of heat without the student having to approach the
experiment, which would have been impractical in a laboratory packed with perhaps
as many as thirty or forty students.

Perhaps the greatest testament to the extent to which the Fahrenheit thermometer
became a regular part of Boerhaave's chemistry was in his suggestion that this thermometer could be used to help regulate heat in chemical operations. The first indication of this came at the conclusion to his lectures on fire, where he discussed the
traditional grades of heat employed by chemists. Recall that in his earliest lectures,
he suggested four degrees of heat, each corresponding to a specific type of operation and heating technology. In the instruments course, Boerhaave expanded these
degrees to six and expressed each category both in terms of the traditional heating
technologies and as a range of measurements made according to Fahrenheit's scale.
So, the first degree of heat, suitable for infusions, extended up to 80° on Fahrenheit's
scale. The second degree, suitable for fermentation and putrefaction, extended from
40° through 94°; the third degree, suitable for distillations, extended from 94° to
212°; and so on.[70]

Following his own recommendations, at some point during the early 1720s, Boerhaave also began to use thermometers in his operations course. It is difficult to determine when he began to do this, since the manuscripts for his operations courses
only list the operations to be discussed and demonstrated without providing further
comment.[71] But when he composed the descriptions of these operations in the "operations" section of the *Elementa Chemiae*, he reported the heat needed for several
operations in term of degrees on Fahrenheit's scale. In the first operation discussed,
the extraction of an essential oil to make a "water" of rosemary, Boerhaave suggested that the process should be conducted at 85°.[72] The significance of this is that
he does not describe the thermometer to be used, or even that a thermometer should
be used; he simply assumes that it is obvious to the reader, who ideally has plowed
through the theory section of the book, what 85° means. He reported the degree
of heat to be used for several other operations as well, most of which involved the
extraction, digestion, or fermentation of plant matter that proceeded at heats below
the boiling point of water. The most likely reason for this was that operations, such
as the calcination of minerals or metals, required degrees of heat that were too intense to be measured practically by a Fahrenheit thermometer. He was most enthusiastic when discussing operations, like fermentation, in which the degree of heat
played a crucial role in the operation. In his "discourse on fermentation," a discussion of this process inserted in the midst of the operations section of the *Elementa*,
he suggested that a heat of 60°–70° was optimal for most fermentations. Allowing
the heat to fall below 36° will stop the fermentation, as will allowing the heat to rise
above 90°.[73]

[70] Ibid., 241–5.
[71] Cf. "Collegium Chemicum" (cit. n. 21), fols. 143r–178v.
[72] Boerhaave, *Elements* (cit. n. 6), 2:9.
[73] Ibid., 123.

Boerhaave's presentation of thermometry in the *Elementa Chemiae* proposed to reshape the way that the chemist understood and controlled the phenomenon of heat. Gone were the days of training the senses to detect the signs and degrees of heat with the hands, sights, and smells. As implied in the *Elementa*, the philosophical chemist now employed the thermometer, and he was required to master the skills and knowledge required for its use, including how to examine the properties and effects of fire and to regulate operations. In effect, Boerhaave had replaced the nuanced, sensory training for chemists with philosophical knowledge—that is, a theory of how the thermometer works, how to make sense of its readings, and how to deploy that knowledge in various experimental and operational situations. According to this construction of the instrument, the Fahrenheit thermometer generated empirical knowledge, just like the senses of the chemist, but was superior to them, because it was more reliable than the unaided senses, and its readings could be openly and unambiguously displayed, reported, and repeated.

CONCLUSION

Boerhaave constructed the thermometer as a chemical instrument by integrating it into a system of theoretical, pedagogical, experimental, and operational practices. As I have argued, his initial aim was to provide empirical evidence for his subtle, elemental fire, and that every new use of the thermometer provided additional evidence. However, for other chemists and natural philosophers, the most enduring part of Boerhaave's thermometry was his pedagogical method for presenting the instrument and its capabilities. For Boerhaave, this included two parts: first, the notion that the expansion of the thermometer's fluid measured the amount of fire and, second, the order of demonstration experiments that empirically validated this notion. Following Boerhaave's method in the *Elementa*, many chemistry instructors presented the thermometer for their courses in the same manner, starting with the expanding bar experiment to establish the principle of expansion caused by heat, followed by the demonstration of thermometers and then experiments to exhibit the heat generated by chemical interactions. Versions of this method can be found, for example, in Peter Shaw's public courses at London and Scarborough and in the courses of William Cullen (1710–90) and Joseph Black (1728–99) at Glasgow and Edinburgh.[74] Like Boerhaave, these chemists saw thermometers as useful in a pedagogical context for revealing small generations of heat or cold and for solving the practical problem of showing the effects of heat to a room full of students. Boerhaave's pedagogical method for the thermometer persisted even through differing interpretations of what the thermometer measured, such as the case of Black's contention that thermometers indicated the intensity, not the quantity, of fire.[75]

[74] See Shaw, *Chemical Lectures* (cit. n. 5), 25–8; Joseph Black, *Lectures on the Elements of Chemistry*, ed. John Robison (Edinburgh, 1803), 1:40–1; Arthur L. Donovan, *Philosophical Chemistry in the Scottish Enlightenment: The Doctrines and Discoveries of William Cullen and Joseph Black* (Edinburgh, 1975). Note that Cullen and Black also conducted research on heat with thermometers; cf. William Cullen, "Of the Cold Produced by Evaporating Fluids, and Some Other Means of Producing Cold," in *Essays and Observations, Philosophical and Literary* (Edinburgh, 1770), 2:159–71; Henry Guerlac, "Joseph Black's Work on Heat," in *Joseph Black: 1728–1799: A Commemorative Symposium*, ed. A. D. C. Simpson (Edinburgh, 1982), 13–22.
[75] Cf. Golinski, "'Fit Instruments'" (cit. n. 7), 196–200.

Many chemists, however, did not accept Boerhaave's program for the thermometer in research or for utilizing the thermometer to control operations. These aspects of thermometry directly challenged the chemist's traditional skills and status. Guillaume-François Rouelle (1703–70), for example, deployed thermometers in his courses at the Jardin du roi to display "the smallest degree of heat produced in bodies at the moment of their mixing and combination" but rejected the use of the thermometer in operations, arguing that the true "chemical artiste" used "his thermometer at the tip of his fingers" to control his operations.[76] This sentiment was echoed by Gabriel-François Venel (1723–75), who argued in his article "Chymie" in the *Encyclopédie*, that chemistry had its own method of understanding the world based in "the immediate exercise of the senses," which distinguished the art from the rationalizations of "physics." Venel suggested, in fact, that the thermometer was an external intrusion into chemistry that could not replace the sensory skills of the chemist, declaring that "thermometers [were] as ridiculous in the apron of a working chemist as in the pocket of a physician visiting the sick."[77]

Venel's remarks about thermometers provide insight into the debates that increasingly gripped both practitioners of eighteenth-century chemistry and its historians over the status and direction of the field. Based in part on Venel's arguments about the demarcation of physics from chemistry, some historians have suggested that physics was the driving force of theoretical change in chemistry, culminating in Lavoisier's "revolution in physics and chemistry." In this interpretation, Boerhaave's theory of fire was characterized as a physical theory, which Venel and the French Stahlians rejected, but then Lavoisier and his collaborators at the Académie royale des sciences revived.[78] In this essay, I have argued seemingly in support of the position that Fahrenheit's thermometer and Boerhaave's pedagogical method derived from practical mathematics and experimental physics, respectively. Yet to say that Boerhaave was really a physicist or was doing physics does not reflect his own understanding of his work. Instead, I suggest that one should examine how the new, philosophical chemistry and its instruments and methods challenged traditional notions of chemical knowledge and practice and, also, what the ramifications of that challenge were. There were many chemical practitioners who were contemporaries of Rouelle and Venel and who embraced Boerhaave's program for chemistry for reasons involving

[76] Quotes from Roberts, "Death of the Sensuous Chemist" (cit. n. 11), 509. On Rouelle's courses, see Christine Lehman, "Mid-Eighteenth Century Chemistry as Seen through Student Notes of the Courses of Gabriel-François Venel and Guillaume-François Rouelle," *Ambix* 56 (2009): 163–89; Lehman, "Innovation in Chemistry Courses in France in the Mid-Eighteenth Century: Experiments and Affinities," *Ambix* 57 (2010): 3–26.

[77] Venel, "Chymie ou chimie," in *Encyclopedie, ou Dictionnaire Raisonné des Sciences, des Arts et des Métiers*, new edition (Geneva, 1777), 2:13–63, on 34. See also Lehman, "Innovation" (cit. n. 76). Note that Jessica Riskin portrays Venel as describing a "culture of chemistry" grounded in specific ideas of language; Riskin, *Science in the Age of Sensibility: The Sentimental Empiricists of the French Enlightenment* (Chicago, 2002), 230–9.

[78] The best summary of this view is Evan M. Melhado, "Chemistry, Physics, and the Chemical Revolution," *Isis* 76 (1985): 195–211. See also Henry Guerlac, "Chemistry as a Branch of Physics: Laplace's Collaboration with Lavoisier," *Hist. Stud. Phys. Sci.* 7 (1976): 193–276; Arthur L. Donovan, "Lavoisier and the Origins of Modern Chemistry," *Osiris* 4 (1988): 214–31; Donovan, *Antoine Lavoisier: Science, Administration and Revolution* (Cambridge, 1993); Siegfried, *From Elements to Atoms* (cit. n. 1). For a critique of this view, see Holmes, *Eighteenth-Century Chemistry* (cit. n. 3), 103–11; Carlton Perrin, "Chemistry as a Peer of Physics: A Response to Donovan and Melhado on Lavoisier," *Isis* 81 (1990): 259–70.

their local situations.[79] Perhaps the question to ask about Venel's strict demarcation between chemistry and physics is, What was it about the French context in which he worked that made his position desirable? There were many proposed plans to reform chemistry in the eighteenth century; Boerhaave and Venel simply proposed two competing programs.

[79] Cf., e.g., the Swedish mineralogists and British brewers; Hjalmar Fors, "Occult Traditions in Enlightened Science: The Swedish Board of Mines as an Intellectual Environment," in *Chymists and Chymistry: Studies in the History of Alchemy and Early Modern Chemistry*, ed. Lawrence M. Principe (Sagamore Beach, Mass., 2007), 239–52; James Sumner, "Michael Combrune, Peter Shaw and Commercial Chemistry: The Boerhaavian Chemical Origins of Brewing Thermometry," *Ambix* 54 (2007): 5–29.

How to See a Diagram:
A Visual Anthropology of Chemical Affinity

by Matthew Daniel Eddy*

ABSTRACT

In 1766, Thomas Cochrane entered the Edinburgh classroom of Joseph Black (1728–99) to learn chemistry for the first time. Cochrane was studying medicine, and, like so many of Black's students, he dutifully recorded several diagrams in his notebooks. These visualizations were not complex. They were, in fact, simple. One of them, reproduced in this essay, was a single "X," a chiasm. Black used it to illustrate ratios of chemical attraction. This diagram is particularly important for the history of chemistry because it is often held to be the first chemical formula, and, as such, historians have endeavored to explain why it was unique and how Black invented it. In this essay, I wish to turn the foregoing premise on its head by arguing that Black's chiasm was neither visually unique nor invented by him. I do this by approaching a number of his diagrams via a visual anthropology that allows me to examine how students learned to attach meaning to patterns that were already familiar to them. In the end, we will see that Black's diagrams were successful because their visual simplicity and familiarity made them ideally suited to represent the chemical theories that he so skillfully attached to them.

The Existence of Chymical Arts is nothing else but the Existence of Chymical Knowledge.   —Joseph Black[1]

Treating diagrams as things in themselves means giving up the notion that they are simply abstractions of reality, stripped down versions of the world of experience.
                    —John Bender and Michael Marrinan[2]

The arrow points only in the application that a living being makes of it.   —Ludwig Wittgenstein[3]

[Visualizations] capitalise on human ability to make rapid inferences about space and the things in it . . . and to perform mental transformations and operations on objects in space.
                    —Barbara Tversky[4]

* Department of Philosophy, 50/51 Old Elvet, Durham University, Durham DH1 3HN, U.K.; m.d.eddy@durham.ac.uk.

[1] Joseph Black, *Notes from Dr. Black's Lectures on Chemistry 1767/8*, Thomas Cochrane (Note-taker), ed. Douglas McKie (Cheshire, 1966). Hereafter cited as Black (1767/1966).

[2] John Bender and Michael Marrinan, *The Culture of Diagram* (Stanford, Calif., 2010), 19.

[3] Ludwig Wittgenstein, *Philosophical Investigations*, trans. G. E. M. Anscombe (Oxford, 1967), 454.

[4] Barbara Tversky, "Communicating with Diagrams and Gestures," *International Conference to Review Research on Science, Technology and Mathematics Education* (2007), 111–23, on 111.

VISUAL ANTHROPOLOGY

The forms and meanings of early modern teaching diagrams have remained largely unexamined by cultural historians. This situation, however, is slowly changing, especially in light of work that treats scientific diagrams as pictures.[5] In addition to this, the growth of digital culture in recent years has slowly begun to erase the division hitherto drawn between timeless works of art and everyday pictures. Instead of being seen as sedentary objects that exist outside time and space, pictures are now seen as objects that were recognized, made, and circulated in many ways that required different modes of access.[6] In this chapter I wish to build on this notion of a picture by focusing on the role played by the visual order of diagrams in the world of early modern chemical knowledge.

My aim is modest in that I merely wish to dig a bit deeper into the design of useful chemical pictures. I am interested in the diagrams inscribed and studied by the hundreds of students who attended Joseph Black's lectures at the University of Edinburgh during the last three decades of the eighteenth century. We know these diagrams were useful because they were copied over and over again by students taking his course. Indeed, they not only wrote them in the rough notebooks that they took in his lectures; they also redrew them in the bound, recopied notebooks that they made at the end of the course.[7]

The history of chemistry has tended to treat the diagrammatic tables and formulas used by Black and his contemporaries as fixed entities. The recent work on the visual anthropology of diagrams used for scientific teaching or research, however, has shown that their designers did not see them as timeless abstractions and that their meaning was strongly influenced by the direct interface between natural knowledge and visual culture. As intimated in the work of the social anthropologist Tim Ingold, the lines of graphic compositions, like diagrams and tables, are shaped by wider epistemic processes and cannot be disentangled from the beliefs of those who used them or the reasons why they were made.[8]

---

[5] Scientific diagrams are treated as pictures in Julia Voss, *Darwin's Pictures: Views of Evolutionary Theory, 1837–1874* (New Haven, Conn., 2010); and Tim Ingold, *Lines: A Brief History* (London, 2007). See also the diagrammatic insights of pictures given in Michael Baxandall, *Painting and Experience in Fifteenth-Century Italy* (Oxford, 1972); Eugene S. Ferguson, *Engineering the Mind's Eye* (Cambridge, Mass., 1992); and Edward R. Tufte, *Visual Explanations: Images and Quantities, Evidence and Narrative* (Cheshire, 1997).

[6] David Freedberg, *The Power of Images: Studies in the History and Theory of Response* (Chicago, 1989); Svetlana Alpers, *The Art of Describing: Dutch Art in the Seventeenth Century* (London, 1983); Martin Kemp, *Visualizations: The Nature Book of Art and Science* (Oxford, 2000).

[7] Black's affinity diagrams occur in most manuscript student notes taken in his lectures from the 1750s to the 1790s. My research is based on the many notes housed in the special collections of the University of Edinburgh, University College London, the Chemical Heritage Foundation in Philadelphia, and the Royal Society of London. Because citing all of these volumes would be impractical, I cite representative sets throughout this essay. The most accurate republication is Black (1767/1966). I cite the sections of these lectures relevant to the topics under consideration. A list of the collections that house manuscript copies of Black's lectures is given in William A. Cole, "Manuscripts of Joseph Black's Lectures on Chemistry," in *Joseph Black 1728–1799: A Commemorative Symposium,* ed. A. D. C. Simpson (Edinburgh, 1982), 53–69.

[8] Ingold, *Lines* (cit. n. 5); see esp. his chapter on evolutionary diagrams, 104–19. See also the visual anthropology of learning as addressed through the publications that emanated from Ingold's "Learning Is Understanding in Practice" project. These are listed on the project's Web site: http://www .abdn.ac.uk/creativityandpractice/ (accessed 1 February 2014). Ingold's work on this topic extends the visual anthropology of Walter J. Ong, *Ramus, Method and the Decay of Dialogue: From the Art of*

In recent years, the visual anthropology of diagrams has been explored by cultural historians of science like David Kaiser, Andrew Warwick, and Hans Jorg Rheinberger.[9] Rather than treating them as ethereal abstractions, these scholars frame diagrams more as a visual genre of representation anchored in both the material and intellectual skills possessed by the community that created or appropriated them. This approach has also been employed by early modern historians of scientific representation such as Sachiko Kusukawa, Sven Dupré, and Barbara Maria Stafford, who research the ways in which early modern diagrams were valued or understood by their users.[10] My view of graphic artifacts extends the work of these authors, especially since I treat diagrams as objects that moved through time and space in a manner that was a knowledge-making process.

In what follows, I present a visual anthropology of the Enlightenment diagrams that Black used to teach his students chemistry. Building on the epistemic concerns of Ingold and Rheinberger, I take the term "diagram" to mean a schematic picture intentionally designed to contain paths of information made from lines or symbols that are meant to represent natural events, objects, or processes. By taking this path, I treat diagrams as linear artifacts, as collections of marks in space, which represent concrete patterns of thought that were assembled and disassembled based on the needs of early modern students as users, and professors as designers.

PICTURING CHEMISTRY

The strong links between early modern chemistry and visual culture have long been recognized by intellectual historians. Studies on this topic tend to approach the chemical arts from either an iconographic or a functional perspective. Both of these traditions have their charms. The tools provided by iconology, for example, extend a rich tradition of motifs and forms that historians have used to examine both the literal and metaphorical nature of chemical imagery, with particular attention being given to the mnemotechnic montages of the hermetic tradition and, more recently, the presence of chemical imagery in portraiture.[11]

Discourse to the Art of Reason (Cambridge, Mass., 1958); and Jack Goody, *The Domestication of the Savage Mind* (Cambridge, 1977).

[9] For experimental diagrams, see Hans-Jörg Rheinberger, *An Epistemology of the Concrete: Twentieth-Century Histories of Life* (Durham, N.C., 2010); for Feynman diagrams, see David Kaiser, *Drawing Theories Apart: The Dispersion of Feynman Diagrams in Postwar Physics* (Chicago, 2005); for Newtonian diagrams, see Andrew Warwick, *Masters of Theory: Cambridge and the Rise of Mathematical Physics* (Chicago, 2003).

[10] Kusukawa, Dupré, and Stafford have numerous publications that are relevant to diagrams. For representative works, see Sachiko Kusukawa, *Picturing the Book of Nature: Image, Text, and Argument in Sixteenth-Century Human Anatomy and Medical Botany* (Chicago, 2012); Sven Dupré, "Visualization in Renaissance Optics: The Function of Geometrical Diagrams and Pictures in the Transmission of Practical Knowledge," in *Transmitting Knowledge: Words, Images and Instruments in Early Modern Europe,* ed. Sachiko Kusukawa and Ian Maclean (Oxford, 2006), 11–39; Barbara Maria Stafford, *Body Criticism: Imaging the Unseen in Enlightenment Art and Medicine* (Cambridge, Mass., 1993). For the epistemology of early modern diagrams, see also Bender and Marrinan, *Culture* (cit. n. 2); Tufte, *Visual Explanations* (cit. n. 5); Lorraine Daston and Peter Galison, *Objectivity* (New York, 2007), esp. chaps. 1 and 2, 17–114.

[11] The iconographic approach to chemical images has been addressed by a number of publications over the past few decades. For the connection between Jungian psychology and alchemical imagery, see William R. Newman, "*Decknamen* or 'Pseudochemical Language'? Eirenaeus Philalethes and Carl Jung," *Rev. Hist. Sci.* 49 (1996): 159–88. For the wider relevance and implications of the iconographic approach, see Lyndy Abraham, *A Dictionary of Alchemical Imagery* (Cambridge, 2001); and

By contrast, the functional approach focuses on the pictorial aspects of chemical materials or graphemes. Studies from this tradition have taught us much about the visual qualities of substances, instruments, and graphic artifacts, like symbols, schemata, and formulae, with the emphasis being placed on how such images were used within experimental settings. Yet, despite their conceptual differences, the temporal and spatial modes of analysis that underpin the iconographic and functional approach are very similar.[12]

Both tend to trace an image through time by attaching it to a specific idea or thinker. Likewise, both tend to bracket the communal conceptions of graphic space that affected how the image was preserved, modified, or valued. This means that, although a visual anthropology of early modern chemical imagery is starting to emerge in the work of scholars like Ursula Klein and Jennifer Rampling, we are only just beginning to understand how the pictures of early modern chemistry were used and iterated, or, more fundamentally, how the skills and routines required to use them were learned.[13]

Bearing its infancy in mind, the visual anthropology of early modern chemical diagrams is probably best seen as a mode of analysis that can be used in addition to, not in place of, the tools offered by iconologists and functionalists. My starting point for early modern chemistry professors like Black is the notion that they treated their pictures as things that were recognized as images, valued as objects, and made through media.[14] Such a perspective transforms early modern chemical diagrams into pictures that were composites of visual concepts, materials, and practices. In this sense, a "compositional" view of pictures is one that provides a way to recover the visual epistemology of Black and his students.[15]

There are many directions in which this compositionalist view could possibly lead us; however, I wish to examine how Black's visualizations were understood and pre-

Christoph Lüthy and Alexis Smets, "Words, Lines, Diagram, Images: Towards a History of Scientific Imagery," *Early Sci. Med.* 14 (2009): 398–439.

[12] Perhaps the most influential exemplar of the functional approach to the representation and meaning of (al)chemical instruments and graphemes is the work of J. R. Partington, esp. *A History of Chemistry* (London, 1970), vols. 1–4. For a functional analysis of diagrams related to the affinity concept, see Alistair Duncan, *Laws and Order in Eighteenth-Century Chemistry* (Oxford, 1996).

[13] Ursula Klein and Wolfgang Lefèvre, *Materials in Eighteenth-Century Science: A Historical Ontology* (Cambridge, Mass., 2007); see esp. the many sections that address the graphic design and meaning of affinity tables. For chemical formulas and nomenclature, see Ursula Klein, *Experiments, Models, Paper Tools: Cultures of Organic Chemistry in the Nineteenth Century* (Stanford, Calif., 2005); its title notwithstanding, it addresses many issues relevant to the graphic representation of eighteenth-century chemistry. See also Jennifer M. Rampling, "Depicting the Medieval Alchemical Cosmos: George Ripley's *Wheel* of Inferior Astronomy," *Early Sci. Med.* 18 (2013): 45–86. A number of the visual issues raised by Klein and Rampling are also addressed in relation to nineteenth-century chemistry in Alan J. Rocke, *Image and Reality: Kekulé, Kopp, and the Scientific Imagination* (Chicago, 2010).

[14] This approach is influenced by the visual compositionalism of the art historian W. J. T. Mitchell, whose use of image, object, and media to define a picture is most clearly addressed in the initial chapters of W. J. T. Mitchell, *What Do Pictures Want?* (Chicago, 2005). His compositionalist approach to pictures is unpacked in Mitchell, *Blake's Composite Art: A Study of the Illuminated Poetry* (Princeton, N.J., 1992). The composite nature of pictures—in terms of both materials and forms—has long been underscored by art historians. See Baxandall, *Painting and Experience* (cit. n. 5); Alpers, *The Art of Describing* (cit. n. 6); Michael Camille, *Image on the Edge: The Margins of Medieval Art* (London, 1992). See also Svetlana Alpers and Michael Baxandall, *Tiepolo and the Pictorial Intelligence* (New Haven, Conn., 1994).

[15] By "visual epistemology" I mean the historical unfolding of the beliefs and values attributed to objects visualized by community. For more on this term, see the first chapter of Daniela Bleichmar, *Visible Empire: Botanical Expeditions and Visual Culture in the Hispanic Enlightenment* (Chicago, 2012). See also John Berger, *Ways of Seeing* (London, 2008).

sented as a set of chemical pictures to university students, that is to say, a specific community of viewers. Following this line of thought allows us to see how treating a picture as a composite "thing" that existed in time and space yields a better understanding of its value and meaning. The pictures under discussion will be three core diagrams used by Professor Joseph Black to represent elective affinity, the influential eighteenth-century theory of material composition that appealed to forces of attraction and repulsion.[16] I focus on the iterations that appear in Black's own lecture notes and in the notes that students made in his lectures, and I highlight the visual skills and tools that were available to students educated in the Scottish primary and secondary school system.

The main forms of visual representation on paper in Scottish schools were pictures composed from either words or lines. Detailed illustrations did play a role, but, due to various economic and cultural factors, the average schoolchild's everyday exposure to visual culture on paper consisted mainly of words plotted as lists and tables. Students interested in learning a trade or going to university were also exposed to simple shapes like squares, circles, and triangles. Thus, when a student first crossed the threshold of a university classroom, a core mnemonic skill he had in his possession was the ability to memorize information that had been plotted along various lines arranged geometrically on the page.[17]

The skills of recognizing, reading, and inscribing visual patterns were practiced every day when Scottish children visualized both words and lines on grids in their school notebooks. The skill was reinforced by the fact that they were concurrently taught that straight lines and geometric shapes were visual examples of how ordered thought ought to work. If the mind's eye was like a lens, then rational thought was like a chain of ideas being paraded in a line across a screen. This metaphor was a common trope in Scotland where it was used to explain cognition, to justify pedagogy, and to chastise immorality. I have treated this topic at length elsewhere, but I am noting it here because I want to underscore the fact that Scottish pedagogy attributed great worth to the visual expediency of words and lines plotted in a simple and straight manner.[18]

Because Scottish sites of learning promoted rectilinearity as a virtue, there were

[16] A number of terms for affinity were used during the early modern period, including: attraction, sympathy, *rapport* (French), and *Affinität* (German). Terms used to describe the act of decomposition were repulsion, antipathy, division, and *partage* (French). The early history of the affinity concept is given in William R. Newman, "Elective Affinity before Geoffroy: Daniel Sennert's Atomistic Explanation of Vinous and Acetous Fermentation," in *Matter and Form in Early Modern Science and Philosophy,* ed. Gideon Manning (Leiden, 2012), 99–124. For the larger context, history, and meaning of the early modern affinity concept, see Ursula Klein, *Verbindung und Affinität: Die Grundlegung der neuzeitlichen Chemie an der Wende vom 17. zum 18. Jahrhundert* (Basel, 1994); Duncan, *Laws and Order* (cit. n. 12); Mi Gyung Kim, *Affinity, That Elusive Dream: A Genealogy of the Chemical Revolution* (Cambridge, Mass., 2003); Georgette N. L. Taylor, "Variations on a Theme: Patterns of Congruence and Divergence among 18th Century Chemical Affinity Theories" (PhD diss., Univ. College London, 2006).

[17] For the print and manuscript sources used to instill graphic knowledge in primary and secondary education in Enlightenment Scotland, see Matthew Daniel Eddy, "The Alphabets of Nature: Children, Books and Natural History in Scotland," *Nuncius* 25 (2010): 1–22; and Eddy, "Natural History, Natural Philosophy and Readership," in *The Edinburgh History of the Book in Scotland,* vol. 2, *Enlightenment and Expansion, 1707–1800,* ed. Stephen Brown and Warren McDougall (Edinburgh, 2012), 297–309.

[18] Matthew Daniel Eddy, "The Shape of Knowledge: Children and the Visual Culture of Literacy and Numeracy," *Sci. Context* 26 (2013): 215–45; Eddy, "The Line of Reason: Hugh Blair, Spatiality and the Progressive Structure of Language," *Notes Rec. Roy. Soc. Lond.* 65 (2011): 9–24. The impor-

many principles of graphic design that were used to arrange words and lines into pictures suitable for university classrooms. The most popular graphic principles of composite diagrams were proximity, symmetry, and contrast. Once words and lines had been arranged into a picture, students were taught to associate various visual relationships with the principle. I will discuss these associations in more detail below; however, at this point it would be prudent to lay out the basic form and importance of Black's affinity diagrams.

DIAGRAMMING AFFINITY

Black used tables and figures as diagrams in his lectures to represent experimental instruments or chemical attraction. His tables contained lists of substances or their properties. Perhaps the most well known is his affinity chart (fig. 1), but there were others as well, including tables that classified the effects of heat, acids, and alkalis. Overall, his figural diagrams came in two varieties.

The first portrayed experimental apparatus, and the second used geometric shapes to depict some sort of chemical reaction. His use of tables and figures is consistent overall with the kinds of visual illustrations that accompanied the chemistry or natural philosophy books cited in his lectures, or which he had studied as a student.[19] Likewise, the practice of using diagrams as teaching aids in university classrooms was common in Scotland. This was especially the case in Edinburgh's Medical School, where Black's colleagues distributed diagrams as handouts, drew on chalkboards, and hung large charts at the front of the classroom.[20]

When Black's diagrams are considered as a corpus, as a group of conceptually related pictures, a striking pattern emerges. Aside from his depictions of experimental apparatus, most of Black's diagrams were used to visualize some aspect of chemical affinity. This pattern is important to note because the affinity concept both explained and predicted the composition of experimental substances and provided the main theoretical underpinning of chemistry during the late eighteenth century. Indeed, for some philosophical chemists, the force of affinity was analogous to the force of gravity.[21]

In his lectures, Black, like most professors of the day, offered little insight into how he created these diagrams. Yet, when they are considered as pictures, it can be seen that he put a great deal of effort into designing three different diagrams that were conceptually unique but visually familiar. Conceptually, each diagram addressed a specific aspect of chemical affinity. Visually, each diagram exhibited a different shape.

---

tance of the graphic metaphor of the mind is raised more generally in Ingold, *Lines* (cit. n. 5) and, with reference to early modernity, in W. J. T. Mitchell, *Iconology: Image, Text, Ideology* (Chicago, 1986).

[19] We lack a definitive study on the chemical texts used by Black to write his lectures, but many of the authors who influenced him and his Edinburgh contemporaries are addressed throughout A. L. Donovan, *Philosophical Chemistry in the Scottish Enlightenment: The Doctrines and Discoveries of William Cullen and Joseph Black* (Edinburgh, 1975); and Matthew Daniel Eddy, *The Language of Mineralogy: John Walker, Chemistry and the Edinburgh Medical School, 1750–1800* (Aldershot, 2008).

[20] The uses of diagrams and figures in early modern Scottish anatomical and botanical teaching are addressed in Joe Rock, "An Important Scottish Anatomical Publication Rediscovered," *Book Collector* 49 (2000): 27–60; in H. J. Noltie, *John Hope (1725–1786): Alan G. Morton's Memoir of a Scottish Naturalist, A New and Revised Edition* (Edinburgh, 2011); and in Noltie, *John Hope, Enlightened Botanist* (Edinburgh, 2011).

[21] This point is intimated throughout Duncan, *Laws and Order* (cit. n. 12); and Donovan, *Philosophical Chemistry* (cit. n. 19).

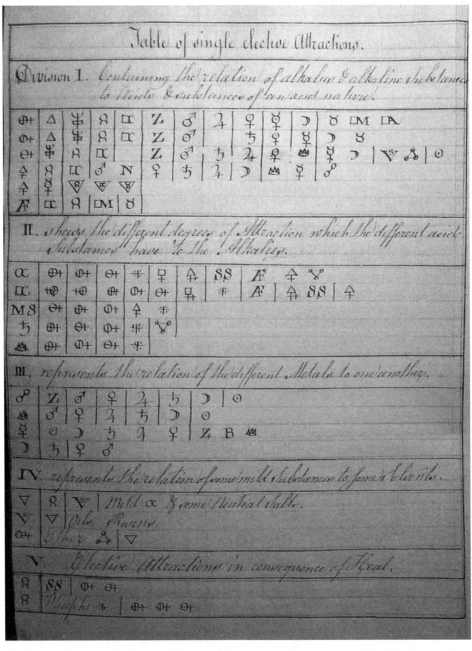

Figure 1. *Affinity table. Black, A Course of Lectures, Volume 3 (cit. n. 36). Black ordered his table so that it read from left to right. To use the table, a student needed to select a substance from the far left column and then read the entries listed in the row to the right. Each substance in the row was listed in descending order of attraction. Reprinted courtesy of the Royal Society of London.*

To achieve this visual differentiation, Black chose an X (fig. 2), a square-shaped table (fig. 1), and a circle (fig. 3). Additionally, he plotted headings within, around, or on these shapes, in a fashion that created a unique flow of information that required a distinctive reading pattern. Such differences required students to draw and use each diagram in slightly different ways, thereby making each one distinct and memorable.

Black fixed the visual patterns of his affinity diagrams during the 1760s, and they remained relatively similar for the rest of his career. In this capacity, they served as stable visual containers of chemical facts that circulated natural knowledge in Edinburgh and the many places where Black's students traveled or immigrated over the next half-century. In this sense, the diagrams provide an excellent example of a representational form that underpinned what Jim Secord once called "knowledge in transit"; that is to say, they were visible objects that circulated scientific information through Scotland's educational community and, more broadly, throughout the British Empire.[22]

Yet while the fixed structure of the lines served as an effective visual container, there was a certain degree of flexibility with the kinds of chemical knowledge that Black associated with the diagram at various points during his long career. Because he lived at a time when "chymistry" was rapidly changing, the conceptual malleability of his diagrams was a great asset because it allowed him to associate and dissociate information as required.

TABLING ATTRACTIONS

Many Scottish professors gave their students lists of facts and books. They also gave them lists of lecture titles called "headings." Next to these simple lists, the table was the most prevalent visual tool used on paper in Scotland's educational settings.[23] From a visual perspective, there were two formats: those with contour lines and those without. This was true for most tables of the time. Lineless tables were often presented in school textbooks and in various handouts used in Scottish universities to supplement lecture courses. Though they were arranged on a grid, their lineless internal structure allowed for the inclusion of full sentences.[24] Lined tables, on the other hand, usually contained more compressed information and placed more restrictions on the directional flow of the heads made available to its viewer.[25]

Black's affinity chart was a lined table (fig. 1). It consisted of an outer square and sets of internal crossed lines. This structure created a series of cells into which the pictographic heads of substances were placed and then read from top to bottom and left to right. Like the logarithmic tables used in natural philosophy courses, Black's chart was effectively a collection of lists that represented ratios of change.

---

[22] James A. Secord, "Knowledge in Transit," *Isis* 95 (2004): 654–72.

[23] See, e.g., the tables used in John Playfair, *Outlines of Natural Philosophy, Being Heads of Lectures Delivered in the University of Edinburgh* (Edinburgh, 1812), 1:160; or Alexander Fraser Tytler, *Plan and Outlines of a Course on Universal History, Ancient and Modern* (Edinburgh, 1782), 223–50. Oftentimes the only way to find copies of tables used by professors is to scour student notebooks.

[24] The graphic format of school textbooks and notebooks is given in Eddy, "The Shape of Knowledge" (cit. n. 18). There is no secondary research on the graphic nature of Scottish lecture heads. For a sample, see the following sets: John Walker's mineralogy lecture heads, *Classes Fossilium: Sive, Characteres Naturales et Chymici Classium et Ordinum in Systemati Minerali* (Edinburgh, 1787); John Hill, *Heads of Philological Lectures, Intended to Illustrate the Latin Classicks,* 2nd ed. (Edinburgh, 1785); Adam Ferguson, *Institutes of Moral Philosophy* (Edinburgh, 1769).

[25] A depiction of one of Black's acid-alkali tables is re-created in Black (1767/1966), 48.

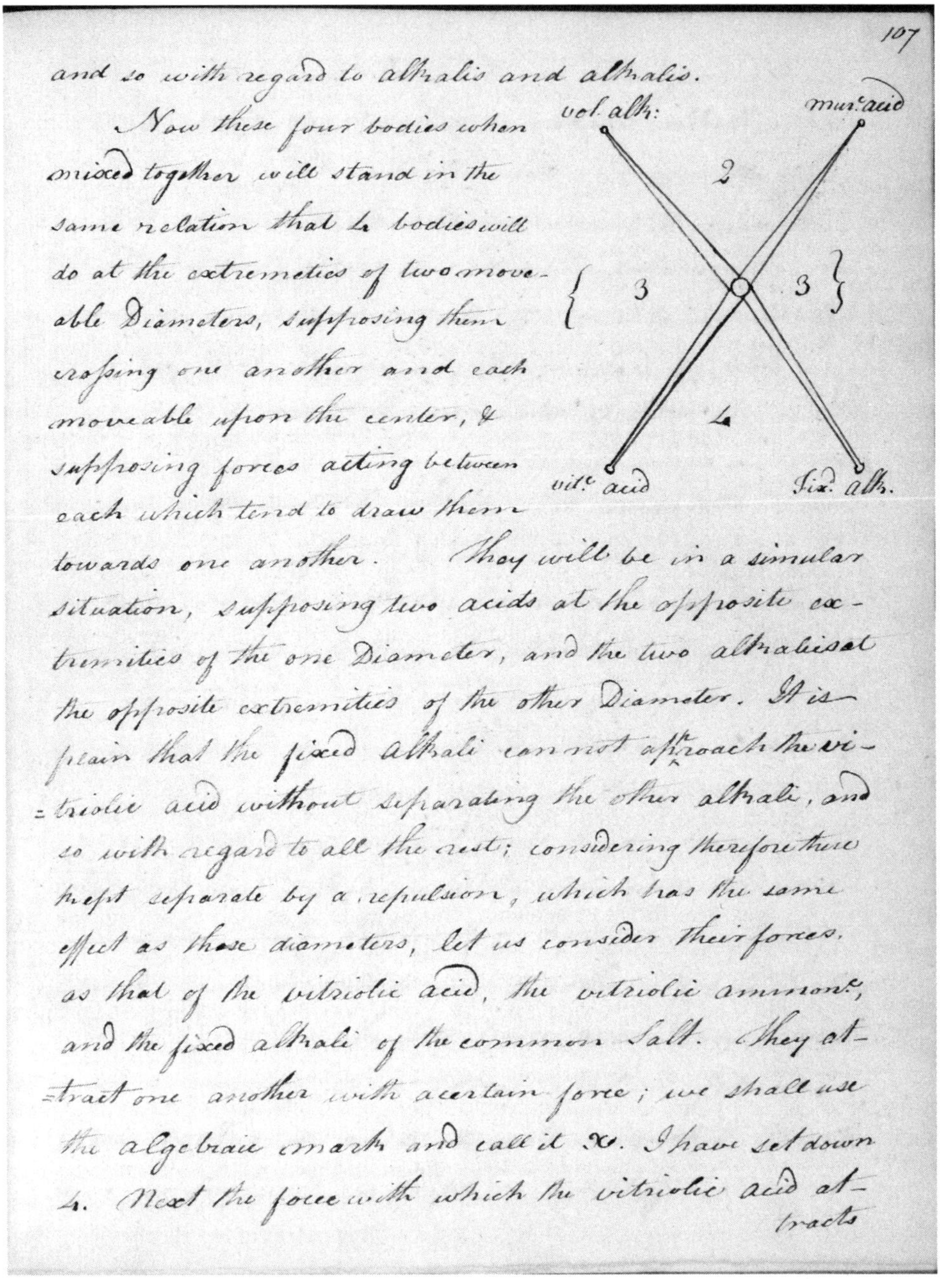

and so with regard to alkalis and alkalis.

Now these four bodies when mixed together will stand in the same relation that 4 bodies will do at the extremities of two moveable Diameters, supposing them crossing one another and each moveable upon the center, & supposing forces acting between each which tend to draw them towards one another. They will be in a similar situation, supposing two acids at the opposite extremities of the one Diameter, and the two alkalis at the opposite extremities of the other Diameter. It is plain that the fixed Alkali can not approach the vitriolic acid without separating the other alkali, and so with regard to all the rest; considering therefore there kept separate by a repulsion, which has the same effect as those diameters, let us consider their forces, as that of the vitriolic acid, the vitriolic ammonc, and the fixed alkali of the common Salt. They attract one another with a certain force; we shall use the algebraic mark and call it ꭗ. I have set down 4. Next the force with which the vitriolic acid attracts

Figure 2. *Chiastic affinity diagram. Joseph Black,* Lectures on Chemistry, *vol. 3 (1778), Paul Panton (Note-Taker), Bound MS, Chemical Heritage Foundation, QD14 .B533 1778, fol. 107. Black used a chiasm to visualize the hypothetical strength of the forces acting in chemical reactions. For example, if a fixed alkali (bottom right) was placed in contact with vitriolic acid (bottom left) and muriatic acid (top right) at the same time, it would elect to combine with the vitriolic acid, the reason being that fixed alkali was "more" (a force represented by a 4) attracted to vitriol and "less" (a force represented by a 3) attracted to muriatic acid. Reprinted courtesy of the Chemical Heritage Foundation.*

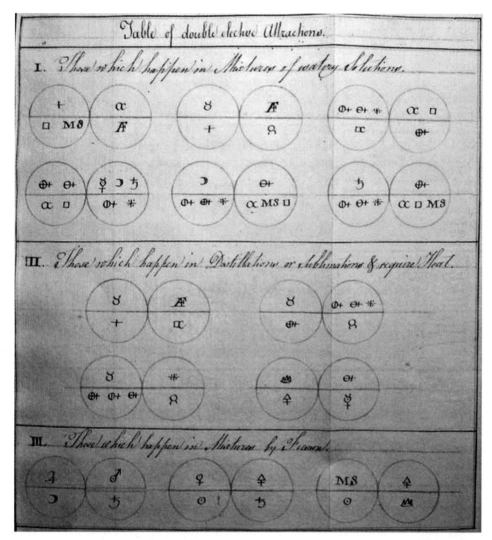

Figure 3. *Circlet diagram. Black,* A Course of Lectures *(cit. n. 36). Black used a double cir-clet diagram to visualize the double elective reactions taking place between the substances in two compounds. Take, for example, the first diagram in Part III of the table above (at the bottom left). The compound in the left circlet is a mixture of tin (on the top) and silver (on the bottom). The compound in the right circlet is a mixture of iron (on the top) and lead (on the bottom). Black explained the double elective attraction between the substances in these compounds in the following manner: "A mixture of tin and silver is melted with [a] mixture of iron and lead. The tin will join the iron, and the silver attract to the lead"; Black (1767/1966), 165. Black only visualized the compounds used at the start of the reaction but not the compounds of the final products. Reprinted courtesy of the Royal Society of London.*

He used it to represent single elective attraction, the form of chemical attraction that was the simplest kind of affinity. The table explained how one substance "elected" to leave a compound and then unite with another substance for which it had a stronger attraction.

Although there are several elements that are unique to Black's chart, the graphic formula of its gridded structure and spatial relationships had played a central role in chemical teaching since the early decades of the eighteenth century, especially in the lectures of his teacher William Cullen (1710–90) in Scotland, and in the chemistry courses of teachers such as France's Gabriel François Venel (1723–75; fig. 4) and Sweden's Torbern Olaf Bergman (1735–84).[26] The diagrammatic nature of the table had existed since Etienne François Geoffroy initially popularized it at the beginning of the century.[27] Within this tradition, the affinities visualized on Black's table were robust representations in that they were modifications based on his own experimental program.

In addition to his affinity table, Black also gave his students smaller tables that represented a select group of chemical affinities or, in the case of temperature, a set of important measurements. The order of the headings in all of these tables (which could be words or numbers) operated on a simple principle of visual proximity. This principle associated nearness with sameness and farness with difference. In affinity tables, for example, nearness represented a stronger attraction and farness represented a weaker attraction. Or, for temperature charts, nearness to one pole of the column represented hotness, and nearness to the other pole represented coldness.

Taking note of this dichotomous principle of visual proximity, moreover, reveals the central role played by the affinity table in the oral component of Black's lectures. Most student notes record him as regularly saying that substances had a weak or strong attraction for each other. These adjectives have traditionally been interpreted merely as qualitative descriptions. When considered in light of the visual relationships depicted on the affinity chart, however, they prove to be ratios that correspond to the dichotomous poles of attraction represented in the column of each substance. He was, therefore, often referring his students to the chemical relationships on the affinity table when he mentioned a strong or weak single attraction during a lecture.[28]

Black's discussions of the theoretical component of single displacement reactions in his lectures, though tersely informative, were diffuse. They occurred as necessary when he wished to point out the theoretical basis of simple chemical reactions. The scattered and brief nature of his comments meant that there was not a specific place in his course where he addressed the affinity concept in a systematic manner. Rather than being an omission, however, the absence of such verbal explanations was compensated for by the presence of the visual relationships depicted on the table. Its col-

---

[26] For the use of affinity tables in Scottish teaching, see Georgette Taylor, "Marking Out a Disciplinary Common Ground: The Role of Chemical Pedagogy in Establishing the Doctrine of Affinity at the Heart of British Chemistry," *Ann. Sci.* 65 (2008): 465–86. For France, see Christine Lehman, "Innovation in Chemistry Courses in France in the Mid-Eighteenth Century: Experiments and Affinities," *Ambix* 57 (2010): 3–26. The pedagogical reaction to the affinity concept in Holland is addressed in John Powers, *Inventing Chemistry: Herman Boerhaave and the Reform of the Chemical Arts* (Chicago, 2012), 163–8.

[27] For a list of the published affinity tables that appeared from the 1720s to the 1790s, see table 4.1 in Duncan, *Laws and Order* (cit. n. 12).

[28] Attractions are discussed in the language of strong and weak in Black (1767/1966), 23, 33–5, 38, 39, 40, 41, 43, 59, 60, 61, 63, 79, 89, 118, 133, and 158.

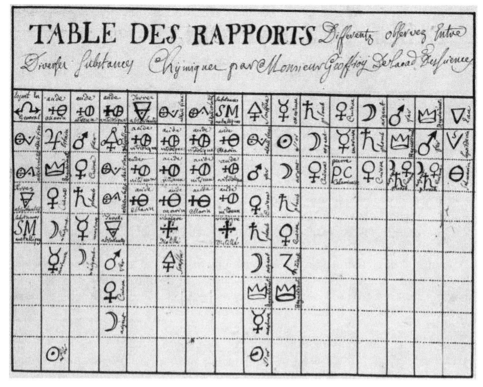

Figure 4. *Gabriel François Venel (1723–75), "Table des Rapports," Cours de Chymie, Wellcome MS 4914. Reprinted courtesy of the Wellcome Library, London.*

umns and rows served as the most accessible and comprehensive representation of single elective attractions known to Black and the other chemists who influenced his thoughts on affinity.[29]

Black's lectures were an introductory course, so he tended not to discuss highly complex compounds or reactions. This explains why a large number of the experiments that he conducted in front of his students were single elective reactions. The single elective attractions depicted on Black's affinity table also constituted the most thorough summary on paper of the affinity concept given to students who took his course. Because of its pictorial nature, the table was not merely a simple reference tool but one of the central documents of the course in that it was the only place where his students could *see* a systematic overview of the affinity concept.[30]

Though explicit definitions of various aspects of affinity were sprinkled throughout his lectures, it was the table that gave his students a constant visual point of reference, and which allowed them to see easily a single elective reaction as one entry in a larger system of knowledge that was based on the theory of affinity. In giving this kind of conceptual priority to a graphic schema, Black was effectively saying that

[29] See the different kinds of affinity tables featured in Duncan, *Laws and Order* (cit. n. 12) and Klein and Lefèvre, *Materials* (cit. n. 13) for comparison.

[30] The same was true for students studying chemistry in eighteenth-century Paris. See Lehman, "Innovation in Chemistry" (cit. n. 26).

a pictorial mode of representation was more practical, and more accessible, than a verbal list of principles or rules of elective attraction. In short, pictures were more effective than words in this case.

CIRCLING COMBINATIONS

Black used two circles set side by side to visualize double elective attraction (fig. 3). Despite the fact that he used such "double circle" diagrams as teaching aids for the bulk of his career, they were not reproduced in the posthumous edition of his lectures that John Robison published in 1803.[31] Because Robison's edition served as the primary reference source for research on Black's ideas over the next two centuries, the meaning and importance of the circlets has remained relatively obscure. Of the studies that actually mention them, it seems that only a few recognize the fact that Black developed them solely to illustrate the concept of double elective attraction.[32] This being the case, it is worth explaining what the visual components of the diagram were supposed to represent.

The use of circlet structures as teaching aids in Scotland was, of course, not unknown. Unlike the ubiquitous presence of tables in Scotland's schoolbooks, freestanding circles were used primarily in geometry, or in subjects that built upon geometry such as gauging, cartography, or architecture.[33] The visual skills required to understand, to access, and to iterate such representations were then expanded at university. At the University of Edinburgh, for example, full circles, semicircles, and quarter circles were often used in mathematics, natural philosophy, and anatomy courses to depict the movement of matter through space (fig. 5).[34] Additionally, students in the arts faculty taking Alexander Tytler Fraser's course on universal history were given world maps that employed the common technique of setting the Eastern and Western Hemispheres side by side in two cartographic circles.[35]

Various approaches to using circles to represent natural knowledge, therefore, were present in the Scottish educational system and provided a good foundation on which Black could begin to build his circlet diagrams. Yet, despite this pedagogical advantage, Black was still faced with a particularly thorny visual problem that plagued early modern chemical knowledge as a whole. Whereas the objects of natural philosophy and anatomy were things like planets and body parts that were readily visible, the objects of chemistry, that is, chemical particles, were entities that had never been seen and had no prospect of being made visible in the near future. The circlets of Black's diagram were heirs to this problem and were not meant to be literal representations of material particles or their movements through time or space.

---

[31] Joseph Black, *Lectures on the Elements of Chemistry* (Edinburgh, 1803).

[32] An exception to this is M. P. Crosland, "The Use of Diagrams as Chemical 'Equations' in the Lecture Notes of William Cullen and Joseph Black," *Ann. Sci.* 15 (1959): 75–90.

[33] Alexander Ewing, *A Synopsis of Practical Mathematics* (Edinburgh, 1771); William Wilson, *Elements of Navigation: Or the Particular Rules of the Art* (Edinburgh, 1773). The graphic elements of these and other books are addressed in Eddy, "The Shape of Knowledge" (cit. n. 18).

[34] For examples of how this linear technique was used in lectures, see the diagrams included at the back of John Playfair's lecture heads, *Outlines of Natural Philosophy*, vol. 2, 2nd ed. (Edinburgh, 1816). See also the anatomical diagrams that Alexander Monro Secundus used in his anatomy course, which are depicted in the tables of his *Observations on the Muscles, and Particularly on the Effects of Their Oblique Fibres* (Edinburgh, 1794).

[35] Tytler, *Plan and Outlines* (cit. n. 23). For the circlet teaching diagrams used in the natural philosophy course of John Playfair, see his *Outlines* (cit. n. 34).

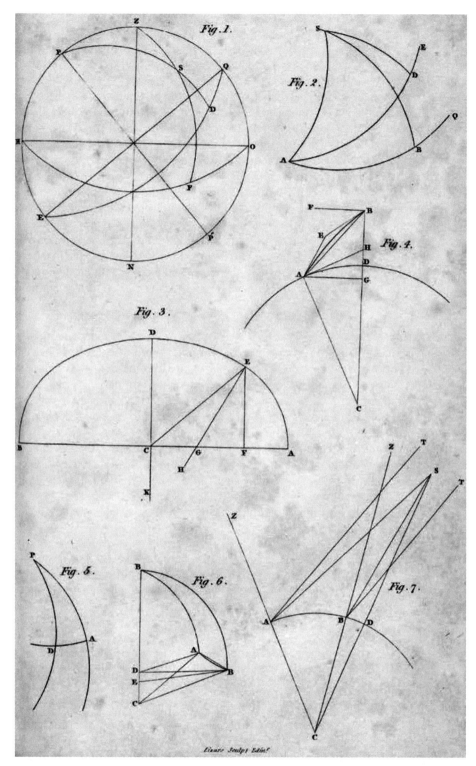

Figure 5. *Playfair,* Outlines of Natural Philosophy *(cit. n. 34), the first of four unnumbered plates that occur at the end of the volume. Huntington Library. Reprinted courtesy of the Huntington Library.*

Each circle represented a compound made of two substances. The 1782 lecture notes that Black read to his students state that the diagram of double elective affinity was "Composed of two circles, each of which is divided by a Horizontal Diameter."[36] Dividing the circles in this manner created four semicircles that Black used to represent four different substances participating in a reaction driven by double elective attraction. The reaction was a "double" attraction because each substance elected not to join another substance on two separate occasions. In this sense, it could also be seen as a double rejection because each substance rejected the other substance in its own circle, as well as another substance in the other circle.

In order to understand more precisely what Black intended his circlets to visualize, we need to ask how the structure and space of the diagram was supposed to be used by his students. Even though he employed circles to represent compounds, those who used the diagram were not meant to read it in a circular manner. Reading the diagram was very much a rectilinear affair. Black instructed his students to perform three visual moves. First, observe the compound in the left circle and then the one in the right circle (fig. 6). Second, associate the two substances in the upper semicircles and then do the same for the substances in the lower semicircles (fig. 7). Third, imagine the two new compounds.[37]

The relatively simple movements of the eye required by Black's diagram were not accidental. Indeed, they were designed. Black arranged the substances in a pattern that was more conducive to a simple reading. In other words, he stacked the visual deck. He did this by arranging the four substances in a symmetric pattern that vertically aligned those that he knew would combine into new compounds in the reaction. This arrangement minimalized the directional possibilities and allowed his students to concentrate on what they were supposed to be learning: the concept of double elective attraction.

What emerges from Black's visual decisions is the fact that he wanted each circlet diagram to be a self-contained picture that was sufficient to illustrate a multistep process. This is why he created one structure that could be read in two ways, depending on which directional path was used. In this sense, it was what anthropologists call a "multistable" image: pictures that offer "different readings in the single image."[38] Black was so keen to keep the design simple that he did not even offer a second diagram that visualized the final products of the reaction. Students simply had to imagine the products in their own mind.[39] This act of imagination was undoubtedly made easier by the fact that Black's circlets were schematic analogies of processes and not figural abstractions of corpuscular entities.

---

[36] This quotation occurs in the set of lecture notes that Black read to his Edinburgh chemistry students in 1782. Joseph Black, *A Course of Lectures on the Theory and Practice of Chemistry*, 3 vols. (1782), Bound MS, Royal Society of London, MS/147. The quotation occurs in the section on elective attractions in vol. 3, lecture 107. His conception of affinity and his use of his circlet diagram remained relatively consistent throughout his career. For an earlier account, see Black (1767/1966), 103, 454, 457.

[37] Steps 1 and 3 represent stasis, and step 2 represents change.

[38] W. J. T. Mitchell, *Picture Theory: Essay on Verbal and Visual Representation* (Chicago, 1994), 45–57. Multistep images that offer two primary readings are called dialectical images. Because Black's diagram has two major directional readings, it is a dialectical image. The anthropological significance of the kind of "multistability" raised by Black's diagrams is addressed in Tsili Doleve-Gandelman and Claude Gandelman, "The Metastability of Primitive Artefacts," *Semiotica* 75 (1989): 191–213.

[39] Following the work of Rocke, I take this historicized notion of imaginative practice to be akin to that which was required to make the "visually imagined microworld" of nineteenth-century chemistry. See Rocke, *Image and Reality* (cit. n. 13), xv.

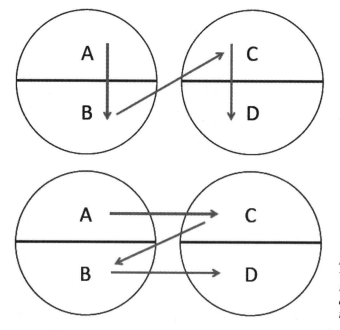

Figure 6. *The first visual pattern required to read Joseph Black's circlet diagrams. Figure by Matthew Daniel Eddy.*

Figure 7. *The second visual pattern required to read Joseph Black's circlet diagrams. Figure by Matthew Daniel Eddy.*

Like the concept of single elective affinity, student notes seldom feature a section where the concept of double elective affinity is explicitly defined. Instead, the concept is mentioned as a matter of course after the introductory lectures. Most sets of student notes, however, contain drawings of circlet diagrams, and this indicates that, in addition to relying on a table to visually represent *single* elective affinity, Black also needed to use another visualisation to depict *double* elective affinity, a core theoretical component of the chemistry course. Unlike the affinity table, however, Black's circlet diagrams were usually accompanied by brief descriptions that explained what kind of double elective attraction was taking place in the picture.[40]

CROSSING RATIOS

Black's chiastic diagram depicted a double chemical reaction as well, but with a twist (fig. 2).[41] Rather than simply illustrating a qualitative change in substances, it also visualized a quantitative relationship between the forces of attraction that held the compounds together. It is this metric aspect of the chiasm that has attracted the attention of historians, particularly because it is seen as one of the first numeric approximations of chemical force expressed in an equation.

[40] More specifically, the three different kinds of double elective reactions represented by the circlet diagrams were: "I. Those which happen in Mixtures of Watery Solutions," "II. Those which happen in distillations or Sublimations & require heat," and "III. Those which happen in mixtures by Fusion." See Black (1767/1966), 164–5. Black addresses double elective attractions in passing on 49, 52, 59, and 67. Black's view of double elective attraction is addressed in Donovan, *Philosophical Chemistry* (cit. n. 19), on 216–8, and the diagrams of this kind of affinity, which preceded and followed him, are summarized in Duncan, *Laws and Order* (cit. n. 12), on 145–53.

[41] Black's chiasm occurs in most sets of student notes and in his notes. See Black (1767/1966), 274–9; Black, *Theory and Practice* (cit. n. 36), in vol. 3, lecture 107. For a 1770s version, see Joseph Black, *Lectures on Chemistry* (1776), transcribed by Paul Panton Jr., vol. 3, Bound MS, Chemical Heritage Foundation, Philadelphia, QD14 B533, fols. 107 and 493.

The chiasm had two visual zones. The outer zone ran around the tips of the chiasm and contained the abbreviated names of the four substances involved in a double elective reaction. The names or pictograms of the substances were inscribed at the end points of the chiasm as headings in the same visual footprint as those in the circlet diagram, and, hence, students could use the same directional path to read them. The inner zone, on the other hand, ran around the angles inside the chiasm. Each angle contained a number. Even if a student did not immediately grasp the meaning of the numbers, their close grouping at the center of the diagram showed that they were somehow related. When read solely as a group of numbers, the inner heads required a diamond line of sight.

Historians seeking to explain the visual origins of Black's chiasm have traditionally pointed to the lectures of his teacher, William Cullen, and Jean Beguin's popular seventeenth-century textbook entitled *Elemens de Chymie*.[42] The chiasms of these chemists were used to represent double attractions, and both positioned names or symbols of substances on the tips of the chiasm. Notably, both were used to teach students, many of whom were adolescents, which may explain why they were so visually simple. When compared to Black's chiasm, however, this simplicity in early modern chemical chiasms points directly to the visual absence of a conceptual piece of information.

Unlike the Cullen and Beguin diagrams, Black positioned numbers on the inner angles of the chiasm. Where did he get this idea? In order to see the origin of Black's inner zone of numbers, we must first remember the intended users of the diagram, most of whom were Scottish students trained in Scottish schools by Scottish tutors and teachers. In Scotland's mathematical tradition, children were taught a mathematical visualization, a trick, called the "Casting of Nines" or "Casting out the Nines." It was a calculation performed to double-check the answers of large arithmetic equations.[43]

The Casting of Nines is not used very often today, but, when it is employed in twenty-first-century classrooms, the numbers of the computation are lined up in a column.[44] Crucially, in early modern Scotland, this calculation was laid out on a chiasm, a practice that most likely originated from its long-standing use in the dichotomous Ramistic tradition of graphic spatialization.[45] An excellent example of this kind of diagram can be found in the marginalia written by the children of the Erskines of Torrie, Scotland, in the books of the family library during the middle of the eigh-

[42] The Cullen connection was addressed in print as early as 1803, when John Robison included a reproduction and description of the diagram in Black's *Lectures* (cit. n. 31), 544–6. The basic conceptual connection between the chiasms of Black, Cullen, and Beguin are addressed in Crosland, "The Use of Diagrams" (cit. n. 32). The graphic context and history of Beguin's chiasm is addressed in Alexis Smets, "Le concept de matière dans l'imagerie des chymistes aux XVIe et XVIIe siècles" (PhD diss., Radboud Univ. Nijmegen, 2014).

[43] The casting of nines was explained in Panton's popular mathematics text used in many Scottish schools. See William Panton, *The Tyro's Guide to Arithmetic and Mensuration* (Edinburgh, 1771), 23–4. The context of its usage is given in Duncan K. Wilson, *The History of Mathematical Teaching in Scotland to the End of the Eighteenth Century* (London, 1935), 2, 31, and 85.

[44] The modern uses of "Casting of Nines" computation are given throughout Isaac Asimov, *Quick and Easy Math* (Boston, 1964).

[45] See the calculation chiasm in Petrus Ramus, *Petrus Ramus, Scholarum Mathematicarum Libri Unus et triginta* (Basel, 1569). The connection between the Ramist chiasm in this text and Beguin's chemical diagrams is addressed in Smets, "Le concept de matière" (cit. n. 42).

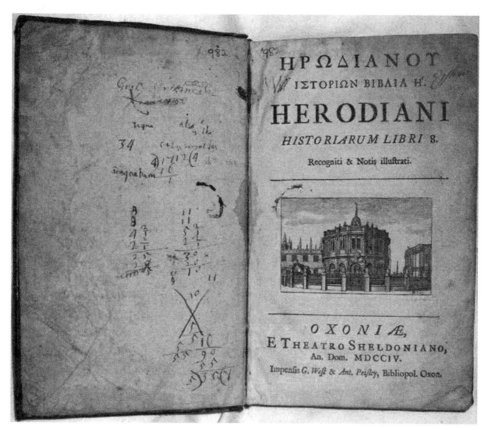

Figure 8. *"Casting of Nines" chiasm used by Scottish children to double-check multistep calculations. Drawn by one of the Erskine children living at Dunimarle Castle during the 1760s. Herodotus,* Herodiani historiarum *(cit. n. 46). Reprinted courtesy of Mrs. Magdalen Sharpe Erskine's Trust.*

teenth century (fig. 8).[46] The early modern version of the calculation consisted of several steps, and the answer to each one was placed on the inner angles of the chiasm. Thus, the visual origin of the inner zone of numbers that Black used for chemical ratios was a simple graphic tool used by schoolchildren that he probably selected because of its familiarity to his university students.

But what, specifically, were Black's numbers supposed to visualize? Stated simply, the numbers in each angle represented the ratio of attraction operating between the two corresponding substances fixed to the pinnacles of its outer zone. To use the diagram, Black instructed his students to pick a compound and then to look to the inner zone for the ratio of attraction between its two substances. If students wanted to read the next compound they simply cast their gaze back out to the tip of the chiasm and

[46] The Erskine chiasm appears on the flyleaf at the front of Herodotus, *Herodiani historiarum libri 8. Recogniti & notis illustrati* (Oxford, 1704), Dunimarle Library, Banff, No. 982. The provenance of the eighteenth-century books from the Erskine family library is addressed in Friends of Duff House, "The Dunimarle Library at Duff House" (Duff House, 2011), leaflet available from the staff of Duff House, Scotland.

then performed the same angular reading that they had used for the previous compound. This kind of reading, moreover, could be performed in either a clockwise or a counterclockwise direction, and Black's discussions of the chiasm reveal that, depending on what he wanted to describe, he read the diagram in both directions during his lectures.

Crucially, Black's ratios did not represent any sort of real unit of measurement. He used them to conceptualize the relative attractions of the substances visualized in the diagram. The use of ratios in this matter was most likely taken from planetary astronomy, where they were employed by natural philosophers to compare the unitless planetary distances and perturbations of orbits. It was not until after astronomers had collected data during the 1761 and 1769 transits of Venus that an accurate distance between the sun and the earth was calculated, thereby allowing ratios to be expressed in known units of measurement (like miles).[47] Because those transits occurred during Black's lifetime, he probably saw his ratios in a similar manner, as formulas simply waiting to be activated with numbers in the event that a viable unit of chemical force was offered or proposed.

CONCLUSION

In this essay I have presented a visual anthropology of Joseph Black's chemistry diagrams with a view to investigating how they preserved and transmitted knowledge. When seen from this perspective, two important points emerge. First, the visual format of Black's diagrams had been used for a long time. This was especially the case for his chiasm, which was used in schools as a calculation tool. Second, Black's diagrams operated collectively as a visual system, an assemblage of pictures that all worked to explain the forces that attracted and repulsed substances. This means that the diagrams were designed to work together, and, as such, they constituted a unified attempt to visualize chemical affinity.

Far from being a unique occurrence, I have shown that Black's diagrams had a pedagogical history. Yet, if the diagrams were not explicitly unique, then what makes them noteworthy? The answer lies in the meaning that Black assigned to them. Whereas it is important to identify the form of simple images, such acts of identification alone offer little insight into how diagrams were used, what they were taken to represent, and how they comprised a unified visual system whose meaning was directly tied to a conceptual system. Likewise, it is very difficult to understand how the meanings of Black's diagrams—or most diagrams for that matter—were learned without taking their uses and iterations into account. So while the diagrams were certainly used to circulate knowledge in Black's classroom, the local use and meaning learned in that setting served as a guide to their global worth when they were taken outside Edinburgh.

[47] The exact ratio was not known in Black's lifetime. The precise distance was contested because of the "drop effect" that occurred when Venus first appeared in front of the sun. See Bradley E. Schaefer, "The Transit of Venus and the Notorious Black Drop Effect," *J. Hist. Astron.* 32 (2001): 325–36.

Between the Workshop
and the Laboratory:

Lavoisier's Network of Instrument Makers

by Marco Beretta*

ABSTRACT

Throughout his career, Lavoisier paid particular attention to the apparatuses he intended to use in his experimental pursuits. Lavoisier engaged many instrument makers in Paris, the French provinces, and abroad, and he made several efforts, more or less successful, to design a new environment for chemical experimentation. In addition to working with famous instrument makers such as Fortin, Mégnié, and Ramsden, Lavoisier had his instruments made by more than seventy other different makers. In this essay, I outline their contributions and make a preliminary attempt to establish their role in the design of Lavoisier's instruments and in the changes that occurred in chemical experimentation.

A DORMANT COLLECTION OF INSTRUMENTS

Lavoisier's collection of instruments represents one of the most interesting examples of a private laboratory because of both its size and the number of apparatuses that have miraculously survived after two centuries of often dramatic vicissitudes.[1] Although the approximately 500 surviving instruments are just a small part of the roughly 7,000 pieces that were briefly cataloged in 1794,[2] they are nevertheless sufficient to illustrate the constant attention paid by the French chemist to updating his laboratory with innovative instruments and machines. It is well known that the meaning of these innovations has been a matter of dispute, and from perusing recent literature,[3] it is not

* Dipartimento di Filosofia e Comunicazione, Università di Bologna, via Zamboni 38, 40126 Bologna, Italy; marco.beretta@unibo.it.

I wish to thank Patrice Bret, Christine Lehman, John Perkins, and the anonymous referees for their helpful suggestions and criticism.

[1] On Lavoisier's collection, see Marco Beretta, "Lavoisier's Collection of Instruments: A Checkered History," in *Musa Musaei: Studies on Scientific Instruments and Collections in Honour of Mara Miniati*, ed. Marco Beretta, Paolo Galluzzi, and Carlo Triarico (Florence, 2003), 313–34.

[2] The documents related to this inventory are published in Beretta, "Lavoisier's Collection of Instruments" (cit. n. 1). At the Archives Nationales, Paris (AN, F17 A. 1337. Dossier 5), some additional descriptions concerning glassware and crucibles are given.

[3] Jan Golinski, *Science as Public Culture: Chemistry and Enlightenment in Britain 1760–1820* (Cambridge, 1992), 137–52; Golinski, "Precision Instruments and the Demonstrative Order of Proof in Lavoisier's Chemistry," *Osiris* 9 (1994): 30–47; Golinski, "'The Nicety of Experiment': Precision of Measurement and Precision of Reasoning in Late Eighteenth-Century Chemistry," in *The Values of Precision*, ed. M. Norton Wise (Princeton, N.J., 1995), 72–91; Frederic L. Holmes, *Antoine Lavoisier,*

clear whether they played a significant role in the design of pioneering experiments or were the results of a subtle strategy to increase the value of precision in chemistry and to exclude those chemists whom Lavoisier did not regard as sufficiently professional to practice it. Because the historiographic issue at stake is of great importance, it is surprising that, after the pioneering studies by Maurice Daumas in the 1950s,[4] the study of Lavoisier's instruments has rapidly declined, and most of the recent disputes on their role and function have principally revolved around the interpretation of texts, and not the instruments per se.[5] The material turn that characterizes the recent trends in the history of science will hopefully inspire a revival of interest in Lavoisier's collection. The data presented in what follows are part of a project that began in 1998, which so far has resulted in a preliminary catalog of Lavoisier's instruments[6] and a survey of his laboratories.[7]

A few words should be said about the cost of Lavoisier's equipment. The richness in quantity and variety of apparatuses in his laboratory has often been regarded as unusual and unnecessarily expensive:[8] as early as 1796 a malicious statement by P. Quenard estimated the cost of the experiments on the decomposition of water at the fabulous sum of half a million *livres*.[9] This kind of accusation became particularly common in the English context, where it was claimed that chemistry did not need costly equipment. However, no comparative studies exist on the laboratories built by Lavoisier's contemporaries and, more generally, on the cost of scientific equipment in other disciplines. Joseph Priestley, who is often portrayed as the emblematic representative of an alternative and more sober vision of the chemical laboratory, owned an expensive apparatus, and after its destruction in 1791 he made a claim to the King for the remarkable sum of £4,000 (including

---

the Next Crucial Year (Princeton, N.J., 1998); Holmes, "The Evolution of Lavoisier's Chemical Apparatus," in *Instruments and Experimentation in the History of Chemistry,* ed. Frederic L. Holmes and Trevor H. Levere (Cambridge, Mass., 2000), 137–52.

[4] Maurice Daumas, "Les appareils d'exérimentation de Lavoisier," *Chymia* 3 (1950): 45–62; Daumas, *Scientific Instruments of the Seventeenth and Eighteenth Centuries and Their Makers* (London, 1989); Daumas, *Lavoisier théoricien et expérimentateur* (Paris, 1955), 112–56.

[5] A relevant exception to this tendency is the study by Peter Heering, "Weighing Heat: The Replication of the Experiments with the Ice-Calorimeter of Lavoisier and Laplace," in *Lavoisier in Perspective,* ed. Marco Beretta (Munich, 2005), 27–41.

[6] Published in Marco Beretta, *Panopticon Lavoisier,* http://moro.imss.fi.it/lavoisier/ (accessed 7 March 2014).

[7] Marco Beretta, "Big Chemistry: Lavoisier's Design and Organisation of his Laboratories," in *Spaces and Collections in the History of Science,* ed. Marta Lourenço and Ana Carneiro (Lisbon, 2009), 65–80. A book on Lavoisier's laboratories is in preparation.

[8] Jan Golinski saw in Lavoisier's instruments nothing more than projections of a view the purpose of which was to make quantification a mere rhetorical argument. See Golinski, "'The Nicety of Experiment'" (cit. n. 3), 72–91. Elsewhere Golinski takes this argument further and declares that "the technical refinements . . . made experiments in this field harder, by raising the stakes in terms of the necessary level of equipment and skills for making contributions in future. Investigators who lacked these resources would no longer need to be taken seriously"; Golinski, *Science as Public Culture* (cit. n. 3), 212.

[9] "Lavoisier n'avait encore que cinquante ans. L'étude lui avait laissé toute sa vigueur: toutes les difficultés étaient vaincus; il avait créé la science; et, comme il jouissait d'une fortune brillante, et il avait la passion de son art, il n'épargnait rien pour découvrir une vérité: il ne calculait ni la peine, ni l'argent, quand il s'agissait d'avancer d'un pas. La seule expérience de la décomposition si simple de l'eau lui avait coûté plus de cinq cents mille livres. C'est sans doute cette passion des recherches qui l'avait rendu avare, insensible même à l'égard de tout ce qui l'approchait"; François Bonneville, *Portraits des personnages célèbres de la révolution* (Paris, 1796), vol. 2, s.v. "Lavoisier," by P. Quenard.

the library). The most significant pieces of equipment were valued at the equally considerable sum of £609, and the makers of the instruments were the most important and expensive ones such as Ramsden, Nairne, Parker, and Wedgwood.[10] In Uppsala, Torbern Bergman's equipment was equally important, and it is no coincidence that most of Carl Wilhelm Scheele's pneumatic experiments leading to the isolation of oxygen were made in Bergman's laboratory—using innovative apparatuses—and not in his apothecary shop.[11] Further evidence that the historiographic myth of Lavoisier's high laboratory expenditures needs reassessment comes from his laboratory assistant Armand Séguin's observation in 1791 that "M: L[avoisier] spends not above £250 on chemistry—that his revenue is £16,000 a year."[12] Another useful comparison to reassess Lavoisier's expenditures on equipment is the cost of astronomical instruments. In 1783 the value of John Dollond's Gregorian telescope was estimated by the English tradesman H. Sykes to be 8,000 *livres*,[13] nearly 1,000 *livres* more than the gasometer made by Mégnié for Lavoisier, which in itself was the most expensive piece of equipment in his laboratory. Mathematical instruments were also quite expensive, decidedly more so than chemical instruments, but they were so widely used in England and France that a small community of specialized instrument makers were able to support themselves by solely catering to this market.

In this essay I will focus on the rich and varied community of craftsmen, inventors, engineers, academic scientists, and laboratory assistants who were employed by Lavoisier and, thanks to his patronage, often managed to embark on successful careers in science. During his lifetime, Lavoisier purchased instruments from more than seventy makers.[14] I have not taken into account the instrument makers employed by Lavoisier on behalf of the Académie Royale des Sciences. One relevant exception is the experiment on the combustion of diamond in 1774 with Trudaine's large burning lens, as it is impossible to draw a sharp line between Lavoisier's contribution and the role of the institution for which he worked.[15]

It has been possible to deduce this impressive figure by analyzing a variety of sources: (1) the signatures on the instruments preserved at the Musée des Arts et Métiers; (2) the explicit references in Lavoisier's *Oeuvres* and correspondence; (3) the numerous references in Lavoisier's laboratory notebooks; and (4) the iconographic sources.

It is also important to add that Lavoisier introduced new instruments within the Régie des Poudres and the Ferme de Tabac, and although we do not have any records of their makers, their use in highly organized manufacture is important in light of Lavoisier's constant concern regarding the standardization of chemical instruments.

---

[10] Douglas McKie, "Priestley's Laboratory and Library and Others of His Effects," *Notes Rec. Roy. Soc. Lond.* 12 (1956): 114–36.

[11] Scheele lived in Uppsala between 1770 and 1775, and during this period he regularly worked in Bergman's laboratory.

[12] Cited in J. A. Chaldecott, "Scientific Activities in Paris in 1791: Evidence from the Diaries of Sir James Hall for 1791, and Other Contemporary Records," *Ann. Sci.* 24 (1968): 21–52, on 34.

[13] Daumas, *Scientific Instruments* (cit. n. 4), 241.

[14] Not all were professional makers. Table A1 in the appendix lists various categories of instrument makers.

[15] On these experiments see Christine Lehman, "Alchemy Revisited by the Mid-Eighteenth Century Chemists in France," *Nuncius* 28 (2013): 165–216.

THE PRODUCTION OF CHEMICAL INSTRUMENTS IN PARIS BEFORE 1770

French instrument making became competitive only after 1760,[16] although it never reached the standards set by English instrument makers. During Lavoisier's lifetime, Paris became an important center for innovation, and the precision developed by French instrument makers equaled those found in English manufacturing.[17] During the second half of the century, a relatively large community of Parisian makers of mathematical, optical, astronomical, and physical instruments owned profitable workshops and furnished their commodities to an expanding clientele of professional scientists and dilettanti. They also worked for institutions such as the Académie Royale des Sciences, the Observatoire, and the colleges of Paris, which either considerably enlarged cabinets and laboratories or created new ones.[18] These makers belonged to different guilds including the founders,[19] *couteliers*, mirror makers, enamellers, and pottery makers, and from the middle of the seventeenth century they were allowed, not without difficulties, to specialize in the production of scientific instruments. Despite their legal rigidity, guilds offered significant guarantees of survival to their members.[20] In order to overcome the constraints of the guilds' regulations, the Académie des Sciences obtained from the King special concessions for instrument makers, which ultimately led to the creation in 1787 of a recognized body of engineers specialized in the construction of all sorts of mathematical and physical instruments.[21]

Considering this background, the absence of specialized makers of chemical apparatuses is conspicuous. In the second half of the century, chemistry became so fashionable that the yearly attendance at the courses given in the French capital reached 3,130 students in 1781.[22] Because the lectures included extensive experimentation, private and public laboratories proliferated at the same pace. Only experimental physics reached the same figures. Unlike mathematical and physical instruments, chemical apparatuses deteriorated easily and needed to be replaced frequently, which would seem to be a good argument for specialization. However, this was not the case for makers of chemical instruments, who were not officially recognized as an independent profession.[23] Who then supplied the Parisian chemical laboratories? Not the makers of the physical and mathematical instruments—their trade catalogs rarely referred to apparatuses that could be used in chemical experiments. An example of one such item was a burning lens with a diameter of 122 cm that was advertised by one of the most successful Parisian instrument makers of the 1760s, Claude Passemant, as a device that could have been productively used in chemical experiments as well

---

[16] Daumas, *Scientific Instruments* (cit. n. 4), 269.

[17] E.g., Etienne Lenoir is commonly known for initiating precision instrument making in France; see A. J. Turner, *From Pleasure and Profit to Science and Security: Etienne Lenoir and the Transformation of Precision Instrument-Making in France 1760–1830* (Cambridge, 1989).

[18] Daumas, *Scientific Instruments* (cit. n. 4), 258–68.

[19] Henri Bouchot wrote that by the end of the eighteenth century, there were 334 *maîtres fondeurs* working in Paris; see Bouchot, *Histoire anectotique des métiers avant 1789* (Paris, 1887), 140.

[20] Daumas, *Scientific Instruments* (cit. n. 4), 91–100.

[21] Maurice Daumas, "Le corps des Ingénieurs brevetés en instruments scientifiques (1787)," *Arch. Int. Hist. Sci.* 5 (1952): 86–96.

[22] John Perkins, "Chemistry Courses, the Parisian Chemical World and the Chemical Revolution, 1770–1790," *Ambix* 57 (2010): 27–47, on 36.

[23] Either recognized by the King or by a patent provided by the Académie des Sciences.

as in architecture and painting.[24] However, it is not a coincidence, I believe, that this lens was not used for chemical experiments and that when the chemistry section of the Académie needed one in 1772–3, they preferred to use the lens of Tschirnhaus.[25]

Between the end of the 1750s and the French Revolution, the market for chemical equipment must have been enormous. Chemists were not the only scientists in need of such instrumentation: druggists, pharmacists, perfume makers, *boulangers* and all the guilds involved in the preparation and preservation of food, wine makers, and distillers were in constant need of furnaces, crucibles, mortars, and glassware of all sorts, in addition to other equipment that was commonly found in chemical laboratories. Paradoxically, it was probably because of this widespread market for "chemical" equipment that specialization was not needed. During the 1750s, this kind of apparatus was easily available and produced in large quantities by various guilds that had met the needs of artisans, pharmacists, and cooks long before professional chemists appeared on the stage. This explains why the laboratories of the chemists relied on very simple apparatuses that changed little over a long period of time. Another factor that prevented chemists from making major changes in their equipment was the fact that most of the chemical courses were held by the guild of Parisian pharmacists, a guild that was subjected to strict regulation and that regarded all the changes introduced by its members with suspicion. Only when the Parisian chemical community began to be composed of a consistent number of practitioners with different educational backgrounds did competition open new paths of enquiry. It took a long time before this happened, however, and as late as 1773 Parisian chemists and pharmacists got their equipment from nonspecialized artisans. Antoine Baumé provides us with a list of the Parisian suppliers of the chemical laboratories of his day:[26] most of the glassware (alembics, retorts, phials, recipients, etc.) could be bought from any *faïencer* (pottery makers) or by glazers [*emailleurs*]; earthenware from the *marchands potiers de terre*; furnaces and crucibles from the *fournalistes* (furnace makers) or the founders. Many useful apparatuses could be obtained from the ironsmiths [*quincailleries*]; balances from the balance makers [*balanciers*]; and copper recipients from the boilermakers [*chaudronniers*]. German crucibles were widely imported by tradesmen in Paris because they were widely regarded as the most resistant. The only device made in the workshops of professional instrument makers was the thermometer, an instrument that, as I shall point out later, was introduced quite late in the chemical laboratory. It is therefore not surprising that the equipment of Parisian chemical laboratories was rarely innovative. The apparatuses listed in the laboratory of the most successful chemistry teacher of the period, Guillaume François Rouelle,[27] was almost identical to that published by Nicolas Lémery in the first part of his *Cours de chymie* (Paris, 1675).

Although glassware was of the greatest importance in chemical experimentations, Parisian chemists were unable to find products with the required resistance to both heat and cold. Macquer remarked:

---

[24] Claude Passemant, *Description et usage des télescopes, microscopes, divers ouvrages et inventions* (Paris, 1764), 64. The mirror was presented to the King in 1757.

[25] See Lehman, "Alchemy Revisited" (cit. n. 15).

[26] Antoine Baumé, *Chymie Expérimentale et Raisonnée*, 3 vols. (Paris, 1773), 1:cxxviii–cxlvi.

[27] A contemporary description of the content of Rouelle's laboratory has been published in Marco Beretta, "Rinman, Diderot, and Lavoisier: New Evidence Regarding Guillaume François Rouelle's Private Laboratory and Chemistry Course," *Nuncius* 26 (2011): 355–79.

> Vessels intended for chemical operations should, to be perfect, be able to bear without breaking the sudden application of great heat and cold, be impenetrable to every substance and inalterable to any solvent, be unvitrifiable and capable of enduring the most violent fire without fusing. But up to the present no vessels are known which combine all of these qualities.[28]

By 1765 French academic chemists and scientists had some interest in glassmaking but did not bother to develop a consistent theory of its composition; in fact, glass was not yet a proper object of chemical inquiry. On the other hand, glassmakers had introduced significant technical innovations, the contents of which rarely went beyond the secrecy carefully guarded by the manufacturers and by the guilds.

Pneumatic discoveries focused renewed attention on the nature of fire and stimulated active research in chemical instruments. Buffon's French translation of Stephen Hales's *Vegetable Staticks* (Paris, 1735), for example, stimulated chemists to repeat Hales's pneumatic experiments and to construct replicas of his instruments. Although Hales was not a chemist, he was inspired by Newton's queries in the *Opticks*,[29] and many of the experiments he presented in his book were explicitly chemical, contributing "to foster the hope, which Hales shared with contemporaries Peter Shaw and John Freind, that chemistry could be made an exact science modeled upon physics."[30] To this aim, Hales devised a piece of equipment to measure the volume of air given off or absorbed in chemical reactions within a closed system, known as a "pedestal apparatus," which set the standard in pneumatic investigation for nearly four decades.

Because of Hales's hybrid background—although not trained as a chemist, he worked on chemical problems—and the revolutionary discovery that air was a constituent of a surprisingly large number of bodies, *Vegetable Staticks* was initially either rejected or ignored by Stahl and Boerhaave, the most authoritative chemists of the period, who continued to consider air as a passive and simple element. It is therefore not surprising that Hales's work was translated into French by Buffon, an enthusiastic follower of the Newtonian philosophy, and was first discussed in Paris not by a chemist but by a natural philosopher. It was in fact Jean Nollet who introduced Hales's work in 1745 to the Parisian public.[31] Interestingly, when surveying the properties of fire, Nollet presented pieces of chemical equipment he used during experiments on fermentation and distillation and on various uses of the burning lens.[32] Nollet's inclusion of devices that could be used by chemists was new in France, but it was probably inspired by Dutch natural philosophers like Petrus van Musschenbroek, who listed in their works instruments such as the pyrometer and the thermometer, which had been designed to be used in Herman Boerhaave's chemical laboratory.[33] Nollet's successful efforts to create a wide public interest in physical apparatuses created the basis for

---

[28] Macquer, *Elémens de chymie théorique* (Paris, 1753), 297.

[29] The Newtonian background of Hales's interest in chemistry is explained in Henry Guerlac's excellent entry, "Stephen Hales," published in the *Dictionary of Scientific Biography*. Here I have used the revised edition, "Stephen Hales: a Newtonian Physiologist," published in Henry Guerlac, *Essays and Papers in the History of Modern Science* (Baltimore, 1977), 170–92.

[30] Henry Guerlac, "The Continental Reputation of Stephen Hales," in Guerlac, *Essays and Papers* (cit. n. 29), 276.

[31] Jean Nollet, *Leçons de physique expérimentale*, 6 vols. (Paris, 1745–70), 3:243–86.

[32] Ibid., 4:228–36, 272–316, 320.

[33] A list of devices for sale was published at the end of the second volume of Musschenbroek's *Essai de physique* (Leiden, 1751). The Latin edition was published in 1741.

a steady market of scientific instruments and machines.[34] It is therefore plausible that Nollet also retailed the chemical apparatus described in his lectures, thus beginning to shake the relatively static environment of Parisian chemical laboratories.[35] Indirect evidence of this can be deduced from the fact that sometime around 1757 Rouelle introduced in his lectures a lengthy description of Hales's pneumatic pedestal, a device that he significantly transformed.[36] Thus, English pneumatic experiments appeared on the scene at about the time Lavoisier was finishing school at the Collège Mazarin. Since Lavoisier attended both Rouelle's and Nollet's lectures we may conclude that he was introduced to Hales's pneumatic chemistry from two different but complementary perspectives. Furthermore he had the opportunity to experience and, eventually, to exploit the favorable conditions created by the competition between the courses by Rouelle and Nollet. Public courses in experimental physics and chemistry became increasingly popular in the late 1760s and early 1770s, and Nollet's inclination to include chemistry and chemical demonstrations in his experimental physics courses was adopted by Nollet's successor, Joseph-Aignan Sigaud de la Fond, and by Jacques Mathurin Brisson. In 1775 Sigaud de la Fond published *Description et usage d'un Cabinet de physique expérimentale,* in which he extensively described the use of several chemical and pneumatic apparatuses that he made (or had made) with the intention of selling them to the students in his lectures. Because pneumatic chemistry was at the center of attention of the Parisian community of scientists, Sigaud devoted special attention to the apparatus made by Priestley and Lavoisier in 1775. Sales of Sigaud's pneumatic instruments increased when his nephew, Rouland, helped him both with the teaching and with the retail of instruments.[37] By 1790 Haering, known officially as *Ingénieur en Instrumens de physique et optique*, published an extensive catalogue of Nollet's instruments and had a specialized section of devices for the analysis of gases.[38]

The entrance of natural philosophers into the production of chemical equipment had a positive effect on the Parisian chemical community. One example is Antoine Baumé's advertisement for a new kind of hydrometer that measured different densities of fluids. The advertisement appeared in the 7 November 1768 issue of the *Avant-Coureur*,[39] the journal launched by Pierre Joseph Macquer's brother, Philippe, which gave detailed reports about inventions and courses developed by Parisian chemists and pharmacists.

So, when Lavoisier was beginning his career as a chemist, pneumatic experiments by Hales had been the subject of public lectures for nearly a decade, and natural

---

[34] See Lewis Pyenson and Jean-François Gauvin, eds., *The Art of Teaching Physics: The Eighteenth-Century Demonstration Apparatus of Jean Nollet* (Sillery, 2002).

[35] As Paolo Brenni points out, "Nollet's action in spreading the culture of scientific instruments prepared the ground for French instrument-makers like Jean Nicolas Fortin and Etienne Lenoir"; Brenni, "Jean Antoine Nollet and Physics Instruments," in ibid., 27.

[36] Guillaume François Rouelle, *Cours de chymie*, 3 vols. (Paris, 1757–80), 1:89–92. Manuscript preserved at the Bibliothèque Interuniversitaire de Médecine (Université René Descartes Paris 5), available online at http://www2.biusante.parisdescartes.fr/livanc/?cote=ms05021_23x01&do=pages (accessed 7 March 2014).

[37] Sigaud de la Fond, *Essai sur différentes espèces d'air fixe . . . Nouvelle édition augmenté par Mr Rouland* (Paris, 1785).

[38] *Catalogue de différentes pièces de physique* (Paris, 1790). A digital copy of the catalog may be found at http://cnum.cnam.fr/redir?8SAR659 (accessed 7 March 2014). I thank John Perkins for bringing this title to my attention.

[39] *Avant-Coureur* 45 (1767): 712–6.

philosophers such as Nollet advertised the sale of chemical apparatuses, combining traditional equipment with physical instruments, such as pyrometers, burning lenses, and pneumatic devices, especially adapted for chemical experimentation. By the end of the 1760s, professional pharmacists such as Antoine Baumé, who were already successful chemical traders, realized that a market for chemical apparatuses was rapidly developing and that chemical laboratories needed to be updated.

LAVOISIER AND THE BIRTH OF PROFESSIONALIZED CHEMICAL INSTRUMENT MAKING

As the demand for chemical instruments increased, Lavoisier became one of the most important private patrons of the Parisian instrument makers. His systematic endorsement of their work contributed to the creation of a group of professional makers of chemical equipment over a twenty-year period from 1764 to 1780. Lavoisier employed both well-known makers of mathematical and physical instruments and, especially at the beginning of his career, workmen belonging to different guilds. Because Lavoisier played such a crucial role in updating and improving chemical apparatuses, it is impossible to separate his contributions from those of the instrument makers themselves.

The first recorded instrument cited by Lavoisier is a microscope used during his experiments on the analysis of gypsum in 1765, but he gives no description of it.[40] Thereafter, references to the use of physical instruments were a constant of his daily work. During his mineralogical travels of 1766–7 with Jean Etienne Guettard, Lavoisier tried several instruments in new ways. An example of this is the barometer that he used in his observations of the geological strata of mountains.[41] Probably inspired by Réaumur's observations,[42] Lavoisier introduced the systematic use of the thermometer into chemistry; during the 1760s he used it in the analysis of mineral waters, and for these purpose he had different ones made by Cappy (in 1761),[43] Claude Simon Passemant, and Gallonde.[44] In his observations of 1767, Lavoisier analyzed mineral waters with three different comparable thermometers,[45] his own hydrometer, and a balance made by Chemin.[46] Thermometers were to play a crucial role for the rest of Lavoisier's career, and he employed the best makers to design them for specific chemical experiments. Lavoisier himself perfected the construction of

---

[40] Lavoisier, Œuvres, 6 vols. (Paris, 1864–93), 3:141 (hereafter LO).

[41] Ibid., 5:109

[42] "Sans les Thermomètres nous n'aurions jamais découvert que certains Sels, en se fondant dans l'eau, la refroidissent"; cited in Arthur Birembaut, "La contribution de Réaumur à la thermométrie," Rev. Hist. Sci. 11 (1958): 302–28, on 311.

[43] Lavoisier, Correspondance, 7 vols. (Paris, 1955–2012), 1:44, 65 (hereafter LC). Musée des Arts et Métiers, Paris, Inventory n. 19934.

[44] Musée des Arts et Métiers, Paris, Inventory n. 19935. Lavoisier used this thermometer during his examination of natural waters in 1767 and during experiments on distilled water ("Pésanteur absolue de l'eau distillé," LO [cit. n. 40], 3:451–4). Comparative data of Gallonde's thermometer with two different thermometers may be found in Lavoisier, Analyses des eaux. Notes diverses, tables, calculs, Archives de l'Académie des Sciences, Paris, Dossier Lavoisier 1376.

[45] In his manuscript notes, Lavoisier mentions a "bleue" thermometer and one "corrigé" that I was not able to identify. See n. 46.

[46] This equipment has been drawn by Lavoisier's manuscript note, Note des articles envoyés a Bourbonnes les Bains le sammedy 13 juin 1767, Archives de l'Académie des Sciences, Paris, Dossier Lavoisier 100-R_6. I found references to Gallonde's thermometer throughout the Dossier Lavoisier 100.

a new type of thermometer that he used during his experiments on cold, which was placed in the cave of the Observatoire and destined to set the standard for many years to come.[47]

Once Lavoisier was interested in a particular instrument, he employed different makers to create instruments for a series of comparisons that enabled him to assess their accuracy and understand how to improve them. Lavoisier used the same approach in the construction and perfection of the barometer, the hydrometers, the balances, the burning lens, the Papin digester, and the pyrometer.

Lavoisier's drive for innovation is particularly evidenced by his early efforts to introduce precision measurement in pneumatic chemistry despite encountering difficulties. As early as 1773, he conceived a recipient to weigh gases by plunging it in water (fig. 1),[48] but it was not easy to get this kind of instrument made. And the delivery of the apparatus for the calcination of lead through the burning lens (also dated March 1773) was delayed because of the slowness of the workmen.[49] It is interesting that Lavoisier refers to the makers as *ouvriers*, thus confirming that by 1773 there were no specialized makers of this kind of apparatus. The situation began to change when he moved to the Arsenal in 1775; he then had access to large facilities and soon built up the richest chemical laboratory of the French capital. In the eulogy of Lavoisier, written after an interview with Madame Lavoisier, Georges Cuvier recalled the laboratory organization of Lavoisier's team in the following:

> In the morning, several enlightened friends whose cooperation he had requested, would gather in his laboratory. He even accepted very young people whose shrewdness he had recognized and certain workers who were especially skilful in producing accurate instruments. He always announced his plans to the assistants with great clarity. Each one made suggestions about their execution; and everything that was considered plausible was soon put to the test.[50]

This statement needs some further clarification. Cuvier seems to be referring to two different groups of people: natural philosophers and skilled artisans. Nicolas Fortin, a talented artisan who had trained in Lavoisier's laboratory and specialized in making chemical instruments, is the foremost individual in the latter group. Lavoisier first recognized Fortin's abilities in June 1779 when he reported at the Académie on his pneumatic machine.[51] Before this presentation Fortin was unknown; Lavoisier must have been quite impressed with his abilities because he soon engaged him for the construction of various sophisticated instruments, for example, a pneumatic machine, a gasometer, a great precision balance, and an apparatus for combusting oils and for fermentation. Fortin also provided Lavoisier and Meusnier de la Place with technical solutions for the assemblage of the machines used during the famous experiments on the decomposition and composition of water.[52] In a letter to Achard dated 6 December 1789, Lavoisier declared that Fortin was the artisan he employed

---

[47] Lavoisier, "Thermomètre des caves de l'Observatoire, Précautions prises pour construire et graduer ce thermomètre" (1776), in LO (cit. n. 40), 3:421–6.

[48] Lavoisier, Registre de laboratoire, Archives de l'Académie des Sciences, Paris, vol. 1, fol. 14r.

[49] Ibid., fol. 18r.

[50] Georges Cuvier, "Lavoisier," *Biographie Universelle Ancienne et Moderne* (Paris, 1819), 23:462.

[51] LO (cit. n. 40), 4:327

[52] Invoice by Fortin, 20 July 1785; LC (cit. n. 43), 4:137–40. This invoice shows that Fortin's role in the construction of the perfected gasometer has been underestimated.

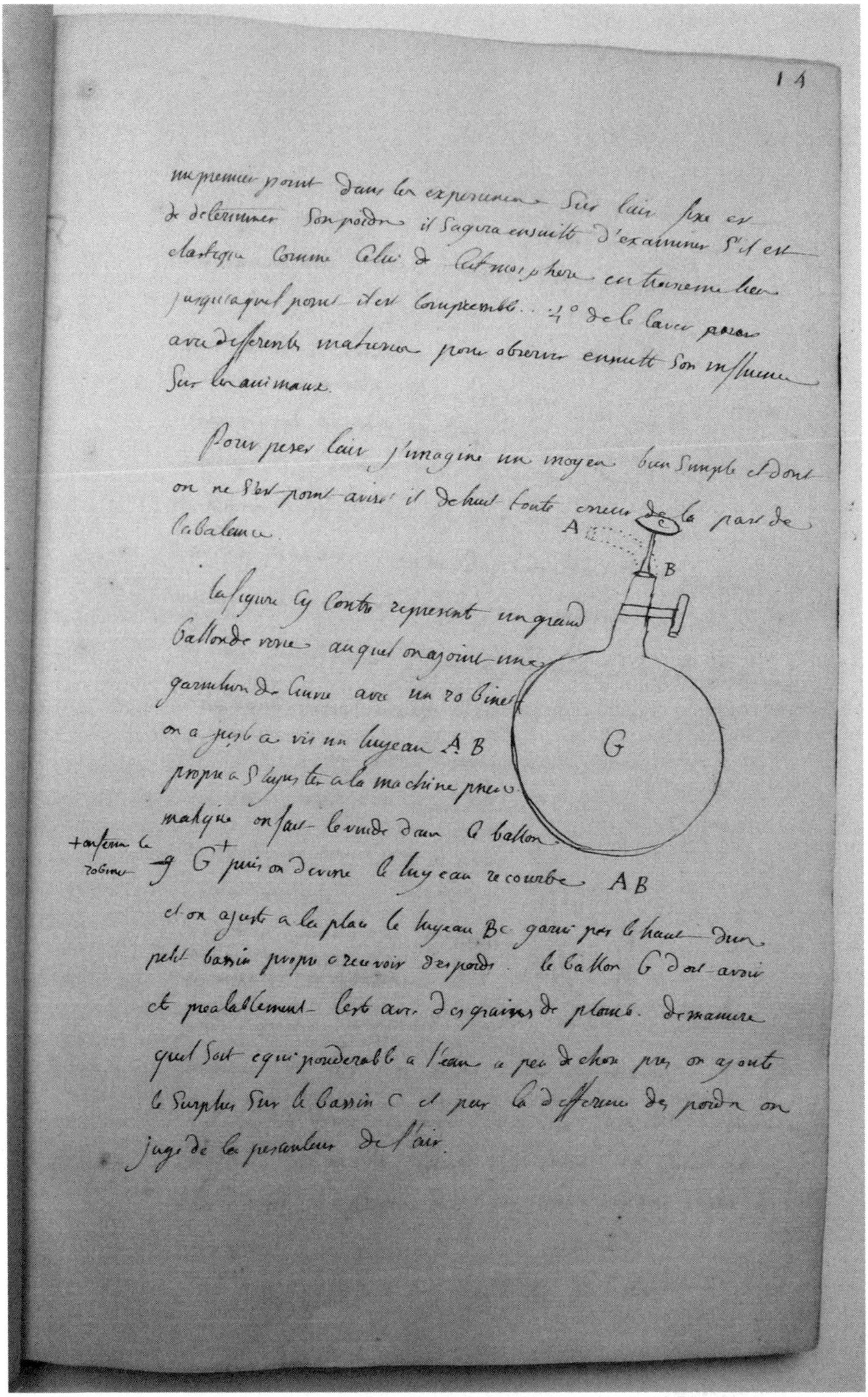

un premier point dans les experiences sur l'air fixe est de determiner son poids il s'agira ensuite d'examiner s'il est elastique comme celui de l'atmosphere en trouver lieu jusqu'a quel point il est compressible. 4° de le laver pour avec differentes matieres pour observer ensuite son influence sur les animaux.

Pour peser l'air j'imagine un moyen bien simple et dont on ne s'est point avisé il dehut toute erreur de la part de la balance.

la figure cy contre represente un grand ballon de verre auquel on a joint une garniture de cuivre avec un robinet on a juste a vis un tuyau AB propre a s'ajuster a la machine pneumatique on fait le vuide dans le ballon

+on ferme le robinet+ g G puis on devisse le tuyau recourbe AB et on ajuste a la place le tuyau BC garni par le haut d'un petit bassin propre a recevoir des poids. le ballon G doit avoir et prealablement l'est avec des grains de plomb. de maniere qu'il soit equipondérable a l'eau a peu de chose près on ajoute le surplus sur le bassin c et par la difference des poids on juge de la pesanteur de l'air.

Figure 1. *Drawing of Lavoisier's recipient to weigh gases (1773). Registre de laboratoire 1 (cit. n 48). Reprinted courtesy of Archives de l'Académie des Sciences, Paris.*

most commonly.[53] It is likely that Fortin worked regularly in Lavoisier's laboratory, at least at an early stage, but sometime during the 1780s he opened his own workshop in the Place de la Sorbonne, where he had only two employees, a striking figure compared to the forty workmen employed by Jesse Ramsden in London. Thanks primarily to Lavoisier, Fortin's reputation rapidly increased, and he began to work for other scientists and, above all, for the Académie des Sciences. Fortin became one of the most authoritative makers used by the Commission des poids et mésures in 1793. Eventually Fortin was officially recognized to have "revolutionize[d] the science of physics and created modern chemistry."[54]

Another Parisian instrument maker used regularly by Lavoisier and who, in time, specialized in the construction of chemical instruments was Pierre Bernard Mégnié. *Ingénieur en instruments mathématiques,* Mégnié was a specialized maker before meeting Lavoisier, and he owned a shop at the Cour de commerce, Faubourg St. Germain in Paris. He was employed extensively by Lavoisier, whom he called his patron [*protecteur*] and benefactor [*bienfaiteur*].[55] The first recorded instrument made for Lavoisier was a barometer used for the meteorological campaign promoted by the French chemist in 1778[56] and was followed by a pair of gasometers in 1787 (costing 7,554 *livres*).[57]

Between the chemist Lavoisier and instrument makers such as Fortin were engineers, including Jean Baptiste Meusnier de La Place, engineer to the King and, since 1776, a corresponding member of the Académie des Sciences de Paris. Lavoisier employed him after he had seen his abilities during the experiments on ballooning promoted by the Académie des Sciences. Although Meusnier and Lavoisier were the principal inventors of the gasometer, another young laboratory assistant with a chemical background, Philippe Gengembre, was also an important member of the team. In the construction of the gasometer, Lavoisier had a special copper pipe made by the chemical Parisian manufacture of the "Périer frères."[58]

Lavoisier owned a vast quantity and variety of glassware (fig. 2). During the 1760s and 1770s, he purchased small glass recipients from the glazers who worked with the glazer lamp to produce test tubes and other refined works; large recipients were provided by pottery makers while crystal glass vases were imported from England.[59] From a very early stage in his career, Lavoisier took an active interest in the study of the properties of glass, and in 1787 he asked the glassmaker Pierre Loysel to write a treatise on glassmaking based on the modern principles of chemistry.[60] In his collected works, the term "glass" occurs 715 times, but "glassmaker" appears only

---

[53] Ibid., 6:90.

[54] "Fortin. Ce dernier, plus spécialement adonné à la construction des instrumens de physique, a secondé, par son talent, les travaux des physiciens français qui ont changé la face de la physique et créé la chimie moderne"; *Rapport du Jury Central sur les produits de l'industrie française* (Paris, 1819), 256.

[55] LC (cit. n. 43), 4 (18 February 1785).

[56] Museé des Arts et Métiers, Paris, Inventory n. 07658; this was followed one year later by another barometer (Inventory n. 08761); see LC (cit. n. 43), 3:623–4.

[57] LC (cit. n. 43), 5:52–5; Musée des Arts et Métiers, Paris, Inventory nos. 7547/1–2.

[58] LC (cit. n. 43), 4:151. On the Périer brothers, see Jacques Payen, *Capital et machine à vapeur au 18e siècle—Les frères Périer et l'introduction en France de la machine à vapeur de Watt* (Paris, 1969).

[59] James Barrelet, *La verrerie en France de l'époque gallo-romaine à nos jours* (Paris, 1953), 116–20.

[60] Marco Beretta, "Unveiling Glass's Mysteries," in *Objects of Chemical Inquiry: The Synergy of New Methods and Old Concepts in Modern Chemistry,* ed. Ursula Klein and Carsten Reinhardt (Sagamore Beach, Mass., 2014), 1–20.

Figure 2. *Samples of Lavoisier's chemical glassware in a photo taken in 1952. Reprinted courtesy of Musée des Arts et Métiers, Paris.*

once.[61] The presence among Lavoisier's instruments of two glass cutters[62] suggests that someone at the Arsenal worked with glass, but it is impossible to say more on the subject.

Another category of people working in Lavoisier's laboratory on the construction of his equipment was young students who, because of their attitude and intelligence, were involved in experimental activities as laboratory assistants. Pierre Adet, Jean Henri Hassenfratz,[63] Subrin,[64] Philippe Gengembre,[65] Eleuthère Iréné Dupont,[66] and Armand Séguin were among Lavoisier's assistants: all began their apprenticeship at the Arsenal, and on several occasions they were asked to propose solutions to improve the existing instruments or to invent new ones. None of them became prominent chemists, but almost all of them had important careers: Adet became ambassador in America, Hassenfratz professor of physics at the École Polytechnique, Dupont founder of the Lavoisier's Mills (eventually renamed Dupont) in Wilmington, Delaware, Gengembre director of La Monnaie (in 1807), and Séguin director of a most profitable tannery. Their work as instrument makers can be reconstructed by taking Séguin as a significant example.

Séguin[67] took his first steps in this capacity in Lavoisier's laboratory during the second campaign of experiments on the decomposition of water in 1785. Apparently impressed by his talent, Lavoisier soon employed him on a permanent basis and by 1791 he and his brothers were living "in the faubourg St Antoine beyond M: Lavoisiers."[68] By then he had made a career, and during one of his numerous visits

[61] LO (cit. n. 40), 6:17.

[62] Musée des Arts et Métiers, Paris, Inventory nos. 20048 and 20069.

[63] Emmanuel Grison, *L'étonnant parcours du républicain J. H. Hassenfratz (1755–1827)* (Paris, 1996), 45–122.

[64] A student at the Ecole des Mines, Subrin was introduced to Lavoisier by Hassenfratz in 1786. See LC (cit. n. 43), 4:209.

[65] Philippe Gengembre (1764–1838) had been a pupil of Lavoisier under the auspices of the Régie des poudres since 1783. See Patrice Bret, "Lavoisier et l'apport de la chimie académique à l'industrie des poudres et salpêtres," *Arch. Int. Hist. Sci.* 46 (1996): 57–74, on 70–2.

[66] René Dujarric de la Rivière, *E.-I. Du Pont de Nemours élève de Lavoisier* (Paris, 1954); and Bret, "Lavoisier" (cit. n. 65), 68–9.

[67] On Séguin, see Pierre Mercier, "Armand François Séguin (1765–1835)," *Bulletin de la Section d'Histoire des Usines Renault* 7 (1976): 218–33. On the collaboration between Lavoisier and Séguin, see Marco Beretta, "Lavoisier and His Last Printed Work: The *Mémoires de physiques et de chimie* (1805)," *Ann. Sci.* 58 (2001): 327–56.

[68] Diary by James Hall of 1791 cited in Chaldecott, "Scientific Activities" (cit. n. 12), 32.

to the Arsenal, Scottish naturalist James Hall reported that Lavoisier entrusted Séguin to show him the machines "for burning charcoal and for oils" as well as an air pump.[69] Séguin is well known for his contributions to the experiments on human breathing made between 1790 and 1792.[70] In 1791 Séguin had devised a new eudiometric method for measuring the salubrity of the air,[71] thus providing Lavoisier with an extremely effective instrument that challenged those used by most of his contemporaries.[72] In the same year he carried out, together with his mentor, a new series of experiments on human perspiration and devised a new apparatus to perform them. Séguin and Lavoisier reported their results to the Académie on 12 February 1792,[73] with the aim of demonstrating the fallacy of Albrecht von Haller's claim that the human body increased in weight when immersed in water (because of the absorption of water). As the combustion of oxygen in lungs and digestion generated heat in the human body, Séguin and Lavoisier wished to measure the mechanism of dissipation of heat that kept the temperature of the body constant. Séguin thought that the excess quantity of caloric passed through the pores of the skin in form of sweat, and to study this phenomenon, Séguin devised a complex experimental procedure (figs. 3, 4). On 21 May 1791, Hall accompanied Séguin to Fortin's workshop to order an air pump for Lavoisier, and a few weeks later Hall reported about a meeting at Charles's Cabinet at the Louvre where Séguin, Lavoisier, and Charles used a large electrical machine made by the latter in order to repeat van Trootstwhyk's and Deiman's experiment on oxygen and hydrogen. Séguin also had a prominent role in the fusion of platinum, which was the experimental challenge of the day. After the disappointing results achieved with the burning lens and with the methods used by other naturalists, Lavoisier asked Séguin "to construct a very simple furnace . . . of very refractory earth . . . to produce a heat greatly more intense than any hitherto known."[74] In order to increase the temperature, Lavoisier suggested that Séguin use his previously mentioned hydrostatic bellows and nourish fire with oxygen only. To the same purpose Lavoisier also tried an "oxy-hydrogen blowpipe" of which we have no further description.[75] The refractory material of Séguin's furnace was supplied by Wedgwood, who also sent Lavoisier two thermometers for measuring the high heat capacity of his clay. Thus, Séguin played a very prominent role in Lavoisier's laboratory as a demonstrator, as a maker and designer of instruments, as a supervisor of other makers (both Parisian and foreign), and, last but not least, as a young scientist. Unfortunately, Lavoisier's early death and Séguin's huge success as an entrepreneur distracted him from science.

---

[69] Ibid., 25.

[70] Marco Beretta, "Imaging the Experiments on Respiration and Transpiration of Lavoisier and Séguin: Two Unknown Drawings by Madame Lavoisier," *Nuncius* 27 (2012): 163–91.

[71] Séguin, "Combustion du phosphore, employé comme méthode eudiométrique" (1790), in Lavoisier, *Mémoires de Physique et de Chimie, with an introduction by Marco Beretta*, 2 vols. (Bristol, 2004), 2:143–53.

[72] See Lavoisier, "De la décomposition de l'air par le soufre, de la formation des acides sulfureux et sulfurique. Et de l'emploi des sulfures dans les expériences eudiométriques" (ca. 1790), in LO (cit. n. 40), 2:720–1.

[73] Séguin and Lavoisier, "Second mémoire sur la transpiration des animaux," in LO (cit. n. 40), 5:379–90, esp. 383–4.

[74] Cited in Chaldecott, "Scientific Activities" (cit. n. 12), 30.

[75] Ibid., 34.

Figure 3. Madame Lavoisier's drawing of a man (Séguin) seated with his head in a glass container during the experiments on respiration (ca. 1790). Reprinted courtesy of Wellcome Library, London.

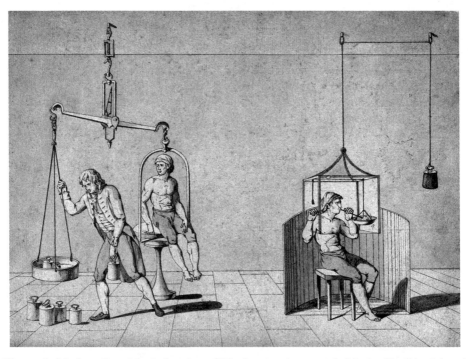

Figure 4. Madame Lavoisier's drawing of Séguin seated on a scale (devised by Séguin) and inside a glass container, during the experiments on respiration (ca. 1790). Reprinted courtesy of Wellcome Library, London.

Figure 5. *Lavoisier's equipment, part of which is now lost, exhibited in 1900 at the Musée Centennial de la Chimie.*

CONCLUSION

When Lavoisier, early in his career, understood that chemical instruments were not of the same quality as those produced for experimental physics, he tried to adapt the skills he acquired during his apprenticeship with Nollet to chemical equipment (fig. 5). This approach was only successful to a limited extent, and it was at this point that Lavoisier realized that only by radically changing experimental practice could he reform chemistry, and use his laboratory as the arena of a new method of apprenticeship. The involvement of so many different instrument makers, young apprentices, and experienced scientists coming from other disciplines indicates a deliberate and consistent effort to change the laboratory practice of a science that in Lavoisier's view was too rudimentary to enable substantial progress. Unlike experienced chemists, young apprentices and instrument makers had, for different reasons, no prejudiced ideas and saw in Lavoisier's laboratory and patronage a formidable opportunity to acquire a public reputation rapidly. Many of them had the chance to present their ideas and results before such an exclusive and selective institution as the Académie at a remarkably young age.

Thus, the eighteenth-century transformation of chemical instrument making did to a large extent take place in the laboratory of the Arsenal. However, the pioneering changes introduced by Lavoisier were more directed at the creation of an innovative site of collegial experimentation than at the construction of costly and sophisticated equipment. The variety of instruments on display at the Arsenal was the fruit of the collaborative efforts of a new group of professionals who demonstrated the crucial importance of practicing instrument makers to chemical research. Through the years,

Fortin, Mégnié, and Séguin came up with technical solutions and suggestions that gave solid experimental support to Lavoisier's theories; at the same time, their technical contributions also served to increase their confidence and, in a few cases, allowed them to become scientific authors. This collaborative network opened up the market for precision instruments that soon became standard features in the main Parisian research laboratories. It is not a coincidence that Balthasard Sage, a fierce opponent of Lavoisier's theory, kept several precision instruments in his laboratory at La Monnaie, including a balance made by Mégnié. Fortin's devices became relatively common and were used by all chemists engaged in research, from Berthollet to Gay-Lussac. The Arsenal, as is well known, was also the laboratory of a new generation of chemists such as Fourcroy, Berthollet, and, in the late 1780s, Guyton de Morveau, all of whom became increasingly engaged in the development of the chemical arts and, in a few cases, in the introduction of new chemical instruments.

Lavoisier's long-term collaboration with different instrument makers and chemists made him appreciate the importance of chemical arts and technology toward the end of his career. Such an increasing awareness led to the undertaking of several projects aimed at the enhancement of chemical artisans and makers of instruments. Lavoisier's founding of the Société pour le Progrès des Sciences et des Arts at the beginning of 1787 and his active role at the Bureau des Consultation des Arts et Métiers and at the Lycée des Arts during the Revolutionary years were all aimed at bridging the gap between theoretical chemistry, chemical arts, and instrument making, a design that inspired the French chemist to draft a general reform of education.[76] All these efforts were directed at making the experience of the Arsenal a model for public institutions, and it is probably not a coincidence that the École Polytechnique embodied most of his ideas on the organization and composition of its chemical laboratories.[77]

The Arsenal then cannot be regarded as a laboratory in which Lavoisier displayed his lavish wealth through the acquisition of exceedingly expensive devices, but, as I have argued in this essay, it became a pioneering site where he advocated the specialized manufacture of chemical instruments. At the same time, he enhanced the scientific skills of instrument makers by involving them in sophisticated scientific and experimental research programs. The shared authorship of Lavoisier's two most important papers on the decomposition of water and on respiration[78] indicate that the construction of sophisticated chemical instruments broke the anonymity that until then surrounded the instrument makers, thus paving the way for a silent revolution that made the roles played by various actors in the laboratory clear.

[76] LO, 4:649–68, 6:516–32; Patrice Bret, "Lavoisier et le Bureau de consultation des arts et métiers," in LC, 7:551–62.

[77] Thomas Bugge thus wrote in 1798: "The Polytechnic School has two very large and fine chemical laboratories, besides two of inferior extent, and some mechanical workshops"; Maurice Crosland, ed., *Science in France in the Revolutionary Era described by Thomas Bugge* (Cambridge, Mass., 1969), 39. Interestingly, Hassenfratz was in charge of the mechanical workshops.

[78] Lavoisier shared the authorship of the report on the experiments on the synthesis of water with Meusnier de la Place (who is duly credited for the construction of parts of the gasometer). In 1790 Lavoisier gave full credit (by putting Séguin's name first) to Séguin's contribution in the construction of the apparatus used during the experiments on human respiration.

Table A1. Makers of Lavoisier's Instruments

Instrument makers, engineers	Laboratory artisans	Laboratory assistants	Academic scientists	Entrepreneurs, traders, apothecaries
Adams, George Jr. (1750–95)	Naudier or Nodier, Pierre	Adet, Pierre Auguste (1763–1834)	Borda, Jean Charles (1733–99)	Alban, Léonard (ca. 1740–1803)
Argand, François-Pierre-Aimé (1750–1803)	Fallot, Jerome	Gengembre, Philippe Joachim (1764–1838)	Bucquet, Jean-Baptiste-Michel (1746–80)	Baumé, Antoine (1728–1805)
Assier-Perrica, Antoine (1730?–1811)		Hassenfratz, Jean-Henri (1755–1827)	Cassini, César-François (1714–84)	Bavière, Jacques[a]
Baradelle, Jacques (1752–94)		Séguin, Armand (1767–1835)	Deluc, Jean-André (1727–1817)	Cadet de Gassicourt, Louis Claude (1731–99)[b]
Baradelle, Nicolas Jacques (?–after 1787)		Subrin	Déparcieux Antoine (1703–68)	De Wandel, François-Ignace (1741–95)[b]
Beringer, David (?–ca. 1780)			Fontana, Felice (1730–1805)	Launoy
Bernière, Claude (?–1783)			Hall, James (1761–1832)	Mitouard, Pierre-François (1733–86)[b]
Bird, John (1709–76)			Laplace, Pierre Simon (1749–1827)	Périer, Jacques-Constantin (1742–1818) and his brother Auguste-Charles Périer
Brander, Georg Friedrich (1713–83)			Volta, Alessandro (1745–1827)	Pluvinet, Jean-Baptiste Charles (fils) (1754–1814)
Canivet, Jacques (1720–74)			Wilcke, Johan Carl (1732–96) (?)	Quinquet, Antoine (1745–1803)
Cappy (?–1775)				Wedgwood, Josiah (1730–95)
Cappy fils				
Carochez, Noël Simon (ca. 1745–1813/14)				
Cartier				
Casbois, Nicolas				
Chabrol de Murol				
Charpentier, François-Philippe (1734–1819)				
Chemin, Nicolas				
Chevalier, Charles				
Clair				
Dellebarre, Louis-François (1726–1805)				
Dinon				
Dumotiez, Pierre François and Louis Joseph (fl. 1780–1815)				
Fahrenheit, Daniel Gabriel (1686–1736?)				
Fortin, Jean Nicolas (1750–1831)				
Fourché, C. (?–1810)				
Gaibert				

(continued)

Table A1. *(continued)*

Instrument makers, engineers	Laboratory artisans	Laboratory assistants	Academic scientists	Entrepreneurs, traders, apothecaries
Gallonde, Louis Charles (1715–71)				
Goubert, Jean Pierre				
Hanin (?–after 1810)				
Janety, Marc Etienne (ca. 1750–ca. 1823)				
Langlois, Claude (1703–56)				
Le Gaux				
Le Maire				
Lennel, Louis-Pierre Florimond (?–ca. 1784)				
Lenoir, Etienne (1744–1830)				
Lepaute, Jean-André (1720–89)				
Magny, Alexis (1712–ca. 1777)				
Mégnié, Pierre Bernard (1751–1807)				
Meusnier de la Place, Jean-Baptiste (1754–93)				
Mossy				
Moth				
Mudge, John (1721–93)				
Nalain				
Nairne, Edward (1726–1806)				
Noël, Nicolas (1712–83)				
Passemant, Claude Paris (1703–63) and Claude-Siméon (1702–69)				
Ramsden, Jesse (1735–1800)				
Richer, Jean François (1743–after 1800)				
Scanegati (or Scanegatty)				

[a] Trader of instruments.
[b] Supplier of chemicals.

An Empire's Extract:
Chemical Manipulations of Cinchona Bark in the Eighteenth-Century Spanish Atlantic World

*by Matthew James Crawford**

ABSTRACT

In 1790, the Spanish Crown sent a "botanist-chemist" to South America to implement production of a chemical extract made from cinchona bark, a botanical medicament from the Andes used throughout the Atlantic World to treat malarial fevers. Even though the botanist-chemist's efforts to produce the extract failed, this episode offers important insight into the role of chemistry in the early modern Atlantic World. Well before the Spanish Crown tried to make it a tool of empire, chemistry provided a vital set of techniques that circulated among a variety of healers, who used such techniques to make botanical medicaments useful and intelligible in new ways.

INTRODUCTION

When Vicente Olmedo arrived in Cádiz in November 1790, he was just one of many officials to pass through Spain's primary Atlantic port on their way to serve the Crown in Spanish America. Yet, Olmedo's appointment was no ordinary one. He was headed for Loja, the southernmost province of the *Audiencia* of Quito, to serve as codirector of an *estanco* (royal reserve) of the trees that produced Loja's most important botanical export: a medicinal tree bark known as *quina*. Healers throughout the eighteenth-century Atlantic World recognized quina as the most effective treatment for intermittent fevers, a prevalent cluster of ailments now associated with the blood-borne disease malaria.[1] As codirector of the royal reserve, Olmedo's primary responsibility was to prepare annual shipments of quina for the Royal Pharmacy in Madrid. While cinchona trees, which gave rise to quina (cinchona bark), were found throughout the Andean forests of South America, quina from Loja was reputed to be

* Department of History, 305 Bowman Hall, Kent State University, Kent, OH 44242; mcrawf11@kent.edu.

Research support for this essay was provided by a Theodore and Mary Herdegen Fellowship in the History of Scientific Information from the Chemical Heritage Foundation. I would also like to acknowledge Dr. Evan Ragland and the anonymous reviewers for their helpful comments and suggestions on earlier drafts of this essay.

[1] J. R. McNeill, *Mosquito Empires: Ecology and War in the Greater Caribbean, 1620–1914* (New York, 2010), 52–7; James L. A. Webb Jr., *Humanity's Burden: A Global History of Malaria* (Cambridge, 2009), 66–91; Saul Jarcho, *Quinine's Predecessor: Francesco Torti and the Early History of Cinchona* (Baltimore, 1993).

of superior quality. In 1751, the Spanish Crown established the *estanco* in order to assert its right to monopolize the very best quina from Loja.[2]

Olmedo's appointment as codirector of this royal reserve is noteworthy not only because of the high-profile product involved but also because of the title and charge of his position. His official title was *Botanico Chimico* (botanist-chemist)—a neologism coined by royal advisers in Madrid.[3] The Crown's original instructions to Olmedo indicated that his duties as a botanist consisted of "harvesting" and "drying" the bark, while "taking all necessary precautions so that [it] does not lose its virtue and good quality."[4] Olmedo learned of his duties as a chemist only after he was on his way to South America. While waiting in Cádiz, Olmedo received a letter from Madrid instructing him to implement production of quina extract in Loja. Olmedo was told that it was "a matter of great interest to His Majesty."[5]

Olmedo's appointment departed from past practice in two ways. First, for much of its existence, the royal reserve had been administered by Creole elites from Loja. In contrast, both Olmedo and the new *corregidor*, with whom he shared directorship of the royal reserve, were appointees from Spain, reflecting an effort by the Crown to reduce local control over quina. Second, whereas previous directors of the *estanco* were career bureaucrats in the colonial government, Olmedo was awarded his position based on his training in botany and chemistry. His appointment was part of a broader effort to integrate science into Spain's imperial enterprise.[6] Since he was responsible for implementing imperial policy as well as serving as an adviser to the Crown, Olmedo was quite literally an "agent of empire."[7]

It should come as no surprise that the Spanish Crown sent a representative of two key sciences of the Enlightenment to oversee this vital imperial project in the Andes. After all, in the eighteenth century, many European rulers sent scores of scientifically minded young men abroad to study and report on the peoples, places, and things found around the world. Such efforts were integral to early modern European commercial and colonial enterprises, including Spain's efforts to know and exploit the natural and human resources of the Americas.[8] In the past few decades, scholarly fo-

---

[2] "Ordén del Rey" [Draft], Madrid, 27 August 1751, Archivo General de Indias (hereafter cited as AGI), Seville, Indiferente General 1552, fols. 343r–348r. For more on the *estanco* of quina, see Matthew Crawford, "'Para Desterrar las Dudas y Adulteraciones': Scientific Expertise and the Attempts to Make a Better Bark for the Royal Monopoly of Quina (1751–1790)," *J. Spanish Cult. Stud.* 8 (2007): 193–212.

[3] "Ynstruccion que han de observer el Corregidor de Loxa y el Botánico Chimico," Madrid, 26 August 1790, Archivo Nacional Histórico de Ecuador-Quito (hereafter cited as ANH/Q), Quito, Fondo Especial, vol. 278, no. 6843, fols. 247r–250r. In the *Diccionario de Autoridades*, an eighteenth-century Spanish dictionary, *chimico* is listed as a noun meaning "someone who professes the art of chemistry and the same as Alchemist"; see Real Academia Española, *Diccionario de la lengua castellana* (Madrid, 1729), 2:319. While this same dictionary lists *botanico* only as an adjective and not as a noun, other sources indicate that the term did exist as a noun for "botanist"; see Real Academia Española, *Diccionario*, 1:659.

[4] "Ynstruccion" (cit. n. 3), fol. 249v.

[5] Vicente Olmedo to Tomás Ruiz de Quevedo, Loja, 7 September 1794, ANH/Q, Quito, Fondo Especial, vol. 316, no. 7644-246, fol. 293v.

[6] Daniela Bleichmar, *Visible Empire: Botanical Expeditions and Visual Culture in the Hispanic Enlightenment* (Chicago, 2012).

[7] David Mackay, "Agents of Empire: The Banksian Collectors and Evaluation of New Lands," in *Visions of Empire: Voyages, Botany, and Representations of Nature*, ed. D. P. Miller and P. H. Reill (Cambridge, 1996), 38–57.

[8] For a recent account of the relationship between science and European expansion in the early modern period, see Harold Cook, *Matters of Exchange: Commerce, Medicine, and Science in the Dutch Golden Age* (New Haven, Conn., 2008).

cus on scientific practitioners as "agents of empire" has provided new insight into the relationship between early modern sciences and empires as part of a general effort to write a more global history of science.[9]

For all the attention to the relationship between early modern science and European expansion, chemistry has been conspicuously absent from the story, even as it has figured prominently in narratives of European science during the Enlightenment.[10] While the story of European botany as a colonial science is well known, much less is known about the roles that chemistry played in European commercial and imperial expansion before 1800.[11] Even the phrase "colonial chemistry" seems a bit awkward in comparison to the more commonplace phrase: "colonial botany." In part, this emphasis in historical scholarship reflects the more prominent role that natural history played in early modern European imperial enterprises. When European academies and states assembled a scientific expedition, they rarely, if ever, included chemical practitioners in such endeavors. For example, while the Spanish Crown supported several Royal Botanical Expeditions in the eighteenth century, there was no Royal Chemical Expedition. It is not that surprising since in early modern Europe, botany was an enterprise that required trips into the field and to markets to collect specimens, while early modern chemistry was an enterprise often performed indoors, in homes and laboratories, even as chemists traveled around Europe to share ideas and techniques.[12] Moreover, the expensive and delicate equipment required to do early modern chemistry did not travel that well. While botanists were not entirely unencumbered in their travels, the botanical library proved more portable than the chemical laboratory.[13]

The appointment of a botanist-chemist to Spain's royal reserve in Loja is striking, then, for its rarity. While many early modern rulers employed chemists at their courts in Europe, they rarely sent chemists overseas.[14] While some recent studies have illu-

[9] Sujit Sivasundaram, ed., "Focus: Global Histories of Science," *Isis* 101 (2010): 95–132; Daniela Bleichmar et al., eds., *Science in the Spanish and Portuguese Empires, 1500–1800* (Stanford, Calif., 2009); James Delbourgo and Nicholas Dew, eds., *Science and Empire in the Atlantic World* (New York, 2008).

[10] Lawrence M. Principe, ed., *New Narratives in Eighteenth-Century Chemistry: Contributions from the First Francis Bacon Workshop* (Dordrecht, 2010); Mi Gyung Kim, *Affinity, That Elusive Dream: A Genealogy of the Chemical Revolution* (Cambridge, Mass., 2003); Marco Beretta, *The Enlightenment of Matter: The Definition of Chemistry from Agricola to Lavoisier* (Canton, Mass., 1993); Frederic L. Holmes, *Eighteenth-Century Chemistry as an Investigative Enterprise* (Berkeley, Calif., 1989). For clarity, I will use the term "chemistry" as an umbrella term for the various ideas and practices associated with the study and manipulations of physical materials in the eighteenth century. Readers should be aware that eighteenth-century "chemistry" was a distinctive enterprise from today's scientific discipline. For more on this important issue, see Lawrence Principe and William R. Newman. "Alchemy vs. Chemistry: The Etymological Origins of a Historiographic Mistake," *Early Sci. & Med.* 3 (1998): 32–65.

[11] Bleichmar et al., *Science* (cit. n. 9); Antonio Barrera-Osorio, *Experiencing Nature: The Spanish American Empire and the Early Scientific Revolution* (Austin, Tex., 2006); Londa Schiebinger and Claudia Swan, eds., *Colonial Botany: Science, Commerce, and Politics in the Early Modern World* (Philadelphia, 2005); Schiebinger, *Plants and Empire: Colonial Bioprospecting in the Atlantic World* (Cambridge, Mass., 2004); John Gascoigne, *Science in the Service of Empire: Joseph Banks, the British State and the Uses of Science in the Age of Revolution* (Cambridge, 1998); Patrick Petitjean, Catherine Jami, and Anne Marie Moulin, eds., *Science and Empires: Historical Studies about Scientific Development and European Expansion* (Dordrecht, 1992).

[12] Pamela Smith, "Laboratories," in *The Cambridge History of Science*, vol. 3, *Early Sciences*, ed. Katherine Park and Lorraine Daston (Cambridge, 2006), 290–305.

[13] Daniela Bleichmar, "Exploration in Print: Books and Botanical Travel from Spain to the Americas in the Late Eighteenth Century," *Huntington Libr. Quart.* 70 (2007): 129–51.

[14] Tara Nummedal, *Alchemy and Authority in the Holy Roman Empire* (Chicago, 2007).

minated chemical practices in colonial contexts of the early modern period, few cases of a chemist directly employed by an early modern European colonial government have come to light. It is significant that "chemist" featured so prominently in Olmedo's title, even if the term was paired with "botanist," a familiar agent of empire in Enlightenment Europe.

This essay uses the case of the Crown's botanist-chemist to discuss the role of chemistry, especially chemical medicine, in the late eighteenth-century Spanish Atlantic World. As recent scholarship has shown, early modern medicine, including pharmacy, made much use of chemical ideas and techniques.[15] In this case, an emphasis on the chemical manipulations of an American medicament illuminates the geography of chemistry beyond Europe in the wider Atlantic World. While there may be a temptation to see the Crown's incorporation of a botanist-chemist into its royal monopoly of quina as anticipatory of the nineteenth century, when chemistry became a full-blown imperial science, this essay argues that this case is best understood with reference to the geography of chemical medicine in the eighteenth-century Atlantic World.[16] This effort to integrate chemistry into the Spanish imperial enterprise was largely unsuccessful. While Olmedo's inability to acquire the equipment needed to produce a quina extract offers a partial explanation for this failure, this essay argues that the real failure was the Crown's inability to recognize and take advantage of a robust tradition of chemical manipulation of quina that pervaded the Atlantic World well before Olmedo's arrival in Loja in 1791.

FROM POWDERS AND INFUSIONS TO TINCTURES AND EXTRACTS

Vicente Olmedo's main duty as imperial chemist was the production of quina extract. Yet, by 1791, when Olmedo departed Madrid, the production of quina extract was already well established throughout the Spanish Atlantic World. Thus, the extract preceded the chemist sent to implement it.

Available evidence strongly suggests that Andean *curanderos* (indigenous healers) used cinchona bark medicinally prior to the arrival of Europeans.[17] After learning about the bark from these *curanderos*, Spanish officials and Jesuit missionaries returning from Peru introduced quina to Europe in the 1630s and 1640s via Seville and Rome. Physicians and pharmacists in Italy, especially near the Jesuit headquarters in Rome, were some of the first in Europe to use quina therapeutically on patients. At the Jesuits' *Colegio Romano*, two pharmacists developed a method for administering the bark in an infusion, which they described in a printed pamphlet known as the *Schedula Romana*. According to a late seventeenth-century manuscript, the Jesuit method involved pouring "two drachms of [powdered bark], very finely ground, and passed through a flour-sieve" into "a flask of strong wine."[18] This infusion was

[15] Ursula Klein and Wolfgang Lefèvre, *Materials in Eighteenth-Century Science: A Historical Ontology* (Cambridge, Mass., 2007); Jonathan Simon, *Chemistry, Pharmacy and Revolution in France, 1777–1809* (Aldershot, 2005).

[16] Daniel R. Headrick, "Botany, Chemistry, and Tropical Development," *J. World Hist.* 7 (1996): 1–20.

[17] Eduardo Estrella, "Ciencia ilustrada y saber popular en el conocimiento de la quina en el siglo XVIII," in *Saberes Andinos: Ciencia y tecnología en Bolivia, Ecuador, y Perú*, ed. Marcos Cueto (Lima, 1995), 37–57.

[18] *Tractatus Simplex de Cortice Peruuiano/A Plain Treatise on the Peruvian Bark ("The Stanizt Manuscript")*, trans. Saul Jarcho (Boston, 1992), 37.

to be administrated "when there were signs of cold, shaking, and rigor," the signs of
the chills that alternated with bouts of fever in intermittent fevers.[19] It is likely that
administering quina in an infusion made the bitter bark easier for patients to ingest.

While some Jesuit pharmacists probably developed their therapeutic method out
of practical experience with patients in hospitals, additional evidence suggests that
other European pharmacists may have derived their method from the practices of
Andean healers. The two earliest accounts of American materia medica that were
printed in Europe described a similar type of infusion used by these healers to administer an unnamed medicinal bark that Fernando Crespo Ortiz has convincingly identified as quina.[20] These descriptions appeared in a work by Nicolas Monardes, a physician in Seville, published in 1571, and another work by Juan Fragoso, physician to
Philip II, published in 1572.[21] Consequently, printed descriptions of quina preceded
the introduction of the bark to Europe by more than a half century. After providing
a physical description of the bark, both Monardes and Fragoso explained in similar
terms how "Indians" administered it therapeutically. For example, Monardes noted
that the Indians took a quantity of bark equal to "a small broad bean and made into
powders" and mixed it "in red wine or in water" in order to treat "the fever or *mal*"
and also diarrhea.[22] The similarities between the Jesuit method from *Schedula Romana* and the methods described by Monardes and Fragoso in the 1570s suggest that
healers on both sides of the Atlantic were making infusions from the bark in the early
seventeenth century.

While infusions remained common, European pharmacists began developing other
chemical methods for manipulating quina in the late seventeenth century.[23] The impetus for developing these new methods derived from various controversies over
quina in the 1650s and the 1660s regarding its medical virtues and therapeutic use.[24]
English and French pharmacists took the lead in developing new ways of using the
bark in response to rising skepticism about the bark's medical utility. By the 1670s
and 1680s, they began developing the method for producing an extract from quina.
An anonymous author in the late seventeenth century wryly commented on the diversity of therapeutic techniques that were emerging: "nowadays [the bark] is taken with
various kinds of ostentation and elaboration, sometimes in the form of a tincture, an
extract, a decoction, and the like."[25]

One way to track the development of these new techniques is through the various
editions of Nicolas Lémery's popular chemical textbook, *Cours de Chymie*, that fo-

[19] Ibid., 35–7. In the late seventeenth century, European physicians and pharmacists classified the
bark as a "hot" medicament on account of its bitter taste. As a result, the humoral theory associated
with Galenic medicine recommended administration of the bark while the patient was experiencing
chills rather than fevers, since the therapeutic goal, according to Galenic humoralism, was to restore
balance in the body between the four humors and their associated properties of hot, cold, wet, and dry.

[20] Fernando Crespo Ortiz, "Fragoso, Monardes and pre-Cinchona Knowledge of Cinchona," *Arch.
Natur. Hist.* 22 (1995): 169–81.

[21] Nicolas Monardes, *Segunda parte del libro de las cosas que se traen nuestras Indias Occiden-
tales, que sirven al uso de medicina* (Seville, 1571); Juan Fragoso, *Discursos de las cosas Aromaticas,
arboles y frutales, y de otras muchas medicinas simples que se traen de la India Oriental, y sirven al
uso de medicina* (Madrid, 1572).

[22] Monardes, *Segunda parte* (cit. n. 21), 117.

[23] Samir Boumediene, "From Tree-Bark to Medicinal Product: The Political Processing of Peruvian
Bark (1640s–1750s)" (unpublished manuscript, 2011). This section owes much to Boumediene's
excellent summary of this complex period in the early history of cinchona bark.

[24] Jarcho, *Quinine's Predecessor* (cit. n. 1), 44–58.

[25] *Tractatus Simplex* (cit. n. 18), 37.

cused on the most common sources of materia medica from animals, plants, and minerals.[26] While the first (self-published) edition made no mention of quina in 1675, Lémery added a chapter on "Quinquina" to the fourth edition of 1681.[27] In the next edition, printed in 1683, Lémery expanded his entry on "Quinquina" to include sections on how to make a "Teinture du Quinquina" and an "Extrait du Quinquina."[28] These additions to the 1683 edition are testament to the increasing prevalence of these techniques, especially since Lémery's book focused on the most common medicaments.

Lémery's descriptions of the "tincture" and the "extract" highlight their differences from an infusion. Whereas the "Tincture of Quina" was "an extraction of the more oily and separable parts of the *Bark* by Spirit of wine," the "Extract of *Quina*" was "a separation of the more substantial parts of the Bark" using "*l'eau du noix.*" While both substances retained the bark's medical virtues, the main difference was the medium used to make them. He explained that quina extract could only be made with *l'eau du noix* and not with "wine or Spirit of wine" because "in the evaporation the Spirit [of wine] carries away with it the more subtle parts of the mixt"—resulting in the extract losing its medical efficacy. Also, whereas the "tincture" was a liquid, the extract was to have the "consistency of thick honey," one of the distinguishing features of extracts.[29] With its inclusion in Lémery's popular textbook and the medical works of other luminaries such as Herman Boerhaave in the early eighteenth century, the extract of quina became an established part of the emerging Atlanticized pharmacopoeia in Europe.

A CHEMICAL TECHNIQUE IN THE SPANISH ATLANTIC WORLD

In 1703, Felix Palacios, Pharmacist of the Court in Madrid, published a Spanish translation of the *Cours de Chymie*, making Lémery's technique for producing the quina extract more accessible to chemists and pharmacists in Spain and its American territories.[30] Evidence that such texts crossed the Atlantic is sparse but suggestive. Using a series of pharmacy inventories from New Spain, Paula De Vos has found that copies of the Spanish translation of Lémery's book appear on the inventories of several pharmacies dating from 1795 to 1818.[31] In addition, anecdotal evidence shows the mobility of Lémery's other works in the eighteenth-century Spanish Atlantic World.[32]

Those who were unable to get a copy of Lémery's work could learn about the extract from Palacios's own pharmaceutical handbook. In 1706, Palacios published the

---

[26] Michel Bougard, *La chimie de Nicolas Lemery* (Turnhout, 1999).

[27] Nicolas Lémery, *Cours de chymie contenant la manière de faire les operations qui sont en usage dans la medicine, par une methode facile* (Paris, 1675); Lémery, *Cours de chymie*, 4th ed. (Paris, 1681).

[28] Nicholas Lémery, *Cours de chymie*, 5th ed. (Paris, 1683), 419–22.

[29] Nicolas Lémery, *A Course of Chymistry: containing an easie method of preparing those chymical medicins which are used in physick*, trans. Walter Harris (London, 1698), 395–7. This edition is the third English edition and the first English translation to appear after Lémery added his entry on quina.

[30] Nicolas Lémery, *Curso chymico del Doctor Nicolas Lemery*, trans. Felix Palacios (Madrid, 1703).

[31] Paula De Vos, "From Herbs to Alchemy: The Introduction of Chemical Medicine to Mexican Pharmacies in the Seventeenth and Eighteenth Centuries," *J. Spanish Cult. Stud.* 8 (2007): 135–68, esp. table 6.

[32] Charles Walker observes that José Llano Zapata, a Creole polymath in Lima, referenced the work of Lémery in his musings on the causes of earthquakes; see Walker, *Shaky Colonialism: The 1746 Earthquake-Tsunami in Lima, Peru, and Its Long Aftermath* (Durham, N.C., 2008), 22.

first of eight editions of his *Palestra Pharmaceutica* that would be released in the eighteenth century.[33] His work also served as the foundation for the first edition of the *Pharmacopoeia Matritensis*.[34] In response to a debate among Spanish physicians over chemical medicine, Palacios embraced a "chemico-Galenic compromise," devoting two sections of his book to each kind of pharmacy: Galenic and chemical.[35]

In his first chapter, "On Pharmacy in general," Palacios defined "Galenic Pharmacy" as "that which only teaches the simple collection and mixture of natural bodies, without speculating on the particles [*particulas*] or substances [out of which] they are composed." He went on to define "Chemical Pharmacy" as "that which teaches us and provides the foundations for knowing and speculating about the parts and substances [out of which] natural bodies are composed."[36] Such definitions characterized both approaches to pharmacy as complementary. While Galenic pharmacy provided the knowledge and skills needed to identify, harvest, and process medicinal plants and other simples, chemical pharmacy provided the knowledge and skills to make more refined medicaments from these materials.

Palacios further defined chemical medicine as "the separation and resolution of the pure parts from the impure and crass parts and how to make more exalted and essential medicaments." Extraction, defined as the "separation of the noble and pure parts from the ignoble and impure," was a key technique.[37] Like his contemporaries in Europe, Palacios understood extraction as a process that, on the one hand, offered deeper knowledge of the medical virtues of plant and animal simples. He also understood it as a process of purification that enhanced the potency and efficacy of medicaments. A desire for both knowledge and purity informed the Crown's decision to send a botanist-chemist to Loja to oversee production of cinchona bark and the quina extract.

Palacios also provided recipes for making extracts from a variety of plants, including one recipe entitled "Extractum antifebrile, Phar. Bateana." It was a treatment for intermittent fevers and was made from a mixture of "the bark of Quinaquinna [*sic*], Gentian root, snakeroot, and the flowers of common centaury." These ingredients were to be infused in "spirit of Wine" and "digested" over "slow heat" in a "well-sealed flask" for two days. After pouring this "tincture" through a strainer and washing the "residue" with more "spirit of Wine," the mixture was to be "digested" and "strained" again. The resulting "tincture" was to be heated in a water bath until it achieved the "consistency of an extract."[38] Through such descriptions, Palacios provided his readers with an account of the basic methods for making plant extracts, including the quina extract. In 1739, the first official pharmacopoeia of the Spanish Empire included Palacios's instructions on how to make an extract from quina.[39]

In addition to pharmaceutical texts, healers in eighteenth-century Spanish America could also learn about extraction and the quina extract from informal interactions with European travelers who visited Spanish America from Europe.[40] While records

[33] Felix Palacios, *Palestra Pharmaceutica, Chymico-Galenica* (Madrid, 1706).
[34] *Pharmacopoeia Matritensis* (Madrid, 1739).
[35] De Vos, "From Herbs to Alchemy" (cit. n. 31), 153.
[36] Palacios, *Palestra* (cit. n. 33), 1.
[37] Ibid., 1 and 32.
[38] Ibid., 400 and 419.
[39] *Pharmacopoeia Matritensis* (cit. n. 34), 188.
[40] E.g., John Gray, William Arrot, and Philip Miller, "An Account of the Peruvian or Jesuits Bark, by Mr. John Gray, F.R.S., now at Cartagena in the Spanish West-Indies; extracted from the Papers given

indicate that such encounters occurred, details on the nature of these encounters are sparse. With regard to the transfer of knowledge of the quina extract, some of the best records come from a joint French and Spanish scientific expedition to the *Audiencia* of Quito in 1735 sponsored by the Royal Academy of Sciences in Paris. The main objective of this expedition was to take measurements at the equator in order to determine the shape of the earth.[41] According to published accounts, Joseph de Jussieu, a member of the expedition who was a physician and naturalist from one of the most important scientific families in eighteenth-century France, visited Loja in 1739 to teach the *corregidor* of Loja and "[the Indians, who worked as bark collectors] the method of making the extract" in addition to teaching them how "to know and distinguish" the different species of cinchona trees according to the methods of European botany.[42]

While European published accounts characterize this encounter as one-way transmissions of knowledge from Jussieu to the "Indians," such interactions are better characterized as exchanges. Before he could teach the Indians anything, Jussieu had to learn from them how and where to find the trees in the first place. Existing evidence suggests that Jussieu worked with Fernando de la Vega, a *curandero*, bark collector, and merchant in Loja, who also assisted Charles Marie de la Condamine during his visits to Loja.[43] Vega was a valuable collaborator, who even took part in the production and testing of the quina extract.[44] In 1752, at the age of eighty, Vega wrote a short report on quina for an official sent to Loja by the Viceroy of New Granada to collect information about the quina trade. In this report, Vega revealed that he used a number of different methods to process the bark, including several different "infusions" of the powdered bark, a tradition among Andean *curanderos*, and also an "extract" and a "salt" of quina.[45] Unfortunately, Vega did not indicate how he learned to make the extract. Nonetheless, his report provides further evidence that, by at least the 1740s, the quina extract was being produced in Loja. When Charles Marie de la Condamine visited Loja in 1743, Vega provided samples of his own quina extract and quina salt to the French explorer.[46]

Additional archival evidence confirms that the quina extract was an established practice in the *Audiencia* of Quito well before the arrival of the Crown's botanist-chemist in 1791. In a report on the quina trade sent to the Viceroy of New Granada in 1776, a bark merchant from Cuenca suggested that quina should no longer be transported as whole bark but as an "extract in paste or in salts" because the extract offered "the convenience of reducing [the bark] to its most noble and spirituous material," a phrase that echoes the language of contemporary chemists and pharmacists in Europe.[47] Similarly, in a 1783 letter to the President of Quito, another bark merchant

---

him by Mr. William Arrot, a Scotch Surgeon, who had gather'd it at the Place where it grows in Peru. Communicated by Mr. Phil. Miller, F.R.S. &c.," *Phil. Trans. Royal Soc. London* 446 (1737): 81–6.

[41] Neil Safier, *Measuring the World: Enlightenment Science and South America* (Chicago, 2008).

[42] Jorge Juan and Antonio Ulloa, *Relacion Historica del Viage a la America Meridional*, 4 vols. (Madrid, 1748), 2:440; see also Charles Marie de la Condamine, *Journal du voyage fait par ordre du roi, a l'équateur* (Paris, 1751), 186.

[43] La Condamine, *Journal* (cit. n. 42), 185–6.

[44] Estrella, "Ciencia ilustrada" (cit. n. 17), 44.

[45] Fernando de la Vega, "Virtudes de la cascarilla, de hojas, cogollos, corteza, polvos y corteza de la raíz," 1752, appendix to: Estrella, "Ciencia ilustrada" (cit. n. 17), 56–7.

[46] La Condamine, *Journal* (cit. n. 42), 185–6.

[47] Marquis de Villa Orellana, "Informe" [copy], Quito, 18 August 1776, ANH/Q, Quito, Cascarilla, box 1, exp. 11, fols. 43r–53r.

from Cuenca, Miguel Perfecto de San Andres, who was seeking a contract as a supplier of quina for the Royal Pharmacy in Madrid, not only mentioned performing "experiments" with quina "salts and extracts" but also proposed using the extract to "supply the Royal Pharmacies."[48]

This cluster of sources illuminates several key features of the early history and historical geography of the quina extract in the eighteenth-century Atlantic World. First, they show that knowledge of making plant extracts was present on both sides of the Atlantic by the mid-eighteenth century and that the circulation of texts and travelers facilitated the spread of this knowledge. Second, published travel accounts and archival evidence show that merchants and bark collectors in Loja and other regions of the *Audiencia* of Quito had been producing quina extract for decades before the Crown sent Vicente Olmedo to Loja. The informal circulation of medical knowledge and chemical practices in this case conforms generally to the contingent itineraries of knowledge highlighted in recent scholarship on science and empire in the Atlantic World.[49] Before the Spanish Crown attempted to control the making of the quina extract, healers on both sides of the Atlantic had already embraced this chemical technique for manipulating the bark.

AN EMPIRE'S EXTRACT

If quina extract was already being produced in Loja, then why did the Crown send a botanist-chemist to implement production in 1791? Both practical and scientific interests played a role. One concern was a looming bark shortage. Starting in the 1770s, reports from several quina-producing regions of Quito noted that trees were becoming increasingly scarce, making it difficult to find bark of sufficient quality for royal purposes.[50] In this context, it became imperative to ensure that all quina reached Madrid in good condition. Tree bark did not travel well. Bark collectors, merchants, physicians, and pharmacists understood that it was essential to protect the bark from humidity in order to preserve its medical efficacy. To this end, many in the quina trade did the best they could using cloth, leather, and wood to construct vessels that would keep the bark dry during its long journey through rain forests, across rivers, and across the Atlantic Ocean.[51] By the late eighteenth century, these efforts had reached their technological limit.

In 1789, the Minister of the Indies asked the Chamberlain of the Royal Household to review a massive dossier on quina and provide recommendations on how to improve the royal monopoly.[52] The Chamberlain was the first to recommend sending a "botanist-chemist" to serve as codirector of the royal reserve in Loja. He may have gotten the idea from Hipólito Ruiz and José Pavón, two Spanish botanists with first-

[48] Miguel Perfecto de San Andres to José García de Leon, Cuenca, 11 November 1783, ANH/Q, Quito, Fondo Especial, vol. 207, no. 142, fol. 167r.

[49] Neil Safier, "Global Knowledge on the Move: Itineraries, Amerindian Narratives, and Deep Histories of Science," *Isis* 101 (2010): 133–45; Delbourgo and Dew, *Science and Empire* (cit. n. 9).

[50] Matthew Crawford, "A 'Reasoned Proposal' against 'Vain Science': Creole Negotiations of an Atlantic Medicament in the Audiencia of Quito (1776–1792)," *Atl. Stud.* 7 (2010): 397–419.

[51] María Luisa de Andrés Turrión and Maria Rosario Terreros Gómez, "Organización administrativa del Ramo de la Quina para la Real Hacienda española en el virreinato de Nueva Granada," in *Medicina y Quina en la España del Siglo XVIII*, ed. Juan Riera Palmero (Salamanca, 1997), 37–43.

[52] Antonio Porlier to Marques de Valdecarzana [Draft], Madrid, 27 July 1789, AGI, Seville, Indiferente General 1555, fols. 285r–296r.

hand experience of the cinchona tree. Ruiz and Pavón had just returned to Madrid in 1788 after a decade of botanizing in South America as codirectors of the Royal Botanical Expedition to Peru and Chile. During their time in South America, they had ample opportunities to study the tree in situ and even contributed to the development of the quina industry in Huánuco in northern Peru.

Ruiz probably gave the Chamberlain a manuscript copy of his new study of the cinchona tree entitled *Quinologia*.[53] In this work, Ruiz devoted an entire chapter to describing how he and Pavón, acting on orders from Casimiro Gómez Ortega, Director of the Royal Botanical Garden in Madrid, successfully introduced the quina extract to Huánuco.[54] In the early months of 1779, after studying several species of cinchona trees found in the region, Ruiz and Pavón began to produce "Extract from fresh barks."[55] The project was a success, according to Ruiz. "Following our example," he wrote, "the natives of this Country have produced [the extract] in great abundance." He reported that "more than 40,000 pounds" of extract were exported to Europe "on various occasions." He also predicted that demand for the quina extract would only increase as the "news of its efficacy" spread. Ruiz supported his predictions with favorable reports of its medical utility from correspondents in Lima, Mexico City, and even "several knowledgeable and circumspect physicians and surgeons" in Spain. Ruiz argued that this evidence challenged European perceptions that quina extract was therapeutically "inferior to the powders of the same bark."[56] The problem, according to Ruiz, was that European pharmacists and physicians lacked sufficient experience with extracts from fresh bark, since the extracts that they made were from bark that had taken months (and sometimes years) to reach Europe.

Ruiz's works also provided the Chamberlain and the Crown with several arguments for the extract's practical utility. First, he argued the extract was not as susceptible to physical degradation as whole bark, the main form in which quina circulated as a commodity in the Atlantic World. As evidence of the extract's durability, Ruiz recounted the experiences of Antoine Laurent de Jussieu, the nephew of Joseph de Jussieu, the French physician and traveler who had visited Loja in 1739. According to results published in the *Mémoires* of the Royal Society of Medicine in Paris, the younger Jussieu had come into possession of some of his uncle's quina extract and found that, even after forty years, the extract still produced "good and marvelous effects" when used therapeutically.[57] It was the perfect example of the extract's durability relative to the bark. In an unpublished work on the quina trade, Ruiz observed,

[53] Hipólito Ruiz, *Quinologia, o Tratado del Arbol de la Quina* (Madrid, 1792). Although Ruiz did not publish this work until 1792, after Olmedo was already in Loja, he did circulate a manuscript copy of the work in Madrid as early as 1789.

[54] [Casimiro Gómez Ortega], "Instrucción a que deberán arreglarse los sugetos destinados por S. M. para pasar a la América meridional en compañía del Médico Don Josef Dombey a fin de reconocer las plantas, y yerbas y de hacer observaciones botánicas en aquellso países," 1776, app. IV in Hipólito Ruiz, *Relación del Viaje*, ed. R. P. A. J. Barreiro, O.S.A. (Madrid, 1931), 365–74; see also Ruiz, *Quinologia* (cit. n. 53), 48.

[55] Ruiz, *Quinologia* (cit. n. 53), 46. After working for several years as an assistant in the Royal Pharmacy at the Royal Palace in Madrid and at the pharmacy of the Royal Palace in San Ildefonso, José Pavón was well acquainted with the chemical manipulation of plants.

[56] Ibid., 46, 48, and 50.

[57] Ibid., 46–8.

"well made extracts retain and conserve the qualities and virtues of the simples, from which they are made, for many years without alteration."[58]

Increased durability relative to the bark was just one of several benefits, according to Ruiz. Another benefit was that the extract offered a "more constant and certain dosage" than using pulverized bark. Ruiz also argued that use of the extract promised "greater exploitation of all [cinchona] bark." While he estimated that "two thirds of *Cascarilla* bark" was "not admitted to commerce" on account of its poor quality at the time of harvesting, Ruiz explained that even inferior cinchona bark could produce an extract of "equal virtue" to high-quality bark. "Water," he wrote, "indifferently takes the extracted part from all [kinds of cinchona bark]."[59] In practical terms, this meant that the process of extraction could render inferior bark, which previously would have been discarded, useful to commerce and to the royal monopoly.

While some argued for its practical benefits, the extract also took on new scientific importance in the second half of the eighteenth century as European chemists became more interested in understanding the "proximate principles of plants" (substances produced by or from plants) rather than in identifying the "ultimate principles" out of which all plants were composed.[60] In this context, chemists and pharmacists, mostly in France, applied themselves anew to the chemical analysis of medicinal plants in order to identify the source of their medical "virtues." Much of this work took place in Paris and at the Royal Academy of Sciences, an epicenter of the chemical analysis of plants since at least the late seventeenth century.[61] Throughout the eighteenth century, Spanish botanists and pharmacists remained engaged with the work of their counterparts in France.

In May 1791, Casimiro Gómez Ortega, Director of the Royal Botanical Garden in Madrid, wrote to the Minister of the Indies suggesting that Vicente Olmedo could make an important contribution to ongoing French efforts to identify the substance that gave quina its medicinal properties. He cited a recent article by French chemist Antoine Fourcroy published in the *Annales de Chimie*. In this article, Fourcroy described a new method for the chemical analysis of plants using a kind of quina from Saint Domingue.[62] Gómez Ortega emphasized that the French chemist had stressed "the importance of repeating the analysis with larger quantities of fresh or recently harvested [quina] in order to obtain more perceptible and certain results."[63] At the time, no one with the appropriate training in the chemical analysis of plants was better positioned to realize Fourcroy's proposal than the Crown's botanist-chemist in Loja. It was a golden opportunity for Olmedo to make a significant contribution to the chemical understanding of the Atlantic World's most important medicament. If successful, Olmedo's findings would have bolstered the Crown's broader effort

[58] Hipólito Ruiz, *Compendio Historico-Medico Comercial de las Quinas*, ed. Eduardo Estrella (Burgos, 1992), 106.

[59] Ruiz, *Quinologia* (cit. n. 53), 46–8.

[60] Klein and Lefèvre, *Materials Science* (cit. n. 15), 211–54.

[61] Lawrence M. Principe, "A Revolution Nobody Noticed? Changes in Early Eighteenth-Century Chymistry," in Principe, *New Narratives* (cit. n. 10), 1–22.

[62] Antoine Fourcroy, "Analyse du Quinquina de Saint-Domingue; Pour servir à celles des matières végétales sèches en general," in *Annales de Chimie, ou Recueil de Mémoires concernant la Chimie et les arts qui en dépendent, et spécialement la pharmacie* (Paris, 1791), 8:113–83, 9:7–29.

[63] Casimiro Gómez Ortega to Antonio Porlier, Marques de Bajamar, Madrid, 26 May 1791, AGI, Seville, Indiferente 1555, fols. 623r–625v.

to establish Spain's reputation as a place of science and Enlightenment as a challenge to prevailing European conceptions of Spain as a place of backwardness and superstition.[64]

A FAILED ENTERPRISE?

Ultimately, Olmedo failed in his efforts to produce quina extract in Loja. One problem was that he lacked the right technology. In a 1794 report, Olmedo noted that he was unable to produce the extract because "some important instruments" had not yet been sent from Spain.[65] Two years later, Olmedo was still waiting; as explained in a letter to the Minister of the Indies, he needed "at least two or three pans of different sizes and two or three stills of different size" made of "glass" or "tin-plated copper."[66] Instruments made out of such materials represented the state of the art in chemistry and pharmacy, especially since many practitioners thought earthenware vessels—the more traditional choice for chemical manipulation of plant parts—compromised the purity of extracts.[67] Producing a pure extract was important. Olmedo wanted to be certain that products of his chemical analysis of quina came from the bark itself and not the vessels used to perform the analysis. He also wanted a pure extract because his quina extract was to be distributed under the auspices of the Crown and the royal monopoly. Just as the original goal of the royal reserve was to acquire the best bark for the Crown, Olmedo wanted to produce the best extract possible, and for that he needed the right technology.

Even if Olmedo had had the instruments he needed, he probably would have had limited success in achieving the Crown's larger goal of making the quina industry as a whole more efficient and effective. As Ruiz suggested, the production of quina extract was one way to make even inferior bark commercially viable and therapeutically useful. Yet, this broader transformation involved a shift in the very ontology of quina from bark to extract. It was a shift that many healers in the Atlantic World, including European pharmacists, had many reasons to resist.

The first thing to recognize is that connoisseurship played a major role in the quina trade. Healers throughout the Atlantic World constituted the most prominent group of connoisseurs of quina and other materia medica. And, in most cases, they served as the gatekeepers to the real consumers of quina: those suffering from intermittent fevers. With the health of the patient and the reputation of the healer on the line, the quality of the quina really mattered. Consequently, potential consumers and users of the bark relied on a variety of tests to assess its medical efficacy. While a trial run in a therapeutic setting was the only way to be certain of any sample of cinchona bark's medical efficacy, many believed that physical characteristics of the bark provided clues as to its quality as a medicament. Although there were techniques to assess the

[64] On Spain and Spanish America in the historiography of the Enlightenment, see Jorge Cañizares-Esguerra, *How to Write the History of the New World: Histories, Epistemologies, and Identities in the Eighteenth-Century Atlantic World* (Stanford, Calif., 2001).

[65] Vicente Olmedo to Tomás Ruiz de Quevedo, Loja, 7 September 1794, ANH/Q, Quito, Fondo Especial, vol. 136, no. 7644-246, fol. 293v.

[66] Tomás Ruiz de Quevedo and Vicente Olmedo to Diego Gardoqui, Loja, 25 November 1796, ANH/Q, Quito, Fondo Especial, vol. 336, no. 8127-160, fol. 199v.

[67] Paula De Vos, "The Art of Pharmacy in Seventeenth- and Eighteenth-Century México" (PhD diss., Univ. of California, Berkeley, 2001), 192–258.

quality of an extract, the techniques for assessing whole bark were much more established in the Atlantic World.

In his *Palestra Pharmaceutica*, Felix Palacios provides some insight into how European pharmacists assessed the quality of quina and other materia medica. He defined such assessment as a vital part of the art of pharmacy. As Palacios explained, the central goal of pharmacy, whether Galenic or chemical, was "to select simples, to give each simple the preparation appropriate to its nature or disposition of parts, and to mix or join them so that they do not lose any essential particles" so as to make medicaments that "produce mild, prompt and certain effects."[68]

The process of producing effective remedies began with "election"—one of three "general operations" of pharmacy along with "preparation and mixtion." Palacios defined election as the knowledge of the "collection, desiccation, reconstitution, and durability" of medicaments.[69] With regard to "the collection and election of vegetables," Palacios instructed pharmacists to make sure to acquire their botanical materia medica from the right "regions" ("hot countries" or "cold countries") and from the "appropriate places" within these regions (forests, plains, gardens, deserts, swamps, etc.), and to collect these materials at the right time of year. Most of these recommendations required the pharmacist to be present for the collection of plant materials.

What about exotic plant materials, like quina, that European pharmacists could not collect themselves? To insure the "good election" of "simples that have already been collected," Palacios provided a list of five characteristics to consider: "substance" (a term referring to a combination of texture and weight), odor, taste, color, and thickness.[70] Pharmacists were also to make sure that exotic medicinal plants came from the regions of the globe reputed to produce the best kinds. If a merchant were offering quina from India or even New Spain, for example, a pharmacist would know that it was either a fake or poor quality because it did not come from South America.

Palacios's discussion of election provides insight into the mindset of pharmacists and other healers in the Atlantic World. Since election was central to the art of pharmacy, pharmacists would not have been too keen on purchasing their quina as an extract rather than as whole bark. The extract had none of the characteristics used for "electing" exotic plant materials.

Such practices were not simply a matter of artisanal connoisseurship. They also played a central role in countering the fraud that was prevalent in the quina trade. Bark collectors and merchants were often paid for their cinchona bark by weight. This system provided an incentive for them to add inferior cinchona barks or the barks of other trees to their shipments of good-quality cinchona bark. And that was what often happened. A discerning buyer, however, could evaluate a given quantity of quina (as whole bark) according to the characteristics outlined by Palacios and other authors. The extract had none of these characteristics, yet was just as easy to adulterate as whole bark. While election was crucial to the art of pharmacy, pharmacists also preferred to trade in whole bark as a safeguard against fraud.

The prevalence of chemical techniques for manipulating plant materials throughout the Atlantic World was yet another obstacle to making the extract the predomi-

[68] Palacios, *Palestra* (cit. n. 33), 1.
[69] Ibid., 2.
[70] Ibid., 4–5.

nant form in which quina traveled as a trade good. Although it is difficult to know exactly how prevalent the practice was, evidence suggests that making an extract of quina was well established in many regions of the Atlantic World. Therefore, most healers, including Andean *curanderos*, did not need to buy quina extract because they could make it themselves. Although some proponents of the extract, like Hipólito Ruiz, argued that extracts made from freshly harvested bark were more potent and retained the potency longer, the quina extract remained a minor medical commodity in comparison to the trade in whole bark.

CONCLUSION

In the nineteenth century, chemistry became a bona fide imperial science for many European states. In 1820, two Parisian pharmacists, building on an existing tradition of chemical analysis, succeeded in isolating and identifying quinine, the alkaloid that gave cinchona bark its febrifugal properties. After 1850, the Dutch government employed chemists to evaluate the quinine content of the barks of different cinchona trees that had been transplanted from South America to Southeast Asia. After cultivating plantations in Indonesia of the cinchona species with the highest concentrations of quinine, the Dutch acquired an effective monopoly of quinine production worldwide. At the same time, the British used cinchona bark and quinine to facilitate their imperial enterprise in Africa.[71]

In the eighteenth century, chemistry's fortunes as an imperial science in the Spanish Atlantic World were quite different. Undoubtedly, the Crown's appointment of a botanist-chemist as codirector of its royal reserve in Loja represents a novel attempt to integrate chemistry into an imperial enterprise. As an agent of empire, Olmedo served scientific and imperial objectives simultaneously. Access to the freshest bark available meant that Olmedo could have contributed significantly to the chemical understanding of quina and made a much broader contribution to knowledge of plant chemistry. In practical terms, use of the extract promised to render even inferior bark useful, to provide a more durable form in which quina could travel, and to act as a medical commodity that offered more consistent dosage and potency relative to pulverized bark. Yet Olmedo achieved neither his scientific nor his practical objectives. There is no evidence to suggest that he ever produced quina extract in Loja.

This episode, then, appears to be a case in which an eighteenth-century science failed empire. So, was eighteenth-century chemistry just poorly suited to being an imperial science? In some ways, yes. But this case is better understood as an instance of conflict between two different modes of integrating chemical practice into an imperial enterprise. On the one hand, there was the Spanish Crown's vision of chemistry as a tool of empire to foster and monopolize production of high-quality quina extract at the point of extraction in Loja. On the other hand, this vision of imperial chemistry was at odds with the informal ways in which chemical techniques had already become a part of the practices of bark collectors, merchants, and healers from the forests of the Andes to the pharmacies of Europe. Undoubtedly, Olmedo was an ineffective agent of empire because he lacked the right technology; but, at the same

---

[71] Mark Honigsbaum, *The Fever Trail: In Search of the Cure for Malaria* (New York, 2001); Daniel Headrick, *The Tools of Empire: Technology and European Imperialism in the Nineteenth Century* (New York, 1981), 58–82.

time, the existing geography of chemical medicine in the Spanish Atlantic World made a shift from bark to extract undesirable, if not unnecessary.

In the early decades of the eighteenth century, knowledge of how to make the quina extract circulated through an informal process of contingent encounters and exchanges between a variety of texts and travelers in the Atlantic World. It was a process that occurred largely outside the control of the Spanish imperial government, although *Protomedicatos* in Lima and Mexico City did attempt to regulate the medical professions in the Spanish colonies by conducting examinations, issuing licenses, and performing inspections of pharmacies and apothecary shops.[72] Second, through the informal circulation of chemical practices, a tradition of making quina extract became established in South America well before the arrival of Vicente Olmedo in Loja in 1791. Finally, healers of all sorts—from European pharmacists to Andean *curanderos*—provided a social milieu that facilitated the movement of such chemical techniques. Although such practitioners did not always contribute to theoretical innovations in chemistry or the chemical analyses of plants, they played an important role in the creation and circulation of chemical medicine well beyond Europe. Even though chemistry was ineffective as an imperial science, it certainly flourished as an Atlantic science by providing healers around the Atlantic with a cluster of techniques that made American materia medica not only intelligible in new ways but also more useful and more potent.[73] Unfortunately for Vicente Olmedo, the Spanish Crown chose to ignore rather than recognize and co-opt this robust, if informal, tradition and community engaged in the chemical manipulations of quina and many other medicinal plants of the Americas. And so, the effort to convert quina from medicinal bark into an imperial extract failed.

[72] John Tate Lanning, *The Royal Protomedicato: The Regulation of the Medical Professions in the Spanish Empire*, ed. John Jay TePaske (Durham, N.C., 1985).
[73] Marcelo Aranda et al., "The History of Atlantic Science: Collective Reflections from the 2009 Harvard Seminar on Atlantic History," *Atl. Stud.* 7 (2010): 493–512.

Elements in the Melting Pot:
Merging Chemistry, Assaying, and Natural History, Ca. 1730–60

by Hjalmar Fors*

ABSTRACT

This essay examines how the modern concept of the chemical element emerged during the eighteenth century. It traces this concept to a group of assayers, mineralogists, and chemists active at the Swedish Bureau of Mines (Bergskollegium). Driven by a deep ontological pragmatism, these "mining chemists" came to regard all inquiries into the component parts of metals as useless speculation. Instead, metals were treated as immutable species that made mineralogical taxonomy possible. Their work was a form of Enlightenment boundary work, which associated chrysopoeia and the pursuit of the components of metals with superstition and disreputable activities such as astrology.

Few stop to consider the deep reductionism that underlies the modern concept of the chemical element. Most would claim that a deep and true knowledge about the nature of a thing entails knowing the elements of its chemical composition. But why do we think this? Do we not also know that elements are not the primary stuff that makes up the universe, but rather a level of composition, signifying the smallest units of homogenous matter that can be bought off the shelf? Using these questions as a starting point, this essay examines the epistemological roots of the modern concept of the chemical element. It traces them to a largely forgotten group of early modern "mineralogical chemists" who, guided by economic concerns, combined chymistry, assaying, and natural history into an assemblage that still influences chemical matter theory.[1] It also shows how a combination of knowledge and skills were assembled into a stable bundle of facts about nature that enabled a new understanding of what constituted worthwhile investigations in chemistry and mineralogy.[2]

The essay's central claim is that, during the first half of the eighteenth century, a new conception of metals as the foundational building blocks of nature emerged.

* Office for History of Science at the Department of History of Science and Ideas, Box 629, Uppsala University, 751 26 Uppsala, Sweden; Hjalmar.Fors@idehist.uu.se.

[1] On the terms "chymistry" and "chemistry," see William Newman and Lawrence Principe, "Alchemy vs. Chemistry: The Etymological Origins of a Historiographic Mistake," *Early Sci. Med.* 3 (1998): 32–65, 38–41; William Newman, "What Have We Learned from the Recent Historiography of Alchemy?" *Isis* 102 (2011): 313–21; Lawrence Principe, "Alchemy Restored," *Isis* 102 (2011): 305–12.

[2] On assemblages and stable bundles, see Bruno Latour, *Reassembling the Social: An Introduction to Actor-Network-Theory* (Oxford, 2007), 1–2, 5.

This epistemological and theoretical innovation became the necessary foundation for the massive changes in chemistry during the second half of the century, including developments in pneumatic chemistry, the so-called Chemical Revolution, and the elevation of composition and decomposition (analysis and synthesis) as central to its pursuit.[3] The idea that metals were simple substances rather than composite objects was, obviously, also of importance for chemists' general abandonment of chryso-poeia (gold making).

To ascertain the full depth of this story, it is necessary to search outside of the alleged centers of eighteenth-century chemical theory as presented in traditional historiography. Instead, I shall focus on Europe's bustling mining regions. Few accounts of eighteenth-century chemistry have paid sufficient attention to the chemical scenes at centers of mining. This is unfortunate. Indeed, the mining arts and mining—or mineralogical—chemistry may be the most important and neglected piece of a puzzle that reveals the larger picture of how chemistry came to be so thoroughly transformed during the century. To eighteenth-century chemical observers, mining regions were not peripheral at all. The essay will focus on one such important location: the Swedish Bergslagen mining region, and its adjoining towns of Uppsala and Stockholm.[4] This area boasted a host of internationally renowned chemists. Texts by Johan Gottschalk Wallerius (1709–85), Axel Fredrik Cronstedt (1722–65), and Torbern Bergman (1735–84), for example, were quickly translated and absorbed into works of natural philosophy and university teaching curricula on the European continent and the British Isles. Indeed, central Sweden was an obligatory stopover for many a young chemist and would-be mining official seeking to perfect his art. Certain skills, such as chemical blow pipe analysis, could be learned only there.[5] As Matthew D. Eddy has remarked, although it had a profound influence, "the impact of this 'Swedish School' has remained relatively unexplored in Anglophone histories of chemistry and/or the nascent earth sciences."[6]

Two local specialities were especially influential and, as we will see, of particular importance to the epistemology of chemical composition. The first was the systematic mineralogy conducted in interaction and dialogue with the botany and zoology of Carl Linnaeus. The second was the assaying, or mineral analysis, which engendered a large number of discoveries of simple substances. As Colin Russell has ob-

---

[3] Precursors of this approach are Robert Siegfried and Betty J. T. Dobbs, "Composition: A Neglected Aspect of the Chemical Revolution," *Ann. Sci.* 24 (1968): 275–93; Theodore M. Porter, "The Promotion of Mining and the Advancement of Science: The Chemical Revolution in Mineralogy," *Ann. Sci.* 38 (1981): 543–70.

[4] For an in-depth discussion of chymistry's and chemistry's importance for early modern mining, see chaps. 1, 3, and 5 in my forthcoming book: Hjalmar Fors, *The Limits of Matter: Chemistry, Mining, and Enlightenment* (Chicago, 2015). Comparable mining regions in the Holy Roman Empire could be found in Hannover/Harz, Saxony, and Austria. It can be noted that similar approaches to those discussed here were also adopted by chemists active in the Holy Roman Empire. See ibid., chap. 5.

[5] Wolfhard Weber, *Innovationen im frühindustriellen deutschen Bergbau und Hüttenwesen: Friedrich Anton von Heynitz* (Göttingen, 1976), 47–57; Brian Dolan, "Transferring Skill: Blowpipe Analysis in Sweden and England, 1750–1850," in *Science Unbound: Geography, Space and Discipline*, ed. B. Dolan (Umeå, 1998), 91–125; Hedvig af Petersen, "Om Torbern Bergmans och C. W. Scheeles franska förbindelser," *Personhistorisk tidskrift* 29 (1928): 190–201.

[6] Matthew D. Eddy, *The Language of Mineralogy: John Walker, Chemistry and the Edinburgh Medical School, 1750–1800* (Burlington, Vt., 2008), 12, 31, 90, 126–7, on 127. See also Marco Beretta, *The Enlightenment of Matter: The Definition of Chemistry from Agricola to Lavoisier* (Canton, Mass., 1993), 89–90, 93.

served, "As late as 1886 no less than 40 per cent of the chemical elements found since the Middle Ages had been discovered in Sweden."[7] Neither of these topics has received sustained attention by historians of chemistry. Their integration into the grand narratives of how chemistry changed in the eighteenth century is, therefore, an important subject of further research for the larger narrative of early modern chemical knowledge.

MATTER AND COMPOSITION

Pure metals are uncommon in nature. Early modern miners and smelters had to engage in a complex sequence of actions in order to obtain any significant quantity of a metal. First, a miner had to extract certain stones, many of which did not look anything like the desired metal, from habitually inhospitable and dark subterranean environments. Next, miners and smelters had to perform various transformative operations on the stones, such as crushing, sifting, washing, roasting, and smelting. This process involved various machines, roasting pits, and furnaces, and, above all, fire—the foremost tool of chemical transformation.[8] For most who had empirical experience with early modern mining and smelting, it would seem a foregone conclusion that metals were manufactured through these processes, much as glass, also an apparently pure and homogenous substance, is manufactured through the application of fire to sand and vegetable ash.

Seventeenth-century theorists of matter generally assumed that metals were composite substances. Some held to the traditional medieval chymical position that they were composed of primal mercury and sulfur. Others proceeded from the Paracelsian *tria prima* of salt, mercury, and sulfur, and some, including Jan Baptista van Helmont, proposed that all substances in the world were made from, and could ultimately be reduced to, water.[9] By the early eighteenth century, the phlogiston theory of Georg Ernst Stahl was embraced widely. It held that metals were composites of, first, a calx or an earth and, second, a flammable substance or phlogiston. From a historiographic perspective, the concept that phlogiston was a component of metal is worth noting: twentieth-century historians often claimed that most eighteenth-century chemists of note prior to Antoine Laurent Lavoisier were strict adherents of the phlogiston theory. This is incorrect. Many influential eighteenth-century chemists—including most of those discussed in this essay—did not proceed from the work of Stahl but from that of Leiden professor Herman Boerhaave. A prominent opponent of Stahl's theories, Boerhaave held that calx and metal were simply two different forms of the same substance. Because he was a proponent of the theory that metals were composed of mercury and sulfur, he held that a true decomposition of metal would result not in earth but in mercury.[10]

---

[7] Colin Russell, "Science on the Fringe of Europe: Eighteenth-Century Sweden," in *The Rise of Scientific Europe 1500–1800,* ed. D. Goodman and C. Russell (London, 1991), 305–32, 323.

[8] See in particular John C. Powers, "Measuring Fire: Herman Boerhaave and the Introduction of Thermometry into Chemistry"; and Christine Lehman, "Pierre-Joseph Macquer: Chemistry in the French Enlightenment," both in this volume.

[9] Allen G. Debus, "Fire Analysis and the Elements in the Sixteenth and Seventeenth Centuries," *Ann. Sci.* 23 (1967): 127–47, on 133–9.

[10] John C. Powers, *Inventing Chemistry: Herman Boerhaave and the Reform of the Chemical Arts* (Chicago, 2012), 158–60, 163, 172, 174. On Boerhaave's critique of transmutation in the 1730s, 180–91.

When considering the early modern views on the chemical composition of stones, one must take into account the empirical knowledge tradition—going back to at least the Middle Ages—that acknowledged the fact that metals often could be recovered intact even when they seemingly disappeared during various processes in the laboratory. This tradition proceeded from what has been called the "negative-empirical concept" of a chemical element, that is to say, the idea that, if chemists no longer could decompose a given substance into further parts, then it was considered and treated as simple and homogenous. As William Newman has put it, "the concept is negative in that it defines an element solely in terms of what chymistry cannot do, and empirical in that it relies on the experience of the laboratory."[11]

There was, finally, a group of early to mid-eighteenth-century chemists who came to completely reject the notion that metals were composites. This rejection was made mostly on theoretical grounds and was promoted by the mineralogical chemists of the Swedish Bureau of Mines (Bergskollegium), particularly Georg Brandt (1694–1768), Henric Teophil Scheffer (1710–59), and Axel Fredrik Cronstedt. Driven by a deep pragmatism combined with skepticism toward the practice of chrysopoeia, this group came to regard all inquiries into the composition of pure metals as useless speculation. This position was a reaction to theories of the foregoing Aristotelian and Paracelsian traditions, which were considered speculative, empirically unfounded, and economically useless. Chemists like Brandt, Scheffer, and Cronstedt held that metals were the end product of the economic production process, and, accordingly, they should also be seen as the end product of chemical analysis. In the hands of these men, the negative-empirical concept was transformed into a heuristic tool that permitted, even encouraged, chemists to systematically seek to discover new substances. Nowhere was this research program more clearly set out than in the final paragraph of Cronstedt's mineralogy of 1758. In a textbook example of Enlightenment boundary work, Cronstedt associated the pursuit of the constituent parts of metals with disreputable activities such as astrology and invoked the general utility of the search for new metals.[12]

> There is no danger attending the increasing the number of metals. Astrological influences are now in no repute among the learned, and we have already more metals than planets within our solar system. It would perhaps be more useful to discover more of these metals, than idly to lose our time in repeating the numberless experiments which have been made, in order to discover the constituent parts of the metals already known. In this persuasion, I have avoided to mention any hypotheses about the principles of the metals,

[11] Newman ultimately traces the concept to the corpus of Aristotle and argues that it was transmitted into eighteenth-century chemistry through Daniel Sennert and Robert Boyle. But he makes no reference to the literature on eighteenth-century developments prior to Lavoisier. However, Cassebaum and Kauffman have argued that Scheele and Bergman were the authors of the concept, and Porter claimed that it was used earlier by Axel Fredrik Cronstedt and Johann Heinrich Pott. See William R. Newman, "The Significance of 'Chymical Atomism,'" in *Evidence and Interpretation in Studies on Early Science and Medicine,* ed. Edith Dudley Sylla and William R. Newman (Boston, 2009), 248–64, on 254; Heinz Cassebaum and George B. Kauffman, "The Analytical Concept of a Chemical Element in the Work of Bergman and Scheele," *Ann. Sci.* 33 (1976): 447–56; Porter, "The Promotion of Mining" (cit. n. 3), 549–50, 557–8. See also Siegfried and Dobbs, "Composition" (cit. n. 3), 280–4, 292–3.

[12] Thomas F. Gieryn, *Cultural Boundaries of Science: Credibility on the Line* (Chicago, 1999), 1–6. On Enlightenment natural philosophy as boundary work, see Fors, *The Limits of Matter* (cit. n. 4), chap. 6.

the processes of mercurification, and other things of the like nature, with which, to tell the truth, I have never troubled myself.[13]

Cronstedt's views present a mature assessment that points to the theoretical and economic implications of the research program he took part in. It is a good example of the larger belief among Swedish mineralogical chemists that natural philosophers should avoid speculating about the (supposed) component parts of substances that were empirically shown to be homogenous. In this view, theory should be limited to that which was empirically known. The economic implication, moreover, was that chemistry should be disassociated from attempts to manufacture metals by artificial means, becoming instead a utilitarian aid in the search for valuable metal in ores.

A NEW CONCEPT OF CHEMICAL SUBSTANCE

Cronstedt's position rested on a thriving local tradition of mineralogical chemistry connected to the Bureau of Mines. Between 1727 and 1768 the bureau's chemical laboratory was led by Georg Brandt. Brandt's chemical credentials were impeccable. His father was a German immigrant ironmaster and apothecary, and Brandt himself had studied physics and chemistry in Leiden in the early 1720s.[14] The Leiden connection is important because Brandt was a student of Boerhaave. Like his professor, Brandt disagreed with Stahl's phlogiston theory and its claim that calxes, not metals, were simple substances. The greatest part of Brandt's work went into the careful analysis of minerals. His first such publication was an investigation of gold and mercury (1731). Two years later he published another paper that carefully outlined the chemical differences between various arsenic compounds and alloys (1733).[15] These were followed by his most famous publication, *Dissertatio de semimetallis* (1735), which analyzed cobalt and characterized it as an independent "semimetal." Brandt also set out the premises for a research program of discovery of simple substances that would be pursued by chemists long into the twentieth century.

He singled out metals (e.g., gold, iron) and semimetals (e.g., zinc, bismuth) as the only known pure substances among all substances in existence. They alone could not be decomposed into further parts.[16] This new grouping of substances laid the ground-

---

[13] [Axel Fredrik Cronstedt], *Försök til Mineralogie, eller mineral-rikets upställning* (Stockholm, 1758). This quotation and the ones that follow are from the English translation, *An essay towards a system of mineralogy: Translated from the original Swedish, with notes by Gustav von Engestrom. To which is added, a treatise on the pocket-laboratory, containing an easy method, used by the author, for trying mineral bodies, written by the translator . . .* (London, 1770), 241. The quotation has been slightly edited for spelling.

[14] Brandt was the laboratory's de facto leader from 1727, although the official title of director was not given to him until 1748. See Sven Odén, "Brandt, Georg," *Svenskt Biografiskt Lexikon* (Stockholm, 1925), 5:784–9; letter from Brandt to the Bureau of Mines, Bergskollegium huvudarkivet. Ink. brev, suppliker, rannsakningar m.m. 1723. 2 E 4:151. Letter 418, registered 2 November 1723, Riksarkivet (National Archives, Sweden).

[15] Georg Brandt, "Experimentum quo probatur dari attractionem mercurii in aurum," *Acta Literaria et Scientiarum Sveciae* (1731), 1–8; Brandt, "De arsenico observationes," *Acta Literaria et Scientiarum Sveciae* (1733), 39–43.

[16] A semimetal should be defined, said Brandt, as a substance with the general form, color, and weight of a metal, but which was not malleable when beaten with a hammer. See Georg Brandt, "Dissertatio de semi-metallis," *Acta Literaria et Scientiarum Sveciae* (1735), 1–12, on 1. Presumably following Brandt, similar definitions could be found in the works of Cronstedt and Bergman. See Cronstedt, *Essay* (cit. n. 13), 235; Torbern Bergman, *Anledning til föreläsningar öfver chemiens beskaffenhet och nytta, samt naturliga kroppars allmännaste skiljaktigheter* (Stockholm, 1779), 45.

work for the main part of the *Dissertatio de semimetallis*. There, Brandt discussed the differences between cobalt and other semimetals. By way of chemical trials, he established that cobalt could not be considered to be the same substance as any other semimetal.[17] From this research it followed that cobalt should be regarded as a pure substance, as defined by the negative-empirical concept of a chemical element.

> It is evident from the definition of semi metal given above, that vitriols, cinnabar, minera, that is, metal bearing veins, as well as earths and glasses [vitra] of that kind in which nothing pure (*gediget*) can be found, cannot rightly be considered to be semimetals or metals.[18]

Brandt's colleagues took up his mode of reasoning and his methods. His paper was to become the first of a series issued from this group of mineralogical chemists, which presented and defined new metals. Of these, the most well known is Cronstedt's paper, *Rön och försök gjorde med en malm-art, från Los kobolt grufvor i Färila Socken och Helsingeland* (1751), in which he presented the discovery of nickel. Scheffer's paper on platinum, *Det hvita gullet, eller sjunde metallen. Kalladt i Spanien Platina del Pinto* (1752), is also well known.[19]

All of these papers were based on the negative-empirical concept and used it in a well-conceived and theoretically articulate way to argue that, first, there was a set number of previously known pure metals and semimetals, and, second, new discoveries had been made that added to the list. These straightforward claims were put forth with the explicit aim of establishing what we now call "scientific priority." For example, Cronstedt stated that there was no known alloy or composition with properties similar to those of nickel. For this reason, nickel "should be regarded as a new semimetal" until someone could present a method to compose it out of the twelve known whole and semimetals.[20]

While Brandt's and Cronstedt's claims to the discoveries of cobalt and nickel were rather straightforward, Scheffer's discovery of platinum is a different matter. It was more of a reinterpretation of a known metallic entity and the integration of this metal into the new epistemic framework of the bureau's chemists. Platinum was known by American native peoples since before the European conquest. Thereafter it was handled by knowledgeable assayers in the Spanish colonies. Finally, in the 1740s, it came to the attention of European scientific practitioners, who began to discuss whether it was another unknown whole metal. It was only in the 1750s that it became subject to chemical analysis. Scheffer's article in the *Transactions of the Swedish*

---

[17] Brandt, "Dissertatio de semi-metallis" (cit. n. 16), 4–6. See also Brandt, *Tal om färg-cobolter, hållit för Kongl. Vet. Acad. Vid praesidii nedläggande den 30 jul. 1760* (Stockholm, 1760), 3.

[18] "Ex definitione semi-metalli supra-memorata patet, vitriola, cinnabarim, minerasque vel venas metalliferas, item terras & vitra hujus generis, in quibus nihil puri (*gediget*) [Swedish: solid or pure (metal)] inest, pro semi-metallis, aut metallis merito haberi non posse"; Brandt, "Dissertatio de semi-metallis" (cit. n. 16), 2. I would like to thank Maria Berggren at Uppsala University library for her help with this translation.

[19] Cronstedt, "Rön och försök gjorde med en malm-art, från Los kobolt grufvor i Färila Socken och Helsingeland," *Kungliga Vetenskapsakademiens Handlingar* (1751) (hereafter cited as *KVAH*); Cronstedt, "Fortsättning af rön och försök, gjorde med en malm-art från Los kobolt-grufvor," *KVAH* (1754); Henric Theophil Scheffer, "Det hvita gullet, eller sjunde metallen. Kalladt i Spanien Platina del Pinto, Pintos små silfver, beskrifvit til sin natur," *KVAH* (1752), 269–75; Scheffer, "Tilläggning om samma metal," *KVAH* (1752), 276–8.

[20] Cronstedt, "Rön och försök" (cit. n. 19), 291–2.

Royal Academy of Sciences is usually considered the first thorough chemical investigation of the metal but cannot be considered a discovery in any normal sense of
the word.[21] Other, less well known papers from this group included Bengt Anders-
son Qvist's paper of 1754, which claimed that a sample of blacklead [*blyerts*] that
he had examined in all likelihood contained a previously unknown metallic entity.
Qvist, however, did not follow through on the investigation and failed to isolate the
new metal. This was done a few decades later by Scheele (1778) and another Bureau
of Mines chemist, Peter Jacob Hjelm, who was the first to obtain molybdenum in a
metallic state (1781).[22]

We have seen how the negative-empirical concept could be used to lay claim to the
discovery of new metals as simple homogenous substances, a practice that seems to
have originated at the Bureau of Mines. For the practically minded chemists, mineralogists, and assayers at the bureau, the primary goal was to identify metals in mineral
samples, not to find underlying chemical principles. This was a choice that derived
from an empirical theory of knowledge (epistemology) taken almost to the extreme:
no principles should be accepted that could not be seen, touched, tasted, or smelled.
Using this theory as their starting point, this group redefined the goal of chemical
analysis in a profound way. Previously, many chymists had had the goal to find, and
learn how to manipulate, the underlying ultimate principles of matter. This goal made
sense from the point of view of chrysopoeians who sought to synthesize metals in the
laboratory. As seen in the Cronstedt quotation above, the Bureau of Mines chemists
sought to discredit this goal. For these mineralogical chemists working in the tradition of assaying, it made much more sense to replace it with another: that of discovering new simple substances as defined through the negative-empirical concept. This
change was in part a consequence of attempts to disassociate the pursuits of the bureau chemists from the increasingly disreputable practice of chrysopoeia.[23]

More importantly for the purposes of this essay, the empirical position of the mineralogical chemists was testimony to the influence of assaying on their chemistry. Indeed, the single-minded pursuit of new substances reduced the goal of chemistry to
that of assaying. More specifically, this pursuit permitted chemistry to become an
economically oriented cameral science [*Kameralwissenschaft*] at the service of the
state, thereby cutting it off from its traditional theoretical ambition to understand
the principles of transmutation.[24] The centrality of this development seems to have

---

[21] L. B. Hunt, "Swedish Contributions to the Discovery of Platinum: The Researches of Scheffer
and Bergman," *Platinum Metals Rev.* 24 (1980): 31–6; A. Galán and R. Moreno, "Platinum in the
Eighteenth Century: A Further Spanish Contribution to an Understanding of Its Discovery and Early
Metallurgy," *Platinum Metals Rev.* 36 (1992): 40–7; Luis Fermin Capitan Vallvey, "The Spanish Monopoly of Platina: Stages in the Development and Implementation of a Policy," *Platinum Metals Rev.*
38 (1994): 22–5.

[22] Bengt Quist, "Rön om bly-erts," *KVAH* (1754); Sten Lindroth, *Kungliga Svenska Vetenskapsakademiens Historia 1739–1818* 1:1 (Stockholm, 1967), 526. On Scheele's cooperation with Swedish mining chemists, see Hjalmar Fors, "Stepping through Science's Door: C. W. Scheele, from Pharmacist's Apprentice to Man of Science," *Ambix* 55 (2008): 29–49.

[23] Fors, *The Limits of Matter* (cit. n. 4), chap. 6.

[24] On cameralist chemistry, Christoph Meinel, "Theory or Practice? The Eighteenth-Century
Debate on the Scientific Status of Chemistry," *Ambix* 30 (1983): 121–32, on 127–9; Andre Wakefield,
"Police Chemistry," *Sci. Context* 12 (1999): 231–67. See also Lisbet Koerner, *Linnaeus: Nature and
Nation* (Cambridge, Mass., 1999) 1–3, 95–7. Koerner extrapolates a species of localism, which she
calls the cameralist ideology of local modernity, from Linnaeus's work to Swedes in general. It can be
noted that this ideology was not shared by eighteenth-century Swedish actors at large; see Koerner,
Linnaeus, 163, 193.

passed under the radar of traditional histories of chemistry. The reason, perhaps, is that assayers were on the social periphery of eighteenth-century natural philosophy and, hence, unsuitable as originators of important innovations in chemistry.

CONTINUITY AND ONTOLOGICAL CHANGE

As the old kind of theory of ultimate principles was cut away, a new, more pragmatically empirical theory of matter developed. The new epistemological framework discussed above was also used in practice to transform and renegotiate perceptions of the ultimate categories of matter (ontology). To illustrate this point, I will use a set of excerpts made by Cronstedt from the manuscripts of Urban Hiärne (1641–1724). Hiärne had formerly been the director of the Bureau of Mines chemical laboratory between 1683 and 1720 and was a Paracelsian chymist and a chrysopoeian. Cronstedt's notes show how he, like many of his eighteenth-century colleagues, drew on the chrysopoetic tradition while simultaneously reinterpreting it. As an *auscultator*, or student/apprentice at the Bureau of Mines during the years 1742–8, Cronstedt spent much time studying old documents on chymistry and mining. One set of documents that he came across consisted of Hiärne's experiments on the chromatic and morphological changes that metals underwent when they were tried with various reagents.[25]

Hiärne had conducted a great number of experiments on earths and metals. For example, in December 1699 he melted *silica alba* with other substances to form a glassy mass, crushed it into a powder, and put it into cold storage. After eight weeks, the powder transformed into a "beautiful clear white thickish liquid," which he called *liqvore silicum*. It was a horribly sharp brew that blistered the tongue when tasted, with the pain continuing for several days afterward. Hiärne then dissolved gold, silver, iron, tin, lead, regulus of antimony, marcasite (or pyrite), and zinc in acids. The metallic substances were separated from the acids and dried out again. Finally, a quantity of each metal was mixed with tartarus Weinstein and placed in the *licvoris silicum* to grow.[26] Hiärne, as excerpted by Cronstedt, then described what happened next:

> And they grew immediately just as little bushes and trees, and one was distinct from the other as to colour, shape and figure, the one better than the other.[27]

Gold gave a yellow tree, lead a white, silver a green, *Reg*(ulus of) *antimon* a yellow, tin a white, marcasita a white, iron a black, zinc a white, and mercury a white. How-

---

[25] Cronstedt, *Utdrag utur framl. Landshöfdingens och Arch: Urban Hiärnes Chymiska Diario och Observationer,* in Vetenskapliga anteckningar och manuskript, vol. 11, 17–36 [F1:9] Axel Fredrik Cronstedts Arkiv in Kungliga Vetenskapsakademiens Arkiv (Archives of the Royal Academy of Sciences, Stockholm; hereafter cited as KVAA). In the cases when chemical signs were used in the original text the common names of substances used during the period are given in italics. Cronstedt's key for translations of some of the signs can be found on 43–4 in the same volume. When he used signs not found in this key, I have used Torbern Bergman's table of signs, as given in Henrik Teophil Scheffer and Torbern Bergman, *Framledne Direct. och Kongl. Vet. Acad. Ledamots Herr H. T. Scheffers chemiske föreläsningar rörande salter, jordarter, vatten, fetmor, metaller och färgning: samlade i ordning stälde och med anmärkningar utgifne* (Uppsala, 1775). Concerning the dating of the text, it was most likely excerpted during 1746, Cronstedt's first year as a student in Georg Brandt's laboratory.

[26] Gold was dissolved in aqua regis; iron in spiritus salis; silver, tin, mercury, lead, regulus of antimony, marcasita, and zinc in aqua fortis. Cronstedt, *Utdrag utur Hiärne* (cit. n. 25), 22–4, on 22.

[27] Ibid., 24.

ever, the trees did not last but fell apart after a few days.[28] In his experiment, Hiärne investigated possible pathways toward augmentation or transmutation of metals. The vegetative patterns and color changes were in all likelihood taken as an indication that there had been a release of the inherent life force that made metals grow.

Cronstedt, however, was not interested in Hiärne's notes because he wanted to investigate the inherent life force of metals, with the further aim of effecting metallic transmutation.[29] He was interested in using the color changes as an indicator to learn more about the chemical characteristics of metals. Hence, Hiärne's investigations of color changes were used *for another purpose*, to indicate the chemical difference between metals. Turning to the table, a representational media often used in early modern chemistry to visualize symbolic forms of meaning,[30] Cronstedt transposed Hiärne's experiments into tabular form based on the negative empirical concept of the substance, that is to say, an ontology of substance that was not used by Hiärne. To this table Cronstedt also added substances that had not been studied by Hiärne (see fig. 1).

Cronstedt made a number of tables based on Hiärne's experiments. His tables and excerpts from Hiärne's notes indicate an instrumental continuity. They show how chemists of the early and mid-eighteenth century could make use of transmutative chymistry and simultaneously transform it into something new. While Hiärne's interest was to learn how to decompose and make metals, Cronstedt wanted to establish what stable species of metals there were and to present the differences between them in tabular form. Cronstedt's appropriation of Hiärne's experiments also shows that there was a strong continuity in laboratory practice.[31] Furthermore, the excerpts indicate that environments like the Bureau of Mines could provide a strong institutional stability, one that connected late seventeenth-century chymistry to eighteenth-century chemistry. What we see here, therefore, is not the demise of alchemy and the birth of chemistry, but how one chemical research program was succeeded by another within a stable institutional environment.[32] Cronstedt's transposition of Hiärne's experiments into tabular form also visualizes an important theoretical break between the transmutative and empirical/pragmatic understanding of a chemical substance. Hiärne's underlying theory of chymical principles and their role in the generation of metals were neatly sidestepped by extracting his chemical operations and presenting them in tables to create an overview of what were considered secure chemical facts.

Cronstedt also incorporated novel facts in his tables based on Hiärne's experiments. In the table reproduced as figure 1, for example, he added cobalt to the bottom of the left column. At the time, as we have seen, cobalt was a new semimetal, the dis-

---

[28] Ibid.

[29] Hiärne stated that he had the experiment from Glauber, and that the latter had used it to substantiate his claim that "*silex is matrix omnium metallorum.*" Hiärne seems to have agreed with Glauber because he held that the *licvoris silicum* was the matrix that permitted the growth of metals. Cronstedt, *Utdrag utur Hiärne* (cit. n. 25), 23–4, quotation about "silex" on 23.

[30] See Matthew Daniel Eddy, "How to See a Diagram: A Visual Anthropology of Early Modern Chemical Affinity," in this volume.

[31] It can be noted that parts of Hiärne's work were of interest to Cronstedt without reinterpretation, e.g., his 1699–1700 investigation of a sample of coal from southern Sweden. See Cronstedt, *Utdrag utur Hiärne* (cit. n. 25), 19–20.

[32] Cf. Beretta, *Enlightenment of Matter* (cit. n. 6), 76–8; although he acknowledges the importance of both the alchemical and metallurgical traditions, Beretta sketches a strong opposition between the two.

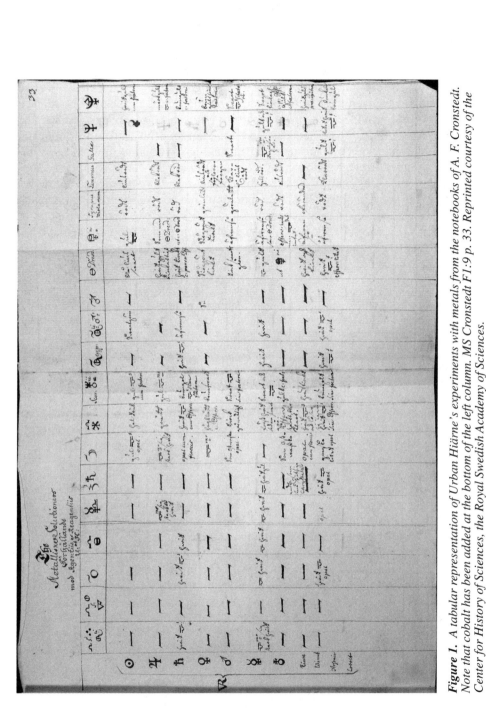

Figure 1. *A tabular representation of Urban Hiärne's experiments with metals from the notebooks of A. F. Cronstedt. Note that cobalt has been added at the bottom of the left column. MS Cronstedt F1:9 p. 33. Reprinted courtesy of the Center for History of Sciences, the Royal Swedish Academy of Sciences.*

covery of which had been published by Brandt in 1735. Cobalt's row was left empty, but it is an empty space that signals the birth of an entirely new research program. The row invited the discovery of new substances. It opened up the table to a potentially endless expansion of new rows and columns. Furthermore, if one conducted the experiments necessary to fill cobalt's row, one also created new knowledge about nature. Finally, if cobalt's reactions with the reagents were found to be different from those of other metals, one had confirmed its identity as a new metal. Hence we see how the table was given the function of defining substances in relationship to each other, and how it provided impetus to look for more substances.[33] This was also the path that Cronstedt and his associates would take in their investigations on nickel, platinum, and other mineral substances. Painstaking attention to difference, as uncovered through chemical analysis and presented in tabular form, permitted Cronstedt and his associates to investigate and map out new chemical territory. Cronstedt's attention to the presentation of chemical operations and new substances in systematic order indicated how one should proceed to find new knowledge. His description of nickel in his *Mineralogy* of 1758 is as good an example as any of this systematic use of difference to define minerals. As he stated about his own discovery,

> I have not, besides the nickel, found any metal or metallic composition, which 1. Becomes green when calcined. 2. Yields a vitriol, whose colcothar also becomes green in the fire. 3. So easily unites with sulphur, and forms with it a regule of such a peculiar nature, as the nickel does in this circumstance; and that 4. Does not unite with silver, but only adheres or sticks close to it, when they have been melted together.[34]

It was an approach very much inspired by another important but neglected influence on eighteenth-century chemistry: that of natural history. In fact, the emphasis Cronstedt and his associates placed on distinguishing individual objects from each other through clear and concise definitions derived as much from natural history as from assaying.

SUBSTANCES AND NATURAL HISTORY

Eighteenth-century natural history was concerned with the systematic ordering of discrete objects. In botany and zoology, these objects were plant and animal specimens. In mineralogy, they were minerals: specific samples of rock and earth, as they could be found in nature. Minerals posed a peculiar challenge to natural history. A great number of minerals were decomposable into constituent parts that were known to be other mineral specimens. Such specimens, therefore, were clearly not species in the same sense as botanical and zoological objects but rather aggregates, or composite substances. How then should they be systematized? In his *Systema Naturae* (1735; 1768), Carl Linnaeus advocated a system analogous to that of botany and zoology. Minerals should be ordered according to their outward characteristics: grouped according to color, structure, and so on. The mineralogical chemists found this solution

[33] This interpretation draws on Kim's analysis of Claude-Joseph Geoffroy's affinity table. Mi Gyung Kim, *Affinity, That Elusive Dream: A Genealogy of the Chemical Revolution* (Cambridge, Mass., 2003), 132–46. See also Lissa Roberts, "Setting the Table: The Disciplinary Development of Eighteenth-Century Chemistry as Read through the Changing Structure of Its Tables," in *The Literary Structure of Scientific Argument: Historical Studies,* ed. P. Dear (Philadelphia, 1991), 99–132.

[34] Cronstedt, *Essay* (cit. n. 13), 240.

problematic and instead opted for a system based on chemical composition.[35] The mineralogical systems of Wallerius (1747), Cronstedt (1758), and Bergman (1782) all integrated chemistry and natural history and emphasized that it was important to use chemical analysis to ascertain the internal composition of minerals.[36] In fact, all three authors felt a certain degree of resentment toward Linnaeus, casting him as something of a megalomaniac. As Wallerius put the matter in a letter to Cronstedt, "Linnaeus is the same as ever, and is presently working on his mineralogical creation; he wants to persuade the whole world that he alone is a Mineralogist as well as a Botanist."[37]

Cronstedt was completely opposed to systems based on external characteristics of minerals. In his view, his mineralogical system was far from complete, but it could function as "a bar or opposition to those . . . who, entirely taken up with the surface of things, think that the Mineral Kingdom may with the same facility be reduced into classes, genera, and species, as animals and vegetables are."[38] This was, of course, a jab at Linnaeus. Nevertheless, Cronstedt as well as his two colleagues relied on *Systema Naturae* as a model and conceptual framework.[39] Wallerius's works, emerging from a didactic and utilitarian enlightenment enterprise very much embedded in the local context of Nordic universities, had much in common with those of Linnaeus.[40] The men were about the same age, sons of countryside vicars, and rivals as professors at Uppsala. Wallerius's mineralogy can be characterized as a midcentury synthesis drawing on textual sources from chymistry and natural history, as well as on direct investigations of minerals. Overall, its guiding principle was utility. Minerals should above all be easy to identify correctly. It mattered less whether this identification was achieved through examination of external characteristics or internal composition.[41]

It would be Cronstedt, a student of both Wallerius and Brandt, who would integrate the concept of the pure substance developed at the Bureau of Mines with the Uppsala tradition of systematic natural history. Cronstedt's taxonomy was completely integrated in his chemical thought. In order to use composition as the main tool to make distinctions between minerals, it was necessary to have a clear idea about what constituted a pure chemical substance. This was provided by the negative-empirical

---

[35] Linnaeus had already published a mineralogical classification in the first edition of *Systema Naturae*. His mature views were published in the third volume of its 12th ed. (Stockholm, 1768). For discussions of his system, see Eddy, *Language of Mineralogy*, 126–33; Beretta, *Enlightenment of Matter*, 50–61 (both cit. n. 6).

[36] Johan Gottschalk Wallerius, *Mineralogia eller mineralriket indelt och beskrifwit* (Stockholm, 1747); [Cronstedt], *Försök til Mineralogie* (cit. n. 13); Torbern Bergman, *Sciagraphia regni mineralis, secundum principia proxima digesti* (Leipzig, 1782). For an overview of eighteenth-century mineralogical systems, see Rachel Laudan, *From Mineralogy to Geology: The Foundations of a Science, 1650–1830* (Chicago, 1987), 23–5.

[37] Johan Gottschalk Wallerius to Axel Fredrik Cronstedt, 17 March 1760, in MS Cronstedt vol. 8, KVAA. My translation of Swedish original.

[38] Cronstedt, *Essay* (cit. n. 13), xii.

[39] Beretta, *Enlightenment of Matter* (cit. n. 6), 95–106.

[40] Lisbet Koerner, "Daedalus Hyperboreus: Baltic Natural History and Mineralogy in the Enlightenment," in *The Sciences in Enlightened Europe*, ed. W. Clark, J. Golinski, and S. Schaffer (Chicago, 1999), 389–422, 389, 398.

[41] Hjalmar Fors, "Vetenskap i alkemins gränsland: Om J. G. Wallerius *Wattu-riket*," *Svenska Linnésällsk. Årssk.* (1996/97), 33–60, 34–8, 42–5; Anders Lundgren, "Bergshantering och kemi i Sverige under 1700-talet," *Med Hammare och Fackla* 29 (1985): 90–124, 103–20. For the international perspective, see Porter, "The Promotion of Mining" (cit. n. 3), 553–5.

concept of an element, and by the list of pure metals and semimetals proposed by the bureau's chemists. According to Cronstedt, the mineral bodies that were found in nature were almost always composites or mixtures of a smaller number of pure substances. It came down to chemical analysis to reveal their true nature, as it must always "remain concealed from every one, however penetrating, who has not employed himself in the compounding or decompounding of such bodies, as far as the present knowledge of these matters will permit."[42] Hence pure substances were assigned elemental status. This meant that the *truth* about mineral bodies—all of interest that there was to know about them—was the composition and proportion of pure substances that they contained. But perhaps more than elements, this ontology treated pure substances as immutable species, as the basic building blocks of the mineral world that made mineralogical taxonomy possible.

Cronstedt discussed only metals (including the semimetals) as pure substances. In contrast, he was vague when discussing the other classes like salts and earths. In his section on earths, he refrained from pointing "out a particular earth for each kind of stone." The reason was that he did not believe that chemical knowledge was advanced enough to permit it.[43] Again we can see how Cronstedt's view of substances invited further research, presenting his system as a sketch or outline that needed to be filled in. Simultaneously, his admission of the incompleteness of the system did not detract from its utility. It was useful precisely because assaying, bolstered by chemical laboratory methods, was at its heart. Hence, its foundation in assaying combined perfectly well with its more overtly theoretical part, that of natural history. With its tabular form of presentation and its painstaking attention to differences between species, natural history provided mineralogical chemistry with an academic pedigree that assaying had not. This was mineralogical chemistry: an assemblage of chemistry, natural history, and assaying that permitted the presentation of metals as individual species, each different from the other, and other minerals as composites of metals, and a number of hitherto unknown other simple substances. The goal of mineralogical chemistry was to define new species, and to arrange them in order to facilitate the economic development of the mining business. The chemical principles of previous generations were rejected along with any pretense to search for the ultimate principles of matter. Instead, it was enough to focus on discovery and systematization and to leave aside questions that could not be resolved through empirical investigation and tabulation.

Stepping back from the mining context, many chemists would soon argue similarly. From about the mid-eighteenth century, stable simple substances were increasingly becoming identified with elements, whether chemists liked it or not. As observed by Lissa Roberts, from the 1760s and a few decades onward, it was considered bad form among chemists to discuss the ultimate constituents of matter. It was enough to describe one's chemical trials thoroughly, to make accounts of credible interpretations of the products that came out of laboratories, and to systematize them in groups. The end products of analysis began to be treated as the only permissible building blocks, from which systems and theories could be built. In a sense, a new ontology had been

[42] Cronstedt, *Essay* (cit. n. 13), xiii.

[43] Cronstedt hazarded a guess: "because we have strong reasons to believe that the calcareous and argillaceous earths are the two principal ones, of which all the rest are compounded, although this cannot yet be perfectly proved by a demonstration"; Cronstedt, *Essay* (cit. n. 13), xvii.

established that depended on an epistemological sleight of hand. It was a reification of entities that most agreed did not have any particular status as elements.[44] As we have seen, these positions were the outcome of historically contingent negotiations and reinterpretations, resulting in a new conception of chemistry as founded on the solid bedrock, as it were, of simple mineral substances.

CONCLUSIONS

This essay has delineated the basic outline of a major epistemic shift, which took place in chemistry during the period ca. 1730–60. Working and thinking essentially like assayers, the Bureau of Mines chemists tied the negative-empirical concept of simple substances to a systematic program for the discovery of unknown metals. This was a new and powerful research program that integrated chymistry with assaying and natural history. Through this assemblage there emerged the possibility of building entire systems and theories, for which these newly defined substances functioned as the foundational building blocks.

Initially metals were the basic species of these systems. They took precedence by virtue of their monetary value, their material solidity, and the painstaking efforts put forth by generations of assayers to identify the differences between them. In the second half of the eighteenth century, the newly discovered components of air would join the swelling ranks of metallic simples, and both groups of substances would eventually be joined by a great number of nonmetallic solids. The innovations outlined in this essay profoundly influenced subsequent chemical and mineralogical systems, including Antoine Laurent Lavoisier's chemical nomenclature and the periodic table of elements, both still in use today.

Why, then, have we not seen this epistemic shift before? The reason, I would say, is twofold. First, the Bureau of Mines tradition of mineralogical chemistry has not hitherto been properly researched. Second, I would agree with Siegfried and Dobbs, who noted in 1968 that an overemphasis on the oxygen theory of combustion has obscured important aspects of what was really going on in eighteenth-century chemistry. Now, forty-six years later, the contribution of Lavoisier has been reevaluated, but the traditional research foci associated with the Chemical Revolution are still given undue explanatory weight.[45] One such focus has been the development of pneumatic chemistry, which resulted in the recognition that air was composed of a number of discrete "airs" (or, to use a modern term, gases). Another is the emergence and eventual overthrow of the phlogiston theory of combustion. A third focus is the development of the theory of chemical affinities, which was elaborated to explain the combination and recombination of chemical substances.[46] In my view, these developments cannot be

---

[44] Lissa Roberts, "Filling the Space of Possibilities: Eighteenth-Century Chemistry's Transition from Art to Science," *Sci. Context* 6 (1993): 511–53, on 513–4, 521–2, 529, 534, 549.

[45] Siegfried and Dobbs, "Composition" (cit. n. 3), 277, 292. Some recent reevaluations of Lavoisier can be found in Kim, *Affinity* (cit. n. 33), 1–16, 439–55; and Kim, "Lavoisier, the Father of Modern Chemistry?" in *Lavoisier in Perspective,* ed. Marco Beretta (Munich, 2005), 167–91; Lawrence M. Principe, "A Revolution Nobody Noticed? Changes in Early Eighteenth-Century Chemistry," in *New Narratives in Eighteenth-Century Chemistry: Contributions from the First Francis Bacon Workshop,* ed. Lawrence M. Principe (Dordrecht, 2007), 1–22; Eddy, *Language of Mineralogy* (cit. n. 6), 5–6.

[46] For some critical evaluations of this emphasis, see Frederic L. Holmes, *Eighteenth-Century Chemistry as an Investigative Enterprise* (Berkeley, Calif., 1989); Powers, *Inventing Chemistry* (cit. n. 10), 11, 155–65.

put in perspective if one does not take into account early eighteenth-century chemists' work on chemical composition, and its importance for the emergence of the concept of chemical elements. Indeed, from the utilitarian and pragmatic point of view of the chemists in northern Europe—which is as privileged a perspective as any—affinity and the phlogiston theory were of little consequence to the development of chemistry. The new chemistry of airs, on the other hand, depended on heuristic tools and concepts developed for the discovery of metallic simple substances. This article draws attention to the importance of the close ties between chemistry and mineralogy for eighteenth-century chemistry and has argued that it was by fusing university learning with artisanal outlooks and skills that chemists laid the foundation for the radical change in research emphasis that occurred during the second half of the century. What is it, then, that we really have inherited from eighteenth-century chemistry? Maybe it is, quite simply, the idea that material reality—in its entirety—can be divided into a finite set of stable commodities that can be sifted out from each other, sold, and bought.

Pierre-Joseph Macquer:
Chemistry in the French Enlightenment

by Christine Lehman*

ABSTRACT

Despite recent studies of chemistry courses and of academic research at the beginning of the eighteenth century, the perception of chemistry in the French Enlightenment has often been overshadowed by Lavoisier's works. This article proposes three specific case studies selected from Pierre Joseph Macquer's (1718–84) rich career to show the continuous evolution of chemistry throughout the century: medicinal chemistry through the application of the Comte de La Garaye's metallic salt solutions, the emergence of industrial chemistry through a few of Macquer's evaluations at the Bureau du Commerce, and finally communal academic research through the experiments on diamonds using Tschirnhaus's lens. These examples attempt to illustrate the innovative, creative, dynamic, multicultural, and multifaceted chemistry of the Enlightenment.

INTRODUCTION

Overshadowed by the new chemistry established by Lavoisier at the end of the century, mid-eighteenth-century chemistry seems to have been neglected by historians until the recent historiography that brought a new vision of the early eighteenth century as well as the study of chemistry courses that showed the infatuation with chemistry in the middle of the Enlightenment.[1] In fact, although it was not marked by spectacular discoveries, mid-eighteenth-century chemistry was anchored in medical applications and integral to industry and academic research.

The purpose of this essay is twofold. The first is to show the diversity and richness of mid-eighteenth-century chemistry exemplified in the exceptional career of Pierre-Joseph Macquer (1718–84). Macquer was not the typical eighteenth-century chemist because his rich and varied career encompassed all of the main facets of

* Paris-Ouest Nanterre University; home address: 7 rue Béranger, F-75003 Paris, France; christine.lehman@wanadoo.fr.

I would particularly like to thank Seymour H. Mauskopf for his encouragement and for his precise review of the final version of this chapter and William R. Newman for his thorough revision of the English. I also thank the anonymous referees for their constructive comments and suggestions.

[1] Frederic Lawrence Holmes, *Eighteenth-Century Chemistry as an Investigative Enterprise* (Berkeley, Calif., 1989); Mi Gyung Kim, *Affinity, That Elusive Dream: A Genealogy of the Chemical Revolution* (Cambridge, Mass., 2003); Bernadette Bensaude-Vincent and Christine Lehman, "Public Lectures of Chemistry in Mid-Eighteenth-Century France," in *New Narratives in Eighteenth-Century Chemistry: Contributions from the First Francis Bacon Workshop*, ed. Lawrence M. Principe (Dordrecht, 2007), 77–96; Lehman, "Innovation in Chemistry Courses in France in the Mid-Eighteenth Century: Experiments and Affinities," *Ambix* 57 (2010): 3–26; John Perkins, "Chemistry Courses, the Parisian Chemical World and the Chemical Revolution, 1770–1790," *Ambix* 57 (2010): 27–47.

eighteenth-century chemistry. Indeed, Macquer's profile, as a physician, teacher of chemistry, academician, and government expert, instantiates the multiple social, institutional, and intellectual spaces that chemistry inhabited in France during the second half of the eighteenth century. My purpose here is not to present a biography of Macquer but rather to illuminate the diversity of chemistry in the Enlightenment through three unpublished case studies selected from this great chemist's activities: the medical application of metallic salt solutions such as the Comte de La Garaye's, the development of the emerging industrial chemistry through a few of his evaluations at the Bureau du Commerce, and, finally, academic research through the experiments on diamonds using Graf Ehrenfried Walther von Tschirnhaus's lens. In the last case, although all the related documents were published in Lavoisier's *Oeuvres*, it appears that the conclusion of the nature of diamond was the result of close communal work within the Académie des Sciences. Furthermore, the chronology of the three episodes provides a view of the evolution of chemistry throughout the whole century from Nicolas Lémery to Lavoisier.

Second, in the examples presented here, the list of Macquer's collaborators shows the variety of actors who were practicing chemistry at a time when the profession of chemist did not yet exist. Indeed, only physicians and apothecaries received a theoretical and practical chemical education, but chemistry was a public science and its practice was not restricted to initiates. The many available publications, such as the *Encyclopédie*, dictionaries, and treatises, in particular those of Macquer, made chemistry accessible to all, from physicians and apothecaries to philosophers and nobles, not to forget craftsmen. The first example concerns the chemistry of metallic solutions that Macquer practiced at La Garaye's castle.

A MEDICINAL CHEMISTRY AT THE KING'S SERVICE: LA GARAYE'S SALTS

After spending his youth attending balls and hunting, Claude Toussaint Marot Comte de La Garaye (1675–1755) decided to devote his entire fortune to helping the poor. He had attended Nicolas Lémery's chemistry courses in Paris.[2] He also had a smattering of the art of the apothecary. Yet he had developed a totally original technique. As an opponent of what he called "pyrotechnics," La Garaye thought that fire altered the principles of the substances exposed to it. In order to extract oil and salts from plants and animal substances, in which the effectiveness of medicines lay, he used a single solvent: water. But dissolution by water alone was a slow process. He had designed and fabricated a hydraulic machine[3] that enabled him to extract oils and salts by means of a long trituration at ambient temperature (fig. 1).

La Garaye praised the simplicity of this new chemistry and said, "This way, one can extract the essential parts without using the violence of fire; I mean salt and oil, which present all the virtues of the mixt."[4] Initially intended for the poor, La Garaye's salts became a medicine for nobles. They were on sale at Versailles's apothecary shops as well as at Antoine Baumé's. Louis 1er Duke of Orléans, grandson of

---

[2] In the seventeenth century, Nicolas Lémery's (1645–1715) chemistry courses taught in his own laboratory attracted a large and curious crowd. His *Cours de chymie* was first published in Paris in 1675 and in many subsequent editions until 1756.

[3] Claude-Toussaint Marot Comte de La Garaye, *Chymie hydraulique pour extraire les sels essentiels des végétaux, animaux, minéraux avec l'eau pure* (Paris, 1745), Planche II.

[4] Ibid., 39.

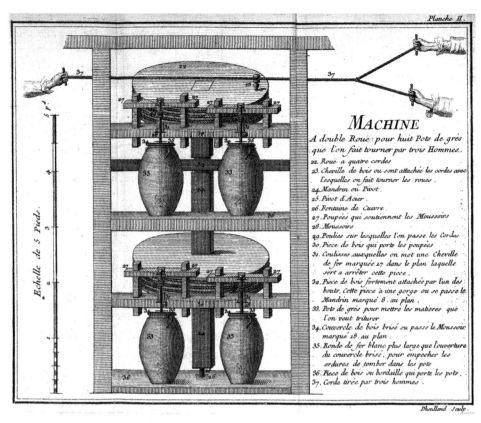

Figure 1. *De La Garaye's hydraulic machine. De La Garaye,* Chymie hydraulique *(cit. n. 3).*

Louis XIV, also owned a hydraulic machine in his laboratory at Sainte Geneviève. It appears that, in addition to physicians and apothecaries, many nobles such as the Comte de La Garaye or Duke Adrien-Maurice de Noailles, Maréchal de France and minister of Louis XV, had very well equipped chemical laboratories.

In December 1751, Macquer received a message from the Duke de Noailles: "He had to talk to him about an affair that could be of interest to him."[5] The affair in question involved Macquer acting as a spy to determine the preparation of La Garaye's metallic salts in order to enable the king to buy the secret. Macquer's status as a physician and his extensive chemical knowledge justified the king's choice for this secret mission.

At his castle, the Comte de La Garaye had developed a new technique for dissolving metals. He was too old to demonstrate it himself to the king as he had done in 1731;[6] it was thus Macquer who had to make this demonstration. The hydraulic machine devoted to extracting salts from animals and plants by using water had been abandoned: this time, the purpose was extracting metallic salts. The "softest neutral

[5] Maréchal de Noailles to Macquer, letter dated 27 December 1751, Bibliothèque nationale de France (hereafter cited as BNF), MS Fr 23226, fol. 1r.

[6] René Richelot, *Éloge historique de Claude-Toussaint Marot Comte de La Garaye grand bienfaiteur des pauvres et du pays de Dinan (1675–1755)* (Dinan, 1955), 9. De La Garaye's castle was located in Brittany.

salts," such as marine, tartar, and ammonia salts, were used as solvents and allowed to react with iron, copper, tin, lead, mercury, gold, and silver. Because salts reacted very slowly with metals, it was necessary not only to accelerate the process by using substances that were finely broken up (e.g., by grinding) but also to wait for their dissolution at ambient temperature: almost one month for mercury, several months for gold and silver. With these salts, La Garaye prepared aqueous or alcoholic solutions that proved to be very effective medicines: with iron one could prepare ferruginous water, a nerve stimulant used in paralysis and numbness cases; lead was used to prepare a sugared water, which was a proven remedy against ophthalmia; copper and mercury solutions were very effective against ulcers and skin diseases, and gold against fatigue.[7] It was an exclusively medicinal chemistry requiring only elementary equipment. In particular, Macquer praised the effectiveness of the copper dissolution in ammonia salt, which, besides being an original chemical preparation, had astonishing virtues. As he reported to Noailles, "It is surprising to see how malignant and inveterate ulcers, sometimes lasting for five or six years, get better within the space of three to four days when tended with this remedy."[8] Indeed, sore legs were very difficult to cure in low-lying and wet regions such as Brittany and Normandy.

Macquer was sent to La Garaye by Noailles's order; then, when he returned from Brittany in June 1752, he immediately reproduced the experiments "before the Marshall's eyes" in the laboratory that Noailles had installed in his residence at St Germain en Laye.[9] These experiments lasted at least until October of the same year. Yet, more interested in chemistry than in medicine, Macquer did not limit himself to repeating medicinal preparations; he changed their operating modes. He distilled the mixtures in glass retorts or used a hotter fire. He then obtained in a few hours what La Garaye got in several weeks. As he wrote to Noailles, studying these metallic dissolutions "invited him to an exciting career in chemical research." This study fully fit into the French tradition of the dissolution chemistry initiated by Simon Boulduc's and Claude-Joseph Geoffroy's works, extending them to metallic dissolutions.[10] The mercury dissolution by ammonia salt particularly caught Macquer's attention. Incidentally, the interpretation of this dissolution formed the basis of a controversy between Macquer and La Garaye. The latter, trained in Lémery's chemistry, interpreted this operation as the decomposition of mercury and based his belief on clinical observation: the mercury dissolution did not make the patients salivate while the ammonia salt was revealed by its urinous smell and continued to react as before. Mac-

[7] "Extrait des lettres de M. le Cte de La Garaye," 4 November 1751, BNF, MS Fr 9151.

[8] "Rapport des expériences réalisées à la Garaye," Macquer to Maréchal de Noailles, BNF, MS Fr 23226, fols. 18–23, on 23r. The surgeon de Raynal, Major des hôpitaux des armées et des troupes de la Marine, who worked at La Garaye's hospital, also attested to the success of these remedies in military hospitals and on warships; BNF, MS Fr 9151, fols. 67r–71v. See also Macquer, "Sur une nouvelle méthode de M. le Comte de La Garaye pour dissoudre les métaux," *Mémoires de l'Académie Royale des Sciences* (hereafter cited as *MARS*) (1755 [1761]), 25–35, on 35.

[9] BNF, MS Fr 23226, fol. 25r. Macquer probably carried out these experiments until 1764 at his home on rue Saint Sauveur; at this time he complained of not having received the yearly pension of 1,200 *livres* that had been granted to him in 1752 in order to "urge him to assiduously work to improve this discovery"; Archives nationales, Paris (hereafter cited as AN), F⁴1961.

[10] At the end of the seventeenth century Simon Boulduc (1652–1729), and later Claude-Joseph Geoffroy (1685–1752), abandoned the analysis of plants by distillation and developed various extraction techniques by means of solvents, mainly water and alcohol; Holmes, *Eighteenth-Century Chemistry* (cit. n. 1), 33–59; Holmes, "Analysis by Fire and Solvent Extractions: The Metamorphosis of a Tradition," *Isis* 62 (1971): 129–48.

quer, who thought that the alkali vapors produced during the operation showed the decomposition of ammonia salt, challenged this interpretation.

His first argument was that of a physician, saying that in order to reach a conclusion from salivation, one would need to have a number of observations from different patients. Then, as a chemist, he put forward that only a part of ammonia salt had been consumed and that this could be demonstrated by weighing the reaction products.

> For this, it would be necessary that, after having made the *teinture*, one could retrieve all the ammonia salt that one had used in the operation, which one will certainly never do, and examining things with exactitude one will on the contrary always find a diminution in the weight of ammonia salt *in proportion* to the quantity of mercury that will have been reduced *en teinture*.[11]

It is worth noting that in the 1750s weighing was used to determine the progress of an operation. For Macquer, mercury was dissolved by the acid of ammonia salt and formed a "mercurial salt composed of marine acid & mercury bound to one another." The difference in chemical composition made the distinction between the *sublimé corrosif*, which was a poison, and La Garaye's salt, containing much less acid, which was an excellent remedy.[12]

Besides, for Macquer, the decomposition of metal mercury by ammonia salt was "an effect deserving much attention because it establishes a new truth in chemistry or at least what one only starts to guess" as it threw into question the first column of the affinity table:

> The affinity of metallic substances for acids is equal to or may even be higher than the one of volatile alkali with the same acids, so that ammonia salts can be decomposed by means of metals with the same ease as metallic salts are by volatile alkalis. . . . Thus the exception [of mercury] to the general rule now becomes a general rule itself.[13]

In this table established in 1718 by Etienne-François Geoffroy (1672–1731), chemical compounds were arranged in columns in descending order of reactivity with the substance at the top of the column.[14] According to the first column, ammonia salt, formed by the union of a volatile alkali and an acid, should not be decomposed by mercury, a metallic substance located at the column bottom, because alkalis were higher and consequently had a stronger affinity for acids, which were at the top of the column.

This controversy about how to interpret the dissolution of mercury by ammonia salt strained the relations between the two men. However, it enabled Macquer to assert, from the very beginning of his career, the superiority of the new chemistry

---

[11] BNF, MS Fr 9134, fol. 73v (emphasis on "in proportion" is mine).

[12] Macquer, "Recherches sur la nature de la teinture mercurielle de M. le Comte de La Garaye," *MARS* (1755 [1761]), 531–46, on 546; "Sur une nouvelle méthode de dissoudre les métaux," *Histoire de l'Académie Royale des Sciences* (hereafter cited as *HARS*) (1755 [1761]), 53–61. The present formula for *sublimé corrosif* is $HgCl_2$, while the one for La Garaye's mercurial salt, calomel, or mild mercury, Hg_2Cl_2, contains much less chlorine.

[13] Macquer, draft of the memoir on La Garaye's salts; BNF, MS Fr 9134, fols. 19r–28v, on 27v–28r.

[14] This table represented the theory accepted at that time. In the first column, titled "*Esprit en général*," i.e., acids, substances are ordered from top to bottom in decreasing affinity order for acids: fixed alkalis, volatile alkalis, absorbent earths, metallic substances. On Etienne-François Geoffroy, see Bernard Joly's essay, "Etienne-François Geoffroy (1672–1731), a Chemist on the Frontiers," in this volume. On the practical use of the table, see Lehman, "Innovation" (cit. n. 1), 20–2.

learned from Rouelle over Lémery's, which was still popular in the middle of the century.[15] The Comte de La Garaye's exotic method caught the attention of the king, who wanted to enable his *sujets* to benefit from it. It obtained its scientific status through Macquer, who brought the analysis of salts to the academic stage when he read his first memoir in May 1754 and the second devoted to mercury dissolution in December 1756, one year after the count's death.[16] After this first secret mission at the king's service, Macquer continued in this role more officially as an expert of the government at the Bureau du Commerce.

A STATE CHEMISTRY AT THE SERVICE OF THE INDUSTRIAL AND ARTISANAL WORLD

Between 1700 and 1791 the Conseil du Commerce, called Bureau du Commerce from 1722 onward, was the royal institution responsible for commerce and industry. It appointed "commissioners" to inspect and evaluate the inventions and technical improvements that came to its attention.[17] Macquer's many reports at the Bureau du Commerce demonstrate the variety of applications that involved chemistry and the state's effort to foster industrial chemistry and national independence.

In 1766, Macquer succeeded Jean Hellot, upon his death, as commissioner of dyeing at the Bureau du Commerce.[18] This position oversaw chemical evaluations of not only dyes and coloring matters, but also chemical products and manufactured products. The purpose of the evaluations was mainly economic in order to reduce imports: crystal and flint glass were generally imported from Bohemia or England, steel from Germany or England, ammonia salt from Egypt, soda from Alicante in Spain, and white lead from Holland. Before the Revolution, France imported one thousand tons per year of white lead or *ceruse*, widely used for painting, and the expenditure amounted to three million five hundred thousand *livres*.[19] Let us focus on this example.

White lead or *ceruse* was obtained by means of a very simple process: lead strips were exposed to vinegar vapors to obtain a white calx. By adding earth mixed with chalk or very white clay in various proportions, one obtained white lead.[20] In April 1779, Jean-Guillaume Laliaud applied to establish a white lead fabrication plant at Rennes near lead mines. Laliaud chose this location not only because of the potential savings but also because of the existence of unemployed unskilled manpower. Another goal was to increase the dynamism of the region:

[15] La Garaye and Macquer had followed the courses of the two great teachers of their respective centuries: La Garaye, those of Nicolas Lémery; and Macquer, those of Guillaume-François Rouelle (1703–70).

[16] Macquer, "Sur une nouvelle méthode" (cit. n. 8); Macquer, "Recherches" (cit. n. 12).

[17] Pierre Bonnassieux, *Conseil de commerce et bureau du commerce 1700–1791. Inventaire analytique des procès verbaux* (Paris, 1900); Harold T. Parker, *An Administrative Bureau during the Old Regime: The Bureau of Commerce and Its Relations to French Industry from May 1781 to November 1783* (Toronto, 1993).

[18] Christine Lehman, "Pierre-Joseph Macquer an Eighteenth-Century Artisanal-Scientific Expert," *Ann. Sci.* 69 (2012): 307–33, on 319–22.

[19] André Guillerme, *La naissance de l'industrie à Paris entre sueurs et vapeurs 1780–1830* (Paris, 2007), 347–8.

[20] Antoine Baumé, "Plomb (Art des préparations du)," in *Dictionnaire portatif des arts et métiers* (Paris, 1767), II, 325–6; Pierre-Joseph Macquer, "Céruse," in *Dictionnaire de chymie*, 2 vols. (Paris, 1766), 1:242–3. White lead is a complex chemical compound containing both lead carbonate and hydroxide. Its present formula is $2PbCO_3 Pb(OH)_2$.

By providing the Kingdom with a considerable new branch of commerce, this plant will not only employ a great number of unfortunate people but will also develop navigation and promote the operation of the lead mines of the Kingdom and will prevent France from paying a heavy tribute to foreign countries due to the amount of white lead it uses.[21]

Macquer's responsibility was to judge the quality of the applicant's white lead and compare it with the Dutch import. Macquer's report dated 7 September 1779 detailed the experiments performed. He compared the respective overall qualities of the samples—fineness, softness, friableness—and their chemical compositions.[22] However, regarding the practical quality of the product, Macquer relied on the opinions of users such as house painters and color merchants; this is why the renowned Jean-Félix Watin was invited to complement Macquer's evaluation.[23]

Moreover, Macquer had to make a selection between competitors because another applicant also asked for permission to install a white lead factory at the same time. As both applicants' white leads had successfully passed the tests, Macquer's report was positive: there was no such manufacturing site in France, demand was increasing, raw materials were available, and the climate was favorable. However, permission was linked to two conditions: the availability of the necessary funding for the investment and the applicant's skill regarding full-scale operation. This shows that Macquer's decisions could obligate the state. For example, Laliaud received 5,500 *livres* as grants from 1781 to 1784.[24] Macquer was accountable for the amounts committed by the state and had to present annual reports on the companies' operation.[25]

When dealing with chemicals, inventors had to reveal their secrets by performing experiments, often in the private laboratory of Macquer, who participated fully in operations.[26] To illustrate this, one can take the example of artificial soda. Dom Malherbe, the first inventor of an industrial production process of soda from marine salt, stated that he performed his process in the presence of Jean-Charles-Philibert Trudaine de Montigny (1733–77) and Macquer in their respective laboratories.[27] It was the same for Malherbe's collaborator, Jacques-Ignace Hollenweger, for the pro-

[21] AN, F[12]1507, Dossier "Céruse."

[22] Macquer, report dated 7 September 1779, AN, F[12]1507. See also the draft of a first report dated 19 August 1779, BNF, MS Fr 9132, fols. 163r–164v.

[23] Watin, *Art du peintre doreur et vernisseur* (Paris, 1772). In the case of manufactured objects made from steel or flint glass, about which Macquer could judge only external qualities, the evaluation was given to professionals such as the cutler Jean-Jacques Perret, author of the *L'Art du coutelier* published in the *Description des arts et métiers* of the *Académie des Sciences*, or the academician Abbé Alexis-Marie de Rochon (1741–1817), who judged the quality of optical glasses. See Macquer's reports: "dossier Mongenet et de Renaucourt" and "dossier Sanche" for steel, AN, F[12]656; and "dossier Lambert et Boyer" for crystals, AN, F[12]*130.

[24] Two out of the five companies that were given royal grants failed. Arrêt royal dated 15 February 1780, AN, F[12]1506; "Réponse des Intendants de Province à la lettre qui leur a été adressée le 24 janvier 1788," AN, F[12]1507.

[25] This was the case for Lambert's and Boyer's crystal making plant though it was under the protection of the Duke of Guignes and of the Duke of Orléans. On 14 January 1783, they obtained a yearly gratification of 5,000 *livres* during ten years together with the authorization of settling in Saint Cloud Park; AN, F[12]1486. Macquer's report dated 7 December 1783 attests to the good health of the company; "Manufacture de cristaux anglois, d'email et de cendre bleue établie dans le parc de St Cloud," 1780, AN, F[12]1489/B.

[26] Macquer, "Demande d'un privilège exclusif de vingt ans pour la fabrication du sel ammoniac par le Sr *Capelle* Me en pharmacie à Falaise," 18 October 1782, AN, F[12]2242.

[27] Dom Malherbe's papers, BNF, MS Fr 9528, fol. 241r.

cess he proposed in 1783.[28] Making artificial soda had great economic importance as France spent more than two million *livres* on importing from Alicante the soda indispensable to glass making,[29] soap making, dyeing, laundering, and so on.[30]

Macquer conscientiously repeated the operations of the various processes, especially when inventors proposed several of them, as was the case for Louis-Bernard Guyton de Morveau (1737–1816) in 1782[31] and for Hollenweger in 1783, or when an operation seemed not to fit into the currently recognized theory, as was the case for the process proposed by Malherbe.

> In 1778 . . . we were charged, the late M^r de *Montigny* and myself, to check Dom Malherbe's process. . . . In fact one could partly alkalize common salt by Dom Malherbe's process and extract some quantity of soda and, although due to some difficulties that we had foreseen we decided not to grant him the exclusive privilege he was requesting, he has nevertheless obtained this privilege since then.[32]

In fact by making iron react with sodium sulfide Malherbe's last operation contradicted the ninth column of the affinity table in which the order—sulfur, fixed alkali, iron—did not permit this action.[33] Besides, as the transformation of marine salt into soda was probably only partial, the operation could not be really profitable; it is probably why Macquer and Montigny expressed reservations, requesting a provision of 120,000 *livres* as capital and asking that full-scale tests be performed in the presence of an inspector.[34]

In order to justify a privilege, the proposed process had above all to be original. When, following Malherbe, Guyton de Morveau wrote to the Bureau du Commerce to request the authorization to open a soda factory, he received the following answer:

> The minister has observed that a privilege had been granted three years ago to start an establishment that still existed today at Rennes, and that before being able to pronounce any judgement it was necessary to check whether your secret was not the same as the one of these entrepreneurs.[35]

Macquer's report on the processes proposed by Guyton laid stress on the following points: they were different from Malherbe's and "much simpler, much safer, much easier and consequently much more practicable and much more economical."[36] Based on this report, in June 1783 Guyton obtained the authorization to sell for fifteen years the artificial soda produced by means of the two proposed methods. The plant had to

---

[28] Macquer's letter dated 20 May 1783 and Macquer's first report dated 22 July 1783, Dossier Hollenweger, AN, F^{12}2242.

[29] "Mémoire du Baron d'Ecrammeville," 6 July 1777, AN, F^{12}2242.

[30] About industrial fabrication of soda between 1777 and 1783, see John Graham Smith, *The Origins and Development of the Heavy Chemical Industry in France* (Oxford, 1979), 196–206; Charles C. Gillispie, "The Discovery of the Leblanc Process," *Isis* 48 (1957): 152–8; Anne-Claire Déré and Jean Dhombres, "Economie portuaire, innovation technique et diffusion restreinte; les fabriques de soude artificielle dans la région nantaise (1777–1815)," *Sci. Tech. Persp.* 22 (1992): 1–176.

[31] Macquer, "Procédés proposés par M^r de *Morveau* pour extraire le sel de soude du sel commun," 31 July 1782, AN, F^{12}1507.

[32] Ibid.

[33] Regarding the affinity table, see n. 14 above.

[34] "Copie du Rapport fait au Comité d'Instruction publique en date du 6 ventôse an III [1795]," AN, F^{12}1508.

[35] M. de Colonna's letter to M. de Morveau, 9 March 1782, AN, F^{12}1507.

[36] Macquer, "Procédés proposés par M^r de *Morveau*" (cit. n. 31).

be located more than ten *lieues* from Nantes, where Malherbe's factory was located. He had no right to oppose sales of soda produced by other methods.

A problem occurred when Hollenweger, supported by the academician Jean-Baptiste Meusnier (1754–93), obtained in his turn the authorization to install a manufactory of artificial soda, vitriolic acid, and marine acid fifteen *lieues* from Nantes. Although the first steps of his process were similar to Malherbe's, Hollenweger had managed to improve a vitriolic acid production apparatus that allowed the recovery and use of by-products such as the marine acid that resulted from the action of vitriolic acid on marine salt.[37] Hollenweger owed his industrial know-how to his experience at the Manufacture des glaces de Saint Gobain and at Malherbe's factory, where he had worked on developing furnaces, and to his collaboration with Meusnier.[38]

Macquer had to justify the respective advantages of the processes of making artificial soda from marine salt in competition at the three sites located in Brittany while taking into account the manufacturers' sensitivities. In spite of similar methods, Hollenweger's process surpassed Malherbe's thanks to its excellent vitriolic acid production apparatus and its more complete and profitable final method.[39] Guyton's method was simpler because it consisted in making calx or litharge react with marine salt; soda was then formed directly in contact with air.[40] As shown by the various manufactories neighboring one another, Macquer did not hesitate to let competition run its course.

> Soda, or soda salt, was a product of so large a consumption that even if there were twelve or fifteen establishments of artificial *soudières* instead of two or three, it is likely that they would all be successful provided that they would be based on good principles and well managed.[41]

However, good principles and good management were not sufficient; a cheap raw material was also necessary. This is why the manufactories were all located in Brittany, where salt was cheap. The good will of the Ferme générale was also required to provide the free circulation of goods.[42]

As he dealt with dyers as commissioner of dyeing,[43] Macquer established close relationships with artisans and entrepreneurs in order to improve processes. Indeed, the many files of the Bureau du Commerce, which Macquer was involved with, reveal the interbreeding of cultures related to chemistry, which required not only practical

---

[37] Macquer, "Addition au rapport des procédés du Sr *Hollen-Weger*," 24 August 1783, AN, F¹²2242.

[38] Déré and Dhombres, "Economie portuaire" (cit. n. 30), 26–8; Parker, *An Administrative Bureau* (cit. n. 17), 76–8.

[39] Macquer, "Addition au rapport des procédés du Sr *Hollen-Weger*" (cit. n. 37).

[40] On 2 May 1782, Guyton de Morveau had filed his process in a sealed note at the Académie des Sciences des Arts et belles lettres of Dijon, "Des vrais procédés économiques de décomposition du sel marin." This sealed note was opened during the session held on 15 July 1818; see *Mémoires de l'Académie de Dijon* (1819), 133–4.

[41] Macquer's report dated 3 November 1783, "Comparaison des méthodes par lesquelles Mr de *Morveau* et le Sr *Hollen-Weger* retirent l'alkali minéral du sel marin," AN, F¹²2242. In contrast, in 1789, Berthollet, Macquer's successor at the Bureau du Commerce, opted for associations. He recommended that Guyton associate with Bouillon de Lagrange and later with Géraud and Carny; AN, F¹²2242 and F¹²1507.

[42] Because Guyton refused to pay these additional taxes, the first three *quintaux* of his soda production sent to a glass factory remained blocked at the Nantes city entrance for three years.

[43] Lehman, "Pierre-Joseph Macquer" (cit. n. 18), 308–22.

know-how and theoretical understanding but also close contact with the industrial and commercial world.

In the eighteenth century, chemists seldom worked alone, and, throughout his career, Macquer often collaborated with other chemists. One can cite his association with Antoine Baumé for the courses he gave from 1757.[44] This collaboration was pursued at Sèvres in the research for making hard porcelain and in writing the *Dictionnaire portatif des Arts et métiers* published in 1766, in which Baumé wrote many articles.[45] Macquer also closely collaborated with Montigny at the Sèvres Manufacture, at the Académie des Sciences when they were commissioners appointed for the examination of dyes, and at the Bureau du Commerce.[46] In addition, it was usual to appoint two or sometimes three commissioners to examine the various memoirs presented to the Académie. The research on the nature of diamond constituted yet more proof of the collaboration among the community of chemists, academicians as well as nonacademicians, in the search for truth.

A COLLEGIAL RESEARCH CHEMISTRY AT THE ACADÉMIE DES SCIENCES

The Diamond Enigma

In the 1770s, chemists began to focus their attention on past experiments on the evaporation of diamonds. In his time Boyle had already observed vapors above heated diamonds. These experiments had been repeated in Florence by order of the Grand Duke of Tuscany between 1694 and 1695 and in Vienna by order of Emperor François I in the 1710s. In the 1770s, Jean D'Arcet (1725–1801), Guillaume-François Rouelle's son-in-law, resumed these experiments, exposing diamonds to very high temperatures:

> Each of these [diamonds] was placed separately in a porcelain crucible. One of them was perfectly closed, and the other one pierced with small holes through its cover: both disappeared like a drop of the purest water would have done.[47]

When exposed to fire, diamonds disappeared, leaving no trace! Then, as quoted by the *Journal des sçavans*, everybody wanted to repeat this surprising experiment. In July 1771, Macquer was the first to reproduce it in his laboratory in the presence of chemists and jewellers as witnesses.[48] On 16 August, Hilaire Marin Rouelle (1718–79)

---

[44] "Contract between Macquer and Baumé for the chemistry course," BNF, MS Fr 9134, fols. 82–4.

[45] Lehman, "Pierre-Joseph Macquer" (cit. n. 18), 322–32; Macquer, *Dictionnaire portatif* (cit. n. 20), ix–xiv. See also Macquer, "Expériences avec Baumé sur la soie du 18 février 1778 au 25 février 1778," Muséum d'histoire naturelle, Paris, MS 283.

[46] E.g., see above the common examination of Dom Malherbe's industrial production process of artificial soda. Christine Lehman, "L'art de la teinture à l'Académie royale des sciences au XVIIIe siècle," *Methodos* 12 (2012), http://methodos.revues.org/2874 (accessed 1 April 2014).

[47] Jean D'Arcet, "Extrait de deux mémoires sur l'action d'un feu égal violent & continu pendant plusieurs jours, sur un grand nombre de pierres, de terres, de chaux métalliques, essayées pour la plupart telles qu'elles sortent du sein de la terre," *Introduction aux observations sur la Physique, sur l'histoire naturelle et les arts* (hereafter cited as *Introduction aux observations*) (August 1771 [1777]), I:108–23, on 119.

[48] "Nouvelles expériences sur les diamans exposés à l'action d'un feu violent," *Journal des sçavans* (June 1772): 435–9; Macquer, "Diamant," in *Dictionnaire de Chymie*, 4 vols. (Paris, 1778), 1:499–524, on 504. On the publication of the second edition of the *Dictionnaire*, D'Arcet took the liberty of criticizing this article; BNF, MS Fr 12305, fols. 249r–255r.

and D'Arcet performed it again in Rouelle's laboratory "in the presence of a very large audience,"[49] and the physician Augustin Roux (1726–76) reproduced it in April of the following year before his students of the Faculty of Medicine in the presence of the Lieutenant général de police, M. de Sartine.[50] Nevertheless, as shown by the works of Antoine-Laurent de Lavoisier (1743–94), Louis-Claude Cadet de Gassicourt (1731–89), and Macquer, who resumed research on diamonds at the beginning of 1772[51] and continued in collaboration with the apothecary Mitouart, diamonds did not disappear under vacuum or when they were well enclosed and sheltered from air.[52] It was the latter observation that put Macquer on the track of the similarity of diamonds with combustible bodies, in spite of the incredulity of jewelers whose ancestral tradition contradicted the volatility of diamond. As a matter of fact, diamond cutters used to expose spotted diamonds to a fire strong enough to reduce or destroy these imperfections. Yet when doing this, they took the precaution of "wrapping their diamonds with a kind of cement containing coal powder and the whole in crucibles closed as exactly as was possible."[53] Macquer then pointed out the very productive collaboration between science and art. Chemists worked with diamond cutters to perform common experiments and learned strictly artisanal know-how from them.[54]

The questions about the behavior of diamonds in fire concerned the whole community of chemists. They did not hesitate to move Macquer's or Cadet's furnaces to the laboratory where the experiments were carried out, they attempted to increase their temperature to the maximum, and they reciprocally invited one another to witness the results. The experiments on the combustion of diamond, reported immediately in *Observations sur la physique*, overthrew the current paradigm. When referring to Stahl in his *Dictionnaire* of 1766, Macquer identified diamond as the representative of the purest vitrifiable earth "able to provide a picture of the primitive elementary earth." Incidentally, Macquer doubted that he would be able to "produce a heat strong enough to melt a very hard material such as diamond, even by using the best burning mirrors."[55] In 1770, he maintained this opinion in his course at the Jardin du Roi,[56] but in the 1778 edition of his *Dictionnaire*, he acknowledged his past

[49] "Procès verbal des expériences faites dans le laboratoire de M. Rouelle sur plusieurs diamans & pierres précieuses; par MM. Darcet et Rouelle," *Introduction aux observations* (January 1772 [1777]), I:480–8.

[50] "Expériences nouvelles sur la destruction du diamant dans les vaisseaux fermés par MM. Darcet et Rouelle," *Journal de Médecine Chirurgie Pharmacie* 39 (1773): 50–86, on 66–7.

[51] "Résultats de quelques expériences faites sur le diamant par MM Macquer, Cadet et Lavoisier lu à la séance publique de l'Académie royale des sciences, le 29 Avril 1772," *Introduction aux observations* (1772 [1777]), II:108–11, and "Expériences & Observations Chymiques sur le diamant par M. Cadet," 401–9; "Expériences nouvelles de messieurs D'Arcet et Rouelle," *Introduction aux observations* (1773), I:17–34; "Nouvelles expériences sur les diamans exposés à l'action d'un feu violent," *Journal des sçavans* (1772), 435–9.

[52] Mitouart, "Résultat des Expériences faites le 30 avril 1772 sur le Diamant et sur plusieurs autres Pierres précieuses, lu à l'Académie Royale des sciences le 2 mai 1772," *Introduction aux observations* (1772 [1777]), II:112–6, and "Résultat des nouvelles Expériences sur le Diamant & le Rubis, faites le 5 mai 1772, & lu à l'Académie Royale des sciences le 9 mai 1772," 197–9 ; "Rapport des commissaires Lavoisier et Macquer lu le 16 décembre 1772," *Académie des sciences pochettes de séance*, 2 mai, 9 mai & 16 décembre 1772. See also Henry Guerlac, *Lavoisier—The Crucial Year: The Background and Origin of His First Experiments on Combustion in 1772* (New York, 1990), 79–90 and 184–7.

[53] Macquer, "Diamant" (cit. n. 48), 507.

[54] Ibid., 508.

[55] Macquer, "Terre," in *Dictionnaire de Chymie*, 2 vols. (1766), 2:567, 571.

[56] Macquer, *Cours du Jardin du Roi*, BNF, MS fr 9133, fol. 28v.

incredulity regarding the nature of diamond.[57] It is worth noting that the revival of interest in diamonds was linked to improvements in furnace design brought about by research on porcelain. D'Arcet worked with the Count of Lauraguais's furnace, and Macquer used a furnace based on the design he had developed at the Sèvres Manufacture[58] and even the hard porcelain kiln of Sèvres, in which diamonds were heated for twenty-four hours. Thus, deepening of knowledge was linked to technical development.

The unexpected behavior of diamonds incited the academicians to bring out Tschirnhaus's beautiful lens, which had lain dormant for more than fifty years (fig. 2). Indeed, it was the same lens that chemists Wilhelm Homberg (1652–1715) and Etienne-François Geoffroy had used between 1702 and 1709, but in 1772 the conditions were totally different. Homberg, physician of the Duke of Orléans, had worked with private equipment in a private laboratory; consequently, information remained confidential in spite of reading memoirs at the Académie des Sciences.[59] In contrast, the renewal of experiments by academicians in 1772 elicited a collective interest far beyond the strict circle of the Académie. The lens was placed in the Jardin de l'Infante, a site well exposed to sunshine and close to the Académie cabinets. Experiments followed a very strict program described by Lavoisier, and, as shown by the notebooks, the observations were carefully noted by the experimenters: Macquer, Lavoisier, Cadet, and the botanist Mathurin-Jacques Brisson (1723–1806). However, the assumptions proposed to explain the behavior of diamonds at high temperature were different: Macquer, the apothecary Mitouart, and the diamond cutter Maillard were convinced that diamonds burned; in contrast, Cadet thought that they vanished by decrepitation. As summarized by Lavoisier:

> Is diamond susceptible of evaporation in free air or simply of division or dislocation? Is it absolutely fixed in closed vessels? Where do the differences observed in function of the ambiances in which it was placed come from? These questions are extremely important.[60]

Only exposure to the focus of a burning mirror seemed to be able to answer this thorny question.

Experiments with Burning Lenses in the Jardin de l'Infante

The behavior of the various substances exposed to Tschirnhaus's burning lens during the summer and autumn of 1772 raised questions. A constant of the observations was the presence of fumes or vapors that escaped, taking away the secret of their composition. Experimenters would have liked to be able to capture and confine them under a bell in order to analyze them. During 1772, little initiative was left to the young Lavoisier, who had to bow to the authority of Macquer, the recognized

[57] Macquer, "Diamant" (cit. n. 48), 500.

[58] In order to check D'Arcet's experiments on the fusibility of earths, Macquer developed a new furnace derived from Pott's but using charcoal instead of wood; see Lehman, "Pierre-Joseph Macquer" (cit. n. 18), 328.

[59] Journal des sçavans (1705), 487–9.

[60] Lavoisier, "Réflexions sur les expériences qu'on peut tenter à l'aide du miroir ardent," in Œuvres, III, 264. The problem had been already raised at the time of the experiments by Macquer, Lavoisier, and Cadet in the latter laboratory; Académie des Sciences, fonds Lavoisier 72 B.

Figure 2. *Tschirnhaus's lens (ca. 1700), diameter 96 cm. Courtesy of the Deutsches Museum, Munich.*

specialist of earths and metals. But everything changed in 1773. Lavoisier, who was then only thirty years old, took the lead in the delicate operations required for exposing diamonds above water or mercury. Why did Macquer let him take on such a responsibility? Was he too busy with his administrative duties, to which the function of assistant director of the Académie had been added that year? Did he feel too old to practice this discipline of new gases, which involved know-how as well as a specific dexterity? Unlike Rouelle and Venel, who had developed Hales's apparatus and used

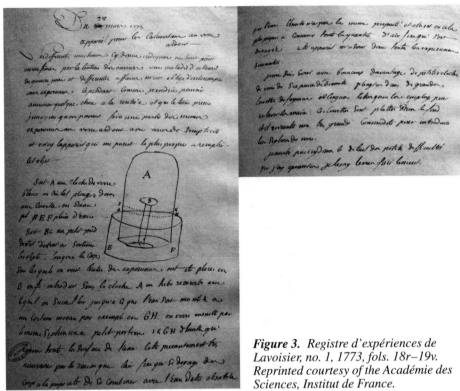

Figure 3. *Registre d'expériences de Lavoisier, no. 1, 1773, fols. 18r–19v. Reprinted courtesy of the Académie des Sciences, Institut de France.*

it in their courses, Macquer had never been interested in the presence of air in natural compounds.[61]

When the academicians began exposing substances to Tschirnhaus's lens once again in March 1773, Lavoisier took the opportunity to assert himself toward his elders. He imagined a device able to measure the air volume absorbed or emitted by calcination operations. He sketched it and described its operating mode in his laboratory notebook on 28 March 1773 (fig. 3).[62] On the following day, he tested the device with equipment used in laboratories at that time: a large glass bell, a crockery wash basin, and a small crystal pedestal for holding diamonds at the lens focus above the water level.[63] The difficulty consisted in reconciling the very high temperature at the focus and the necessity of maintaining the water level while allowing sunlight to get in. Holding samples at a sufficient distance from the water required a small diameter pedestal made of crystal in order to withstand the high temperature. The bell was

[61] Christine Lehman, *Gabriel François Venel (1723–1775). Sa place dans la chimie française du XVIIIe siècle* (Lille, 2008), 400–5.

[62] Académie des Sciences, fonds Lavoisier R1, fols. 18r–19v. This drawing was recently published by Marco Beretta in the context of the study of respiration; Beretta, "Imaging the Experiments on Respiration and Transpiration of Lavoisier and Séguin: Two Unknown Drawings by Madame Lavoisier," *Nuncius* 27 (2012): 163–91.

[63] Or a crystal column used to present fruits on sale at crockery shops; Lavoisier, "Opuscules physiques et Chimiques," in *Œuvres* (1844), I:599, n. 1. When exposing diamond to the burning lens, the crystal bell was first turned over on distilled water on which an oil layer was floating or on mercury in order to prevent gas dissolution during calcinations.

made of white glass or crystal in order to be as transparent as possible for the light beam; its diameter was six inches (about fifteen centimeters), large enough to let the beam pass through before being too concentrated. This true innovation, which looks common at first sight, reveals the sharp analysis of the problem by Lavoisier.

These experiments on diamonds with Tschirnhaus's burning lens show an unusual type of collaboration between Macquer and Lavoisier. Lavoisier was probably the operator because the experiments were performed in a closed environment, a device that he had been testing since April, and Macquer logged the observations. The analysis of the report of the seventh experiment (13 August 1773) using a glass bell over a water basin in Lavoisier's second memoir on the destruction of diamond enables us to grasp their respective roles (fig. 4).[64] Considering the words "we," "us," "one," and "I," it seems possible to distribute the respective tasks. In the first part headed "Experiment preparation," Lavoisier used "I"; he was thus manipulating. He gave the bell dimensions and the sample arrangement "on a crystal support." He turned over the bell filled with water and raised the water level by sucking. After putting the device in place, it was again he who oriented the lens and focused it on the samples. The style changes when dealing with collective observation, headed "Effect"; Lavoisier's use of "one" suggests the presence of other participants. The roles were inverted when coming to the microscope examination of diamonds that Lavoisier explicitly attributed to Macquer, transcribing the terms of Macquer's report in the common experiment notebooks.[65] The latter reported his observations as naturally as possible, keeping the strictly impersonal style of any experiment report. It is the same for the "Thoughts" that follow: they are expressed in a strictly factual style and were probably the fruit of shared discussions on the observations in the frame of common hard work. One should remember that handling the lens was hardly feasible for one person and that the light was dazzling. Besides, regardless of the need for publicity in order to validate the observations, it was generally better not to be alone to collect them.

Macquer's presence was recorded on 16 October 1773 for having observed a smell "similar to marine acid."[66] But from 20 October 1773 on, when coming to expose diamonds in fixed air under a bell above mercury, a complex device that required all Lavoisier's know-how, Macquer confessed his incompetency.

> It is easy to feel that it was a type of experiment full of difficulties; but they neither frightened nor rebuffed M *Lavoisier* to whom we should be indebted for these new experiments as he performed a large part of them, alone and at his own expense.[67]

Thus, leading the experiments became Lavoisier's responsibility.[68] He took upon himself sizing and ordering the glassware while bearing the expenses.[69] He also decided on the sequence of the experiments to be performed.

It is worth noting that the two academicians were not alone during the experiments.

---

[64] Lavoisier, "Second mémoire sur la destruction du diamant," *MARS* (1772 [1776]), II:591–616, on 596–8, or *Œuvres*, I:338–40.

[65] The list of the diamonds sent to Macquer shows this collaboration; Académie des Sciences, fonds Lavoisier 72 F.

[66] Académie des Sciences, fonds Lavoisier R2, fol. 27r.

[67] Macquer, "Diamant" (cit. n. 48), 519.

[68] Ibid., 521.

[69] Lavoisier, "Second mémoire" (cit. n. 64), 594.

SEPTIÈME EXPÉRIENCE.

Évaporation du diamant fous une cloche de verre plongée dans de l'eau.

Préparation de l'expérience.

J'ai mis fur un teffon de porcelaine très-réfractaire creufé convenablement, neuf diamans du poids de 11 grains $\frac{3}{16}$; le teffon a été placé fur un fupport de criftal, lequel a été lui-même affujetti au milieu d'une jatte de faïence émaillée, remplie d'eau diftillée ; l'appareil a été recouvert avec une cloche de criftal de fix pouces & demi de diamètre ; enfin, j'ai fucé l'air avec un tube de verre recourbé, pour faire monter l'eau à une hauteur convenable, & j'ai fait tomber fur les diamans, au travers de la cloche, le foyer du verre brûlant.

EFFET.

On n'a obfervé dans cette expérience, ni vapeur ni fumée fenfible ; mais on a remarqué très-diftinctement que le diamant [...]

Lorfque l'appareil a été fuffifamment refroidi, nous avons levé la cloche avec précaution, & nous n'avons remarqué aucune odeur fenfible ; les gouttes de liqueur qui s'étoient attachées aux parois de la cloche, pendant le refroidiffement, ne nous ont pas paru avoir aucun autre goût que celui de l'eau diftillée ; mais pour nous affurer plus particulièrement de leur nature, nous avons rincé cette cloche avec environ une demi-once d'eau diftillée, que nous avons mis foigneufement à part : nous avons de même raffemblé toute l'eau qui étoit dans la jatte ou cuvette, & nous avons réfervé le tout pour en faire un examen fcrupuleux.

Les huit diamans reftans ne fe font plus trouvés pefer que 7 grains $\frac{1}{8}$, poids de marc, au lieu de 11 grains $\frac{3}{16}$, quelques-uns étoient de couleur noire ; d'autres étoient brunâtres ; quelques-uns enfin étoient grifâtres, & avoient confervé une demi-tranfparence ; tous étoient fpongieux & caverneux comme des pierres de meulières & des pierres-ponces, & [...]

M. Macquer ayant eu la complaifance de fe charger de les examiner au microfcope, en porta lui-même le rapport fur notre journal d'expériences, & je vais le tranfcrire ici.

« Ces diamans vus au microfcope avec une lentille foible, « d'un pouce de foyer, paroiffoient fingulièrement altérés, & « comme détruits en grande partie ; la plupart étoient caverneux « comme des pains de fleur d'orange : un d'entr'eux paroiffoit « feuilleté comme un fpath ; un autre étoit creufé dans fon « [...]

RÉFLEXIONS.

Il réfulte de ces dernières obfervations, 1.° qu'il s'eft détaché du diamant, pendant fon évaporation, de petites parcelles qui ont fauté à quelque diftance ; 2.° qu'il eft probable que ces petites parcelles de diamans ont fervi de fondans à la porcelaine, qu'elles en ont procuré la vitrification & la fufion ; puifque la porcelaine feule, & dans les endroits où elle n'avoit pas eu le contact des parcelles de diamans, n'a donné aucun figne de vitrification, & eft demeurée dans le même état qu'elle étoit avant d'avoir été préfentée au foyer.

SEVENTH EXPERIMENT

Evaporation of diamond under a glass bell immersed in water

Experiment preparation

I put 9 diamonds weighing 11 grains 3/16 on a suitably carved very refractory porcelain potsherd. The potsherd was placed on a crystal support secured at the middle of a glazed crockery jar filled with distilled water; the apparatus was covered with a crystal bell of six and a half inch diameter; Last **I sucked** the air using a bended tube to make the water come up to suitable height, & **I let** the focus of the burning glass **fall** on the diamonds through the bell.

EFFECT

In this experiment **one** observed neither vapour nor smoke...
[...]
When the apparatus was enough cooled, **we** pulled the bell up with precaution, & **we** observed no appreciable smell; the liquor drops that fixed themselves on the bell sides during the cooling down seemed to **us** to have no other taste than the one of distilled water. [...]|
We rinsed the bell [...] and **we** kept the whole in order to make a scrupulous examination.

The remaining eight diamonds were found to weigh 7 grains 1/8 instead of 11 grains 3/16; some of them had a black colour. All were spongy, cavernous....
[...]

M. Macquer, who was so kind to charge himself with examining them under the microscope, wrote the report on our experiment journal and I will transcribe it here: 'these diamonds seen under the microscope with a weak lens...' [...]

THOUGHTS

It follows from the last observations 1° that small fragments have come off from diamond during its evaporation and have jumped at some distance 2°that its probable that these diamond fragments served as flux to porcelain and provided its vitrification & fusion; as porcelain alone, and in the locations where it had no contact with the diamond fragments, has given no sign of vitrification & has remained in the same state as it was before being presented to the focus.

Figure 4. *Exposure of diamond to Tschirnhaus's lens on 13 August 1773. Lavoisier, "Second mémoire" (cit. n. 64), 596–8.*

Macquer's reports imply that other people attended part of the experiments on diamonds, most likely the other two commissioners, Cadet and Brisson, and, according to Lavoisier, other witnesses such as Baumé and du Fourmi de Villiers.[70] In addition, the Baron de Sickingen—the Palatine elector—and the ambassador of Spain attended the first optical tests of Trudaine's lens, which was used in the experiments that continued in 1774.

[70] Ibid., 591.

What Is the Nature of Diamond?

According to Lavoisier's second memoir, which is limited to the year 1773,[71] and to Macquer's article, "Diamant," in his *Dictionary*, written in September 1774,[72] the answer was not categorical: "The diamond is a combustible body having an equal fixity to the one of coal." During its combustion, "it was reduced to a sort of gas, which precipitates lime water and has much similarity to the gas released from effervescences, fermentations, and metallic reductions."[73] Could the diamond (i.e., Stahl's and Macquer's "elementary earth") be mere common coal?

The similarity of coal and diamonds continued to excite the curiosity of chemists. In 1776, Rouelle and D'Arcet established that coal and diamonds did not have the same chemical properties since diamonds did not detonate in burning niter.[74] The question of the nature of diamond seems to have been finally answered by Guyton de Morveau's experiments before his students of the École Polytechnique in 1798 (*an VI*). When placed in an oxygen stream and at the focus of Tschirnhaus's lens, a diamond burned completely, and the product of the combustion, or of its combination with oxygen up to saturation, was carbonic acid with no residue.[75] Finally, as Brisson stated one year later, "The *diamond* is *pure carbon.*"[76]

CONCLUSION

This original itinerary through Macquer's activities provides glimpses of the important legacies of late seventeenth- and early eighteenth-century chemistry: for instance, the developments of Lémery's medicinal chemistry by a noble for a charitable purpose, Homberg's experiments on metals with the help of Tschirnhaus's lens, a technological research jewel of that time, and the diamond enigma raised by Boyle. It shows the impact of the affinity table on the progress of chemical theory and its role in the selection of processes at the birth of the chemical industry, the state's place in the economy, and the emergence of laissez-faire. Finally, it presents the collegial research conducted by the Académie des Sciences at the end of the century, notably, the capture of combustion gases at very high temperatures by the young Lavoisier and the abandonment of Aristotle's fourth element, Earth.

In brief, these few examples bring to light the multifarious cultures related to eighteenth-century chemistry: a medicinal chemistry, a state chemistry, a chemistry practiced not only by apothecaries and academicians but also by nobles, craftsmen, and industrialists, a public chemistry and a communal science.

---

[71] Académie des Sciences, fonds Lavoisier 72 J.

[72] Macquer, "Diamant" (cit. n. 48), 518 n. (a); Trudaine de Montigny, Macquer, Cadet, Lavoisier, and Brisson, "Premier essai du grand verre ardent de M. Trudaine," *MARS* (1774 [1778]), 62–72, and *HARS* (1774 [1778]), 1–5. This memoir was read by Brisson during the public session of 12 November 1774, Lavoisier, *Œuvres*, III, 274–83.

[73] Macquer "Diamant" (cit. n. 48), 523; Lavoisier, "Second mémoire" (cit. n. 64), on 613 and 615.

[74] "Expériences de Mrs Rouelle et Darcet sur des diamants du Brésil lues par M. Macquer," BNF, MS naf 5153, fols. 57r–v.

[75] Guyton de Morveau, "Extrait du procès-verbal des expériences faites à l'école polytechnique dans les années V et VI sur la combustion du diamant," *Annales de chimie* 31 (an VII): 72–112. Read at the session of the first class of *Institut national*, on 26 prairial an VII. *Procès verbaux Institut*, I, 16 fructidor an VI, 456 and I, 26 prairial an VII, 587.

[76] Mathurin-Jacques Brisson, *Elémens ou principes physico-chymiques destinés à servir de suite aux principes de physique à l'usage des écoles centrales* (Paris, An VIII), 175; emphasis in original.

Chemical Expertise:
Chemistry in the Royal Prussian Porcelain Manufactory

by Ursula Klein*

ABSTRACT

Eighteenth-century chemists defined chemistry as both a "science and an art." By "chemical art" they meant not merely experimentation but also parts of certain arts and crafts. This raises the question of how to identify the "chemical parts" of the arts and crafts in eighteenth-century Europe. In this essay I tackle this question with respect to porcelain manufacture. My essay begins with a brief discussion of historiographical problems related to this question. It then analyzes practices involved in porcelain manufacture that can be reasonably identified as chemical practices or a chemical art. My analysis yields evidence for the argument that chemical experts and expertise fulfilled distinct technical functions in porcelain manufacture and, by extension, in eighteenth-century "big industry," along with its system of division of labor.

HISTORIOGRAPHIC CONSIDERATIONS

The relation between the early modern "sciences" and the "arts and crafts" (or "technology" from the late eighteenth century) has long been a problematic issue for the history of early modern sciences.[1] In the 1930s and 1940s historians such as Robert K. Merton, Boris Hessen, Henryk Grossman, and Edgar Zilsel argued that there were significant interconnections between the early modern sciences, technology, and economy.[2] After a period of fierce attacks against these historians in the 1950s

---

* Max Planck Institute for the History of Science, 14195 Berlin, Boltzmannstrasse 22, Germany; klein@mpiwg-berlin.mpg.de.

[1] I use the term "science" as a shorthand for methodically gained, systematized bodies of knowledge and the activities yielding that knowledge; using this term for a period before the establishment of the modern disciplinary structure of sciences in the nineteenth century, I neither presuppose absolute continuity nor a radical rupture between the early modern and modern sciences. The term "technology" [*Technologie*], in the sense of the arts and crafts and knowledge about them taken collectively, was introduced by Göttingen professor Johann Beckmann: Beckmann, *Anleitung zur Technologie, oder zur Erkenntnis der Handwerke, Fabriken und Manufakturen, vornehmlich derer, die mit Landwirtschaft, Polizey und Cameralwissenschaft in nächster Beziehung stehn. Nebst Beiträgen zur Kunstgeschichte* (Göttingen, 1777).

[2] Robert K. Merton, *Science, Technology and Society in Seventeenth-Century England* (New York, 1970; first published in *Osiris,* 1938); Edgar Zilsel, *The Social Origins of Modern Science*, ed. Diederick Raven, Wolfgang Krohn, and Roberts S. Cohen (Dordrecht, 2000); Gideon Freudenthal and Peter McLaughlin, eds., *The Social and Economic Roots of the Scientific Revolution*, texts by Boris Hessen and Henryk Grossmann (Dordrecht, 2009). During the Cold War, the part of Merton's book that dealt with the influence of the Protestant ethos on early modern science was well received, whereas

and 1960s, their empirical studies and arguments were simply ignored.[3] During the Cold War, most historians took for granted that the early modern sciences and technology were clearly separate enterprises. In the last two decades, however, many historians have argued, based on a plethora of detailed empirical studies, that early modern natural knowledge often intersected with manufacturing and technological inquiry.[4] These historians have shown that in certain practical areas learned knowledge and manual labor were interconnected and that chemistry presents convincing examples of this interface.[5] For example, Pamela Long has argued that there were distinct "trading zones" between Renaissance humanists and practitioners, which existed mostly in urban centers and at large-scale technological sites (e.g., sites of construction of cathedrals, arsenals, mining). Harold Cook, Sven Dupré, and others have scrutinized "social networks" of early modern artisans, merchants, and savants evolving in centers of colonial trade and manufacture of luxury goods.[6] A third way to study the interface between the early modern sciences and technology highlights "experts" and "expertise," and this is the approach taken in this essay.[7]

But many questions concerning the more concrete relationship, epistemological and social, between early modern learned men, engineers, and other types of practitioners are still open to debate. While it has been broadly accepted that empirical-minded humanists and eighteenth-century savants, respectively, interacted with practitioners and took up their instruments, materials, and techniques, far less attention has been

the larger part of the book, on the interconnection of the early modern sciences with technology and capitalist economy, was ignored.

[3] A leading figure in these attacks was Rupert Hall; see Rupert A. Hall, "The Scholar and the Craftsman in the Scientific Revolution," in *Critical Problems in the History of Science*, ed. Marshall Clagett (Madison, Wis., 1969), 3–23; Hall, "Merton Revisited or Science and Society in the Seventeenth Century," *Hist. Sci.* 2 (1963): 1–16.

[4] The literature on this issue has grown quickly in the last several years. The most recent publications include Eric H. Ash, ed., "Expertise: Practical Knowledge and the Early Modern State," *Osiris* 25 (2010); Bruno Belhoste, *Paris Savant, Parcours et rencontres au temps des Lumières* (Paris, 2011); Pamela O. Long, *Artisan/Practitioners and the Rise of the New Sciences, 1400–1600* (Corvallis, Ore., 2011); Sven Dupré and Christoph Lüthy, eds., *Silent Messengers: The Circulation of Material Objects of Knowledge in the Early Modern Low Countries* (New Brunswick, N.J., 2011); Celina Fox, *The Arts of Industry in the Age of Enlightenment* (New Haven, Conn., 2010); Ursula Klein and E. C. Spary, eds., *Materials and Expertise in Early Modern Europe: Between Market and Laboratory* (Chicago, 2010); Isabelle Laboulais, *La maison des mines: La Genèse révolutionnaire d'un corps d'ingénieurs civils (1794–1814)* (Rennes, 2012); Lissa Roberts, Simon Schaffer, and Peter Dear, eds., *The Mindful Hand: Inquiry and Invention from the Late Renaissance to Early Industrialization* (Amsterdam, 2007); Matteo Valleriani, *Galileo Engineer* (Dordrecht, 2010); Sacha Tomic, *Aux origines de la chimie organique: méthodes et pratiques des pharmaciens et des chimistes (1785–1835)* (Rennes, 2010); Leslie Tomory, *Progressive Enlightenment: The Origins of the Gaslight Industry, 1780–1820* (Cambridge, Mass., 2012); Andre Wakefield, *The Disordered Police State: German Cameralism as Science and Practice* (Chicago, 2009); Simon Werrett, *Fireworks: Pyrotechnic Arts and Sciences in European History* (Chicago, 2010).

[5] See, for example, Matthew Daniel Eddy, *The Language of Mineralogy: Chemistry, John Walker and the Edinburgh Medical School* (Farnham, 2008); Ursula Klein, "The Chemical Workshop Tradition and the Experimental Practice: Discontinuities within Continuities," *Sci. Context* 9 (1996): 251–87; Ursula Klein and Wolfgang Lefèvre, *Materials in Eighteenth-Century Science: A Historical Ontology* (Cambridge, Mass., 2007); Tara Nummedal, *Alchemy and Authority in the Holy Roman Empire* (Chicago, 2007); Pamela H. Smith, *The Business of Alchemy: Science and Culture in the Holy Roman Empire* (Princeton, N.J., 1994).

[6] Long, *Artisan/Practitioners*; Dupré and Lüthy, *Silent Messengers* (both cit. n. 4); Harold J. Cook, *Matters of Exchange: Commerce, Medicine, and Science in the Dutch Golden Age* (New Haven, Conn., 2007).

[7] Ash, "Expertise" (cit. n. 4); Ursula Klein, "Introduction," in "Artisanal-Scientific Experts in Eighteenth-Century France and Germany," ed. Ursula Klein, special issue, *Ann. Sci.* 69 (2012): 303–6.

given to the question of whether certain types of practitioners actively created connections with the learned world.[8] Furthermore, in stark contrast to the argument that the early modern sciences developed in close contact with technology, the question of whether technology implemented scientific knowledge and methods is often answered in the negative.[9] In this essay I focus on the latter issue, leaving aside the inverse question of how chemical science was conditioned by the arts and crafts or technology.

The social and political nature of the relationship between early modern savants and practitioners is a related issue that is currently the subject of controversy. Historian of technology Cynthia Koepp, for example, has argued that Diderot's circle aimed to dominate artisans and craftsmen through scientific knowledge and ultimately contributed to their de-skilling in the Industrial Revolution.[10] In a fascinating recent essay on eighteenth-century English programs to improve naval architecture, Simon Schaffer has highlighted, on the one hand, conflicts and differences between practitioners (dockyard workers and shipwrights) and mathematicians allied with administrative reformers. But, on the other hand, he has also described certain common interests and collaborations between these men.[11] In the same collection of essays, Peter Dear has reinforced the emphasis on conflicts between practitioners and savants, but with reference to Enlightenment France. Dear presents the French philosophe and mathematician D'Alembert as the typical Enlightenment savant who conversed with artisans but also insisted that proper "knowledge" must be codified linguistically. As artisans often kept their knowledge secret and did not communi-

---

[8] Among the historians who have long studied the latter question are Pamela O. Long and Pamela H. Smith; see Long, *Artisan/Practitioners* (cit. n. 4); Pamela H. Smith, *The Body of the Artisan: Art and Experience in the Scientific Revolution* (Chicago, 2004); and Smith, "In a Sixteenth-Century Goldsmith's Workshop" in Roberts, Schaffer, and Dear, *Mindful Hand* (cit. n. 4), 33–58.

[9] For example, in their introduction to *The Social and Economic Roots of the Scientific Revolution*, Freudenthal and McLaughlin have emphasized that Hessen and Grossmann argued that theoretical mechanics developed in the study of technology, and that this argument is "diametrically opposed" to the wrong argument, often ascribed to them, that "theoretical mechanics was pursued in order to apply it in practice," or that existing technology could be improved by science; Gideon Freudenthal and Peter McLaughlin, "Classical Marxist Historiography of Science: The Hessen-Grossman-Thesis," in *The Social and Economic Roots of the Scientific Revolution* (cit. n. 2), 11. Likewise, Dym and Wakefield have recently argued that eighteenth-century German savant officials' (or "cameralists'") attempts to improve technology were a failure; Warren A. Dym, *Divining Science: Treasure Hunting and Earth Science in Early Modern Germany* (Leiden, 2011); Wakefield, *The Disordered Police State* (cit. n. 4). The vast majority of historians who argue instead that the eighteenth-century sciences had an impact on technology have been concerned with the early stage of the Industrial Revolution in Great Britain. See, for example, Ian Inkster, *Science and Technology in History: An Approach to Industrial Development* (New Brunswick, N.J., 1991); Margaret C. Jacob and Larry Stewart, *Practical Matter: Newton's Science in the Service of Industry and Empire, 1687–1851* (Cambridge, Mass., 2003); Peter M. Jones, *Industrial Enlightenment: Science, Technology and Culture in Birmingham and the West Midlands, 1760–1820* (Manchester, 2008); Joel Mokyr, *The Enlightened Economy: An Economic History of Britain, 1700–1850* (New Haven, Conn., 2009).

[10] Cynthia J. Koepp, "The Alphabetical Order: Work in Diderot's Encyclopédie," in *Work in France: Representations, Meaning, Organization, and Practice*, eds. Steven L. Kaplan and Cynthia J. Koepp (Ithaca, N.Y., 1985), 229–57. In the same book, W. H. Sewell presents a similar argument: William H. Sewell Jr., "Visions of Labor: Illustrations of the Mechanical Arts before, in, and after Diderot's Encyclopédie," in *Work in France,* 258–6.

[11] Simon Schaffer, "The Charter'd Thames: Naval Architecture and Experimental Spaces in Georgian Britain," in Roberts, Schaffer, and Dear, *Mindful Hand* (cit. n. 4), 279–305. With respect to the German (Saxon) context, W. A. Dym argues even more radically, and far less convincingly, that enlightened savants aimed to wholesale replace artisanal traditions by a scientific approach; see Dym, *Divining Science* (cit. n. 9).

cate it in written form, D'Alembert and (according to Dear) more or less all enlightened savants questioned whether artisans possessed "knowledge." Dear interprets D'Alembert's "Preliminary Discourse" to the *Encyclopédie* as a stark example of a text that actively created a categorical distinction between science and technology. "This active creation of difference between domains claimed to be categorically distinct from one another," he further states, "is why the now-popular term 'technoscience' has such limited usefulness in historical work, whatever its validity in the metaphysical absolute."[12] Thus, based on his interpretation of D'Alembert, he goes a step further in his argumentation by questioning the methodological validity of historical studies that explore the overlapping fields of early modern scientific and technological expertise.

By contrast, several historians have effectively argued that there were many eighteenth-century French savants, among them many chemists, who had an entirely different attitude toward craftsmanship and technical expertise than D'Alembert presented in his works. As Christine Lehmann has recently shown, one of the leading eighteenth-century French chemists, Pierre-Joseph Macquer, held significant practical positions in royal manufactories for many years.[13] Even the well-to-do academician Antoine-Laurent Lavoisier closely collaborated with instrument makers and artisans.[14] If we approach the issue from the other end of the spectrum, the Paris distiller Déjean, the apothecary Antoine Baumé, and the carpenter André-Jacob Roubo, recently studied by E. C. Spary, and Bruno Belhoste, were artisans who actively created contacts with the Paris Academy of Sciences and published their knowledge.[15] Add to this the fact that half of the French chemists were apothecaries, and, like other members of the Paris Academy of Sciences, many of them were consultants of the State.

THE GERMAN CONTEXT

These recent studies have shown that the traditional "scholar-craftsman" dichotomy is an oversimplification that ignores the broad range and diversity of experts that had developed since the late Middle Ages. From the mid-eighteenth century, the term "expert" [German: *Sachverständiger* or *Sachkundiger*] was frequently used to denote a distinct group of practitioners who were hired, or encouraged, by the state and possessed outstanding skills as well as some mathematical, natural, and technical knowl-

---

[12] Peter Dear, "Towards a Genealogy of Modern Science," in Roberts, Schaffer, and Dear, *Mindful Hand* (cit. n. 4), 431–41, on 435.

[13] Christine Lehman, "Pierre Joseph Macquer an Eighteenth-Century Artisanal-Scientific Expert," in Klein, "Artisanal-Scientific Experts" (cit. n. 7), 307–33.

[14] Belhoste, *Paris Savant* (cit. n. 4). For more on Lavoisier and instrument makers, see Marco Beretta's essay, "Between the Workshop and the Laboratory: Lavoisier's Network of Instrument Makers," in this volume.

[15] The largely unknown Paris distiller [Antoine Hornot?] Déjean, who defined himself as a chemical expert, was the author of a *Traité raisonné de la distillation* (1759), while Antoine Baumé is well known in the history of chemistry as a member of the Academy of Sciences, collaborator of Macquer, and author of several chemical and pharmaceutical treatises and essays; see E. C. Spary, *Eating the Enlightenment: Food and the Sciences in Paris, 1670–1760* (Chicago, 2012), 146–94. Roubo contributed a long essay titled "Art du menuisier" to the Academy's *Descriptions des arts et métiers*; see Bruno Belhoste, "A Parisian Craftsman among the Savants: The Jointer André-Jacob Roubo (1739–1791) and His Works," in Klein, "Artisanal-Scientific Experts" (cit. n. 7), 395–412; Belhoste, *Paris Savant* (cit. n. 4), 174–9.

edge.[16] "Experts" was a generic term that comprised figures such as engineers, architects, mathematical practitioners, mining officials, arcanists, chemists, mineralogists, botanists, physicians, and so on. As some of these experts were also recognized as savants [German: *Naturforscher*], studies of the knowledge of experts, or "expertise," are a promising way to gain insights into the interconnections between early modern science and technology.

I have shown elsewhere that eighteenth-century German chemistry had close connections with pharmacy as well as with mining and metallurgy and that many eighteenth-century German chemists operating in these contexts were "artisanal-scientific experts," or "savant experts," working as consultants, or officials, of the state.[17] Some of these hybrid figures had studied at universities (mostly medicine) and were recognized both as *Naturforscher* and as mining officials doing technical and administrative work. Others had received apprenticeship training and had gone on to become practicing apothecaries as well as chemists, botanists, and members of scientific academies.[18] The German apothecary chemists were recognized as *Naturforscher*, but they were also directly involved in trade and production processes. Likewise, the savant mining officials were both members of scientific academies and technical experts; they were not merely administrators but also carried out technical labor that was placed in a highly developed system of division of labor.[19] These hybrid figures implemented chemical knowledge about materials and analytical methods in labor, and they further performed exploratory technological experiments. At the same time they contributed significantly to the development of chemistry, mineralogy, and geology.

There were many material preconditions and social incentives for eighteenth-century German apothecaries to study chemistry, the most notable of which were pharmaceutical laboratories, chemical periodicals, and new forms of education and training, ranging from the Collegium medico-chirurgicum in Berlin to the private chemical-pharmaceutical boarding schools. We now have concrete examples of rather unknown apothecaries who began to write down and publish their experimental trials in the *Annalen der Chemie* or in Göttling's *Almanach oder Taschenbuch für Scheidekünstler und Apotheker*. From here it was only a small step to contribute more regularly to these journals and then to broaden chemical studies to new objects of inquiry beyond chemical pharmacy.[20] I have also followed the ways in which

---

[16] See Ash, "Expertise" (cit. n. 4); Stephen Johnston, "The Identity of the Mathematical Practitioner in 16th-Century England," in *Der „mathematicus": Zur Entwicklung und Bedeutung einer neuen Berufsgruppe in der Zeit Gerhard Mercators*, ed. Irmgard Hantsche (Bochum, 1996), 93–120; Ursula Klein, "Savant Officials in the Prussian Mining Administration," in Klein, "Artisanal-Scientific Experts" (cit. n. 7), 349–74; Klein, "The Prussian Mining Official Alexander von Humboldt," *Ann. Sci.* 69 (2012): 27–68.

[17] See Klein, "Introduction" (cit. n. 7); Ursula Klein and E. C. Spary, "Introduction: Why Materials?" in Klein and Spary, *Materials and Expertise* (cit. n. 4), 1–23.

[18] Ursula Klein, "Apothecary's Shops, Laboratories and Chemical Manufacture in Eighteenth-Century Germany," in Roberts, Schaffer, and Dear, *Mindful Hand* (cit. n. 4), 246–76; Klein, "Blending Technical Innovation and Learned Natural Knowledge: The Making of Ethers," in Klein and Spary, *Materials and Expertise* (cit. n. 4), 125–57; Klein, "Apothecary-Chemists in Eighteenth-Century Germany," in *New Narratives in Eighteenth-Century Chemistry*, ed. Lawrence M. Principe (Dordrecht, 2007), 97–137.

[19] Klein, "Prussian Mining Official"; Klein, "Savant Officials" (both cit. n. 16); for a broader overview of technoscientific, or artisanal-scientific, activities in eighteenth-century chemistry, see Ursula Klein, "Technoscience avant la lettre," *Perspect. Sci.* 13 (2005): 226–66.

[20] Klein, "Blending Technical Innovation" (cit. n. 18). It should be noted that my studies have also shown that there were some differences between a leading apothecary chemist like Martin Hein-

university-educated men such as Alexander v. Humboldt (1769–1859) and Carl Abraham Gerhard (1738–1821) became acquainted with mining and metallurgy, appropriated knowledge and techniques of mining experts, inspected and improved mines and metallurgic factories, performed technological experiments, and contributed to the establishment of new teaching institutions for miners and mining officials.

In this essay, I want to further extend this approach by considering the relationship between porcelain manufacture and chemical expertise. Concentrating on the Royal Prussian Porcelain Manufactory in Berlin and its various types of chemical experts, I take a close look at "laboratory workers" [*Laboranten*], "arcanists" [*Arcanisten*], and external savant experts such as the mineralogist and mining official Dietrich Ludwig Gustav Karsten (1768–1810), the apothecary chemist Sigismund Hermbstaedt (1760–1833), and the apothecary chemist Martin Heinrich Klaproth (1743–1817). From 1786 onward, the board of the Royal Prussian Porcelain Manufactory pursued a new policy of recruiting laboratory workers and arcanists and organizing their more formal chemical education and training. It also promoted the collaboration between internal experts and external savant experts, for example, in the context of committees of inspection. And it further promoted the improvement of the manufactory's laboratory.[21] All of this entailed refinement of chemical expertise and analytical methods involved in porcelain manufacture. Based on archival material, I trace these changes. I will focus, in particular, on the collaboration between the laboratory worker, and later arcanist, Friedrich Bergling and the apothecary chemist and member of the Royal Prussian Academy of Sciences Martin Heinrich Klaproth.[22]

THE ROYAL PRUSSIAN PORCELAIN MANUFACTORY

Production of porcelain was a novel art in eighteenth-century Europe. In the sixteenth century, demand for hard-paste porcelain, imported from China, had increased significantly. A century later, china became the great passion of princes and upper classes, who also supported efforts to imitate it. The first European hard-paste porcelain, which was almost identical to the imported china, was invented in 1708 by the German alchemist and apothecary Johann Friedrich Böttger (1682–1719) and the mathematician and chemist Ehrenfried Walther von Tschirnhaus (1651–1708). This was done in close collaboration with the Saxon mining official Gottfried Papst von Ohain (1656–1729). In 1710, Friedrich August I founded the first porcelain manufactory in Meissen, located in his castle, Albrechtsburg.

---

rich Klaproth and an apothecary occasionally contributing to chemical periodicals. Unlike Hufbauer, however, I argue that there was no huge gap between the German apothecaries and the chemists; see Karl Hufbauer, *The Formation of the German Chemical Community (1720–1795)* (Berkeley and Los Angeles, 1982).

[21] As Christine Lehman has recently shown, similar developments took place in France. The Royal Porcelain Manufactory of Sèvres employed *chimistes académiciens*, of whom Pierre-Joseph Macquer was one of the most famous; Lehman, "Pierre Joseph Macquer" (cit. n. 13). From 1731, the Royal Porcelain Manufactory of Meissen employed four *Arcanisten*, two of whom were university-educated physicians knowledgeable in chemistry; see Otto Walcha, *Meißner Porzellan* (Gütersloh, 1975), 85, 128. A comparison between these three manufactories is included in Ursula Klein, "Depersonalizing the Arcanum," *Tech. & Cult.* (forthcoming).

[22] In this essay, my focus is on Klaproth alone. For a more elaborate analysis of the activities of Karsten and of Humboldt, see Ursula Klein, "Chemical Experts at the Royal Prussian Porcelain Manufactory," in "Sites of Chemistry in the Eighteenth Century," ed. John Perkins, special issue, *Ambix* 60 (2013): 99–121.

Figure 1. *Drawing of the Royal Prussian Porcelain Factory from 1818 by Eduard Gaertner. Reprinted courtesy of the Stiftung Stadtmuseum Berlin.*

The Royal Prussian Porcelain Manufactory of Berlin was founded in 1763 at the end of the Seven Years' War.[23] It comprised several buildings (fig. 1), and in the 1780s it employed an average of 400 workers [*ouvriers*], among them children and women. Like state-directed mining, the building of warship fleets, drainage of swamps, construction of canals, and so on, the eighteenth-century royal porcelain manufactories belonged to "big industry" of the time. Such industry was marked by a high degree of division of labor, a complex organization, and the deployment of outstanding expertise. The following data may suffice to provide an overview of the system of division of labor in the Berlin Manufactory.[24]

Around 41% of the manufactory's workers, called *ouvriers*, were highly qualified artisans and craftsmen who had been apprenticed to a master of the manufactory for six to seven years and were relatively well paid. To this group belonged the painters and the artistic workers who fashioned the porcelain ware, the masters responsible for tools and machines, and the laboratory workers [*Laboranten*]. Around 36% of the manufactory's workers were less qualified people in charge of firing, polishing, and so on. The remaining 23% were workers without much previous training who performed simple, but often physically demanding, labor in the mill, furnace house, stable, and so on. The manufacturing process was managed and supervised by a corps

[23] Gustav Kolbe, *Geschichte der Königl. Porzellanmanufaktur zu Berlin* (Berlin, 1863); Erich Köllmann and Margarete Jarchow, *Berliner Porzellan*, Textband (Munich, 1987); Arnulf Siebeneicker, *Offizianten und Ouvriers: Sozialgeschichte der Königlichen Porzellan-Manufaktur und der Königlichen Gesundheitsgeschirr-Manufaktur in Berlin, 1763–1880* (Berlin, 2002).

[24] The following paragraph is based on Siebeneicker, *Offizianten und Ouvriers* (cit. n. 23), 87–8, 236–40. For a more detailed description of the various technical steps involved in porcelain manufacture, see Klein, "Chemical Experts" (cit. n. 22).

of officials [*Offizianten*] who numbered more than thirty in the 1780s. This group consisted of the manufactory's two directors, accountants, and other specialists for management and commerce (ca. 60% in the 1780s) as well as artistic and technical experts and their assistants (ca. 40%).

The most important technical officials, studied in detail in this essay, were the manufactory's two arcanists [*Arcanisten*]. They were the experts who knew the secret [Latin: *arcanum*] of how to make porcelain. As we will see below in more detail, they were knowledgeable in chemistry, one of them being a university-educated man who was explicitly recognized as a chemist. The arcanists were responsible for a kind of quality control that required them to chemically analyze the ingredients for the porcelain paste and the glaze, calculate the proportions for mixing the paste and the glaze, and supervise the groups of workers who actually made the paste and the glaze and who fired the porcelain.[25] The arcanists also performed technological experiments for improving all of these items.

A second group of technical experts were the manufactory's "laboratory workers" (usually two to three men). They worked in the "pigment laboratory" [*Farbenlaboratorium*], where they prepared pigments, fluxes, and oils used by the painters for overglaze ornamenting and subsequent enameling of the porcelain ware (fig. 2). They knew the second part of the manufactory's *arcanum:* the preparation of pigments and enameling. These experts also carried out experiments for improving pigments and extending the spectrum of colors. When the manufactory was reorganized in 1786–7—a crucial step in the development of chemical expertise—they were promoted to the group of officials.

Generally speaking, the division of labor in the manufactory was organized along three different axes: first, type of work (hard physical labor, administrative, technical, and artistic work); second, work that required long, short, or almost no previous training; third, division into workers [*ouvriers*] and state officials [*Offizianten*]. The manufactory was hierarchically structured, the king and the director, or the board (from 1786), being at the top of the hierarchy, the officials below them, and the *ouvriers* below the officials. Moreover, the officials and highly qualified workers had assistants.

In addition to the laboratory workers and arcanists, the Prussian State administration also invited savants, mostly members of the Royal Prussian Academy of Sciences, who were recognized as chemists or mineralogists, to become teachers of the manufactory's experts as well as members of committees to inspect porcelain manufacture and promote improvements. It was this practice that led the well-known chemist Martin Heinrich Klaproth to become a member of an inspecting committee. These external chemical experts brought new chemical knowledge into the manufactory. Even if they were teachers of would-be laboratory workers and arcanists from 1787, they also collaborated with the latter, as we will see below.

It is perhaps tempting to regard the division between the *ouvriers* and the technical officials, who collaborated with external savants, as a continuation of the ancient distinction between "hand and mind." This conclusion, however, would be a simplification that obscured other important features. While such a move would help to illuminate social hierarchy and distribution of power, it would be a hindrance to an

---

[25] Siebeneicker, *Offizianten und Ouvriers* (cit. n. 23), 101, 236–48. A more detailed story about the manufactory's arcanists and external chemical experts before 1786 can be found in Klein, "Depersonalizing the Arcanum" (cit. n. 21).

Figure 2. *The left part depicts grinding of pigments, painting, and enameling in a muffle fur-*
nace. Nicolas-Christiern de Thy, Comte de Milly, Die Kunst das ächte Porcellan zu verferti-
gen, *translated and annotated by Daniel Gottfried Scherber (Kaliningrad, 1774), plate VIII.*

appropriate historical understanding of the role played by the state and its officials in
the evolution of technical expertise. I argue that the high degree of division of labor
in the big industry of the eighteenth century also meant the employment of new types
of "experts" and that these experts' activities and knowledge ("expertise") resist the
simple hand-mind or craftsmen-scholar distinction. A careful analysis shows that all
of the manufactory's experts actually combined hand and mind.

First, the artistic and technical officials—the master painters and arcanists—
worked with their hands as well. This kind of handiwork was not related to the immediate production of goods but, rather, to the preparatory technical work comprising
the creation of plaster models, design, quality tests, and chemical analysis of materials. With respect to the arcanists, it also included calculating the proportions of the
ingredients of the paste and glaze. The latter was based on the chemical analysis of
ingredients, which varied according to their specific local provenance. Clearly, "hand
and mind" were joined in their case. Regarded from the perspective of the overall
system of division of labor, these officials participated in the production process,
though in a more mediated way than the *ouvriers*. Their expertise and work had an
important technical function in the production of high-quality, luxury porcelain. Second, I would further argue that the arcanists' expertise was not categorically different
from that of the highly qualified technical *ouvriers*, the laboratory workers. This can

be seen if we scrutinize the arcanists' and laboratory workers' activities and expertise (see below). It is further evinced by the careers of laboratory workers who became arcanists as well as by the fact that the laboratory workers were promoted to the group of officials in the course of the manufactory's reorganization in 1786–7. The archival material about the laboratory worker, and later arcanist, Bergling, presented below, will illuminate the role played by these experts in experimentation, the writing of reports, and the making of chemical knowledge.

INSPECTION COMMITTEES

Shortly after Friedrich Wilhelm II had ascended to the throne in August 1786, the manufactory was reorganized in several respects.[26] In April 1787, the influential Minister Friedrich Anton von Heinitz (1725–1802), who had long been in charge of the administration of the Prussian mining and smelting industry, convinced the new king that the manufactory's directors must be replaced by an administrative board (the so-called *Porzellanmanufacktur-Kommission*).[27] The new board consisted of the former two directors and a mining councillor named Friedrich Philipp Rosenstiel (1754–1832), who had worked in Heinitz's Mining and Smelting Department. Heinitz served as the committee's president. This reorganization linked the manufactory to the general Prussian State administration in Berlin, the *Generaldirektorium*, thereby enabling Heinitz to pursue the same policy of innovation that he had long sought to establish in Prussian mining and metal production.

Minister Heinitz was an adherent of cameralism and of the Enlightenment, but he was also a practical man, long experienced in the technicalities of mining and manufacture. He fully endorsed the high value that the Enlightenment attributed to knowledge in a manner that did not equate "knowledge" with scientific theory. Heinitz's understanding of technical innovation was instead accompanied by attempts to find new ways of combining technical expertise with knowledge produced by mathematicians and empirically minded savants such as chemists and mineralogists. Instead of wholesale transformation of artisanal traditions, he was looking for experts who combined their knowledge with insights about things and processes [*Sachwissen*] stemming from chemistry, mineralogy, mathematics, and other "useful sciences."[28] In Heinitz's reorganization of the manufactory, the significance of chemistry changed through the establishment of long-term inspections that organized the collaboration between internal and external chemical experts. These changes also included making improvements in the pigment laboratory, writing experimental reports, and requiring more formal chemical training and education of the laboratory workers and arcanists.

On 4 July 1787, the new board formally ordered the establishment of a committee for the inspection of the manufactory's pigment laboratory and further defined its goals.[29] It called for:

[26] Siebeneicker, *Offizianten und Ouvriers* (cit. n. 23), 24, 30–3.

[27] The minister's name was first written "Heynitz," but during his stay in Prussia it changed to "Heinitz." On Heinitz, see Wolfhard Weber, *Innovationen im frühindustriellen deutschen Bergbau und Hüttenwesen: Friedrich Anton von Heynitz* (Göttingen, 1976).

[28] "Useful sciences" is a contemporary term; see Ursula Klein, "Ein Bergrat, zwei Minister und sechs Lehrende: Versuche der Gründung einer Bergakademie in Berlin um 1770," *NTM* 18 (2010): 437–68.

[29] Archive of the Königliche Porzellanmanufaktur (hereafter cited as KPM), XVII 12 (Acta die Untersuchung des Farben-Laboratorii . . . betreffend), fols. 9 [3]–10 [4].

an exact examination of the pigment laboratory of the Royal Manufactory [to be] under-
taken, in order to get information about the state of this laboratory with respect to the
quantity and quality of raw materials used for the preparation of the various pigments as
well as about the laboratory workers' techniques for preparing the individual pigments
and the fluxes.

The committee was also charged with checking the quality of the laboratory instruments and furnaces, and with observing whether the laboratory workers "worked
with exactitude and cleanliness and according to the good principles of chemical
science." In addition, it had to examine the quality of the finished pigments and to
answer the question of "whether all of them possessed the required *egalité*, so that it
was certain that they always yielded the same effects when used in painting."[30] The
latter item was, in modern terms, a quest for standardization. Furthermore, the order
requested that the inspection committee make proposals for improvements and begin
"to perform chemical experiments of its own."[31]

Hence the overall goal of the inspection was twofold. The first was to observe and
gather information about a part of the *arcanum* that had long been personal knowledge of the laboratory workers. This goal was part of Heinitz's policy to eliminate
the secrecy of the arcanists and laboratory workers and transform it into a property of
the State Manufactory.[32] The second goal was technical improvement and invention.
As figure 2 shows, the preparation of pigments, painting on porcelain, and enameling were done on a small scale. The notorious problem of scaling up technological
experiments, or models, did not occur here. Hence, apart from economic considerations, exploratory technological experiments could have direct consequences for
manufacturing.

The inspection committee consisted of mining councillor Rosenstiel, the manufactory's university-educated arcanist Dr. Schopp, and the chemist Martin Heinrich
Klaproth. The latter was the most famous German chemist of the late eighteenth
century. His academic reputation derived from his public lectures, analytic rigor,
and the sensational discovery of new substances between 1789 and 1803.[33] Klaproth
was of humble social origin (his father was a tailor), and he was an apothecary, having received his training outside the university system via an ordinary pharmaceutical apprenticeship. In 1780 he had bought A. S. Marggraf's apothecary's shop in
Berlin, and he ran this shop until 1800, when he became the director of the laboratory of the Royal Prussian Academy of Sciences. Between the 1780s and 1800, his
pharmaceutical laboratory was a site both of production and of frequent scientific
experimentation.[34]

In 1788 the apothecary Klaproth had become a member of the Royal Prussian
Academy of Sciences and was thus formally recognized as a savant [*Naturforscher*].
Around this time he was also one of Heinitz's most reliable scientific experts. In 1784

[30] Ibid., fol. 9 [3].

[31] Ibid., fol. 10 [4]. For more details, see Klein, "Chemical Experts" (cit. n. 22).

[32] For this particular issue, which also concerns the fate of alchemy in the eighteenth century, see
Klein, "Depersonalizing the Arcanum" (cit. n. 21).

[33] On Klaproth, see Georg Edmund Dann, *Martin Heinrich Klaproth, 1743–1817* (Berlin, 1958).
On Klaproth's discovery of uranium in 1789, see Ursula Klein, "Klaproth's Discovery of Uranium,"
in *Objects of Chemical Inquiry*, ed. Ursula Klein and Carsten Reinhardt (Sagamore Beach, Mass.,
2014), 21–46.

[34] On this dual function of early modern laboratories, see Ursula Klein, "The Laboratory Challenge:
Some Revisions of the Standard View of Early Modern Experimentation," *Isis* 99 (2008): 769–82.

Minister Heinitz had hired him for teaching at the so-called Mining Academy in Berlin.[35] From that point forward his chemical expertise was constantly in demand. Thus his participation in the inspection committee of the Porcelain Manufactory was just one of the apothecary chemist's contributions to Heinitz's policy of innovation.

The manufactory's inspection committee was expected to do its work during the next three months and to finish its report in October at the very latest.[36] However, things turned out somewhat differently. The university-educated chemist and arcanist Dr. Schopp was apparently too old or not willing to cooperate. In the first report of the committee, which did not appear until 22 June 1789, Dr. Schopp was no longer mentioned. Instead, the name of another man, Friedrich Bergling, who soon became the major experimenter for the committee, turned up. His arrival signaled the rising importance of laboratory workers and the refinement of chemical expertise involved in porcelain manufacture.

FORMAL CHEMICAL INSTRUCTION OF LABORATORY WORKERS

Friedrich Bergling (unknown–1797) had originally been an apothecary living in Berlin. In January 1788, the board of the manufactory hired him as a "laboratory assistant" to be further trained as a true "laboratory worker." In fact, he was the manufactory's first laboratory worker, and later an arcanist, to receive some formal chemical training. For the latter purpose Heinitz involved Klaproth.

As a first step, Heinitz asked Klaproth to examine Bergling's "chemical knowledge" and send him an evaluating report.[37] On 10 January 1788, Klaproth wrote back that he had examined Bergling the same day. He also included a list of his questions along with Bergling's answers. All eight questions on his list concerned knowledge about material substances, including knowledge of their chemical identification, classification, composition, reactions, and techniques of preparation. For example, Klaproth asked Bergling how many species of earth were known and what the main differences between them were. He also asked Bergling to name the components of clay and to explain the characteristics of a metal. Klaproth concluded that "although he does not lack skill and knowledge," Bergling "is still backward concerning the proper scientific part of chemistry." He added that Bergling was "more familiar with the part of chemistry belonging to the pharmaceutical discipline." Concerning his further education and training, he recommended that Bergling should attend his own chemical course, combining it with "diligent reading of good textbooks." With respect to his practical training in the manufactory he further recommended keeping a notebook. He pointed out that in the notebook he should "not only report the work itself but also pay attention, in accordance with chemical principles, to the rationale behind procedures, their success, and the observed phenomena."[38]

---

[35] The Mining Academy of Berlin was founded in 1770. However, compared with the mining academies of Freiberg and of Schemnitz, the Berlin institution was not a true school or academy, but rather a series of lectures, organized by the Mining and Smelting Department; see Ursula Klein, "Ein Bergrat" (cit. n. 28).

[36] KPM, XVII 12, fol. 10 [4].

[37] KPM, II 1, vol. 1 (Acta die Etablirung der Königlichen Porzellan Manufaktur-Commission . . . betreffend), fol. 28 [17|; KPM, V 6 (Acta die Anstellung des Friedrich Bergling beim Laboratorio der Königlichen Porzellan Manufaktur betreffend), fol. 2.

[38] KPM, V 6, fol. 3.

Heinitz immediately followed Klaproth's recommendation to combine Bergling's practical training in the manufactory with formal instruction through lectures and textbooks. Notably, his practical training differed from traditional forms of apprenticeship in one significant aspect: the use of a notebook. When Klaproth emphasized that the notebook would help Bergling to make observations and think about chemical procedures "in accordance with chemical principles," he pointed to the essentials of the discipline of chemistry and not to first causes or principles. Here the notion of "chemical principles" referred to the ensemble of chemical techniques, concepts, and values that were shared in the chemical community.

Thus on 29 January 1788, Bergling became a laboratory assistant, and it appears that he soon convinced his superiors that he did a good job. In spring 1789 he moved on to a position as a laboratory worker, with an annual salary of 200 taler. One of his first tasks was writing a report on the pigment laboratory. On 29 May 1789, he received the order to "apply the chemical knowledge and experience hitherto acquired to the perfection of the preparation of pigments, the invention of new pigments and fluxes and the most advantageous use of enamel fire." In addition, he had to perform all preparations done in the laboratory and "clearly explain the reasons behind the applied techniques."[39]

A few weeks later, on 16 June 1789, Bergling completed the foregoing report. It was very critical.[40] The water used for preparing the pigments, the young laboratory worker pointed out, was not pure enough, and he also observed additional sources of impurity. Furthermore, materials were not stored appropriately, and the laboratory needed new pots, glass vessels, and another furnace. For a true chemical laboratory, purity of materials, clean instruments and vessels, and well-ordered storage of materials were essential. The manufactory's board accepted almost all of Bergling's suggestions. Accordingly, the manuscript includes comments at the margins such as "this shall be done for the sake of order and cleanliness."[41]

CHEMICAL IMPROVEMENT OF THE LABORATORY

Until the end of 1791, the inspection committee was preoccupied with laboratory visits, the examination of materials, and the reconstruction of pigment recipes that had long been a preserve of the laboratory workers. On 22 June 1789, a few days after Bergling had completed his own report, Rosenstiel presented the report of the committee, now consisting of only himself and Klaproth.[42] In the accompanying letter he remarked that his and Klaproth's own views "often coincide with Bergling's ideas."[43] This remark, like many similar ones, is indicative of the committee's collaborative style of working. The high mining official Rosenstiel and the savant Klaproth supported Bergling and collaborated with him rather than exploiting his handiwork for their own purposes.

The committee's report was critical, too. Like Bergling, the committee noted a deficit of good furnaces, balances, vessels, cabinets for storing materials, and many tools,

[39] Ibid., fol. 10.
[40] KPM, XVII 12, fols. 15 [9]–16 [10]. For more details concerning this report, see Klein, "Chemical Experts" (cit. n. 22).
[41] KPM, XVII 12, fol. 16 [10].
[42] Ibid., fols. 18 [12]–32 [26].
[43] Ibid., fol. 17 [11].

"whose lack is utterly unpleasant for a meticulous chemist" [*reinlicher Chemist*].[44] They even observed that the laboratory workers often used their bare hands instead of ladles to take substances out of vessels. The laboratory workers also did not weigh the ingredients for preparing pigments and fluxes precisely, and many of the materials were impure and not well ordered on shelves. Likewise, there was a problem with the standardization of ingredients. Although they did not use the term "standardization," Rosenstiel and Klaproth's insistence on "unchanging equality of quality" [*beständig gleiche Güte*] of materials used for the preparation of pigments and fluxes meant they highlighted the need for the regularized quality of raw materials.[45] The quality of the finished pigments frequently varied, as commercial raw materials were often impure. Thus, the two commissioners stated, "In our opinion the preparation of invariably good pigments for the Royal Manufactory is too important to be always dependent on the accidental quality of their ingredients, based merely on what a merchant has in stock."[46] They recommended quality tests of all materials and the construction of a storeroom to stock tested ingredients for at least a whole year.

The report also contained a long list of ingredients for pigments and fluxes as well as recipes for their preparation. The inclusion of this material stemmed from the committee's desire to shift craft secrecy away from individual laboratory workers and toward the board of the manufactory.[47] The recipes were short, resembling the type of recipes presented in pharmacopoeias. They first presented a list of ingredients, along with information about their quantities, and then gave some basic information about the techniques and tools to be used. Clearly, a lot of additional knowledge and skill were involved in the actual making of a pigment.

Klaproth added personal information for each material about the best source for obtaining it in a pure form.[48] He also offered his help to get access to pure copper and tin from overseas. This was one way in which the experienced apothecary chemist, who participated in an international commercial network, sought to improve manufacture. But for a manufactory that needed high-quality materials for its production of luxury goods, there was yet another way to achieve its goal: to buy pure materials from a chemist. Thus Klaproth began to experiment with *Magisterium plumbi*, prepared in his own pharmaceutical laboratory, to be used as a substitute for minium (used as a flux). He also sold pure nitric acid to the manufactory.[49] And he recommended the purchase of sal ammoniac, also used as an ingredient for fluxes, from the chemist Friedrich A. C. Gren, professor at the University of Halle, who "prepared and sold an excellent sal ammoniac."[50] Only a month later, the laboratory workers Bergling and Riedel experimented with samples of Gren's sal ammoniac, which the professor had sent on 14 July (1789).[51] Based on their trials, they concluded, however, that Gren's sal ammoniac was not better than the material hitherto used, although it was more expensive.

The committee's report thus presented many suggestions for improving extant

---

[44] Ibid., fol. 24 [18].
[45] Ibid., fol. 19 [13].
[46] Ibid.
[47] See Klein, "Depersonalizing the Arcanum" (cit. n. 21).
[48] KPM, XVII 12, fols. 20 [14]–23 [17].
[49] Ibid., fols. 20 [14] and 38 [32].
[50] Ibid., fol. 20 [14].
[51] Ibid., fols. 33 [27]–34 [28].

practices. In addition to the improvement of existing technologies, it also mentioned the possible utility of an invention made by Klaproth on the use of "Plantina" (platinum) for ornamentation. Klaproth had reported this invention to the Royal Prussian Academy of Sciences and had included a demonstration of samples of porcelain ornamented with platinum. He described his technique of preparing and using platinum for ornamenting porcelain and recommended that it be used as a substitute for silver, as silver did not sufficiently coat the porcelain and quickly lost its luster.[52] While Rosenstiel apparently failed to notice this invention, a comment by Klaproth in the margins of the committee's report reveals that it could have important implications for the manufactory. There he stated, "*Plantina* should be mentioned, which has been in use since the completion of the draft [of the report]."[53]

A number of the suggestions made by Bergling and the committee were soon put into practice. As early as 20 August 1789, Grieninger (Junior), who was an assistant on the manufactory's board, reported the enlargement of the laboratory (a change necessary for installing a distillation retort), improvements of storage devices as well as the purchase of many new barrels, glass vessels, and tools.[54]

COLLABORATION BETWEEN LABORATORY WORKERS AND CHEMISTS

In April 1791, Bergling finished another long report on the preparation of pigments and fluxes.[55] Around this time, he had become the most important experimenter for the inspection committee. The style of this report, especially its presentation of recipes, was similar to that of Rosenstiel and Klaproth's earlier report. Bergling presented 40 recipes altogether for different pigments, fluxes, and oils that were based on numerous experimental trials and repetitions of trials in which he had altered the proportions of ingredients or the techniques.

Many of his recipes were considerably longer than Rosenstiel and Klaproth's because they included more detailed practical information as well as explanations of observations. A good example of his explanations occurs in his observations on the preparation of a flux to be used for ornamenting with gold. During the preparation he noticed a strong smell. The ingredients used for the flux were equal proportions (1 *Loth*) of pure sal ammoniac, pure saltpeter, and calcinated borax. Bergling explained the smell by reconstructing the chemical reaction taking place after the mixture of the substances. Borax, he explained, was oversaturated with mineral alkali. When it was mixed with the other ingredients, the mineral alkali "displaces the volatile alkali of the sal ammoniac and combines itself with its salt acid [later hydrochloric acid]." The result of this reaction was the smelly volatile alkali, sal digestivum (understood as a combination of salt acid and mineral alkali) and borax glass. In a similar way, he explained other observations with displacements of, or combinations with, phlogiston.[56]

---

[52] Martin Heinrich Klaproth, "Über die Anwendbarkeit der Platina zu Verzierungen auf Porcelan," *Mém. Acad. Roy. Sci. Belles-Lett.* (1788–9), 12–15. On the history of platinum, see Bernhard Neumann, *Die Metalle: Geschichte, Vorkommen und Gewinnung nebst ausführlicher Produktions- und Preis-Statistik* (Halle, 1904), 353–64.

[53] KPM, XVII 12, fol. 21 [15].

[54] Ibid., fols. 11 [5]–14 [8]. The expenses for the enlargement and improvement of the laboratory, listed by an accountant, amounted to 95 taler; ibid., fol. 35 [29].

[55] Ibid., fols. 36 [30]–49 [43].

[56] Ibid., fols. 39 [33], 40 [34], 41 [35], 42 [36], 43 [37].

Bergling's explanations and comments about phlogiston clearly show that he shared the chemists' collective understanding of chemical reactions and movements of invisible substances such as phlogiston (or Lavoisier's "caloric"). Well into the 1790s, most German chemists, including Bergling's teacher Klaproth, kept the phlogiston theory. Klaproth, the precision analyst and discoverer of several substances, hesitated to quickly overthrow phlogiston, which was a useful concept in chemistry. From early on, he had accepted oxygen (or "dephlogistated air") as a major substance involved in combustion and other processes yielding phlogiston, but he also argued in 1789 that Lavoisier's rejection of phlogiston relied "on weak reasons."[57]

Six months later, on 19 September 1791, Rosenstiel sent a second report of the inspection committee to the board of the manufactory, the opening paragraph of which reads as follows:

> Following the order of your Excellency [von Heinitz] I had a meeting with Professor Klaproth and Herr Bergling, and we have studied in fine detail Bergling's essays about the materials used in the laboratory of the Royal Porcelain Manufactory and about the preparation of fluxes for gold, silver and so on and of the pigments.[58]

Again, this statement demonstrates that Klaproth and Rosenstiel regarded Bergling, who was a laboratory worker, as their collaborator. Far from enforcing abstract scientific ideas of Enlightenment savants onto the artisan, they first read his essay and thought about his ideas before they added their own. Although Bergling had a lower social status than the professor and the mining official, there is no indication that either saw his report as anything other than a respectable contribution to chemical knowledge. Add to this the fact that, from 1791 on, Klaproth and Bergling continually exchanged ideas about improvements of existing pigments. Klaproth made comments on Bergling's suggestions, written on the margins of the latter's report or in the committee's own report. Likewise, Bergling commented on Klaproth's suggestions for improvement. All of this took place in a context in which the two men technically had a student-teacher relationship.

Beginning in 1791, Bergling and Klaproth performed a larger number of inventive experiments. It had long been a goal of the directors of the Porcelain Manufactory to extend the spectrum of pigments so that new shades of colors could be made. To this goal, Klaproth had suggested testing entirely new pigments containing scheelite, lapis lazuli, and uranium. Among these substances, uranium was a material product stemming from systematic chemical analysis. In his analysis of uranium ores, Klap-

---

[57] Martin Heinrich Klaproth, *Vorlesungen über die Experimental-Chemie nach einer Abschrift aus dem Jahre 1789,* ed. Rüdiger Stolz, Peter Lange, and Rita Schwertner (Berlin, 1993), 7. As Karl Hufbauer has rightly observed, before ca. 1792, "German chemists still dismissed his [Lavoisier's] system as a passing French fad"; Hufbauer, *The Formation* (cit. n. 20), 96. With respect to Klaproth, it is questionable whether the reason for this "phlogistic loyalty" was "cultural nationalism" (ibid., 104). According to Hermbstaedt, Klaproth accepted Lavoisier's system in 1792 (ibid., 103). However, Klaproth did not publish on the issue. On the contrary, as Dann has argued, "he was never engaged in a written controversy (*literarischer Meinungsstreit*) with Lavoisier"; Dann, *Klaproth* (cit. n. 33), 78. Given the fact that Klaproth was an empiricist with a critical attitude toward any kind of comprehensive chemical theory, or "system," the question of what exactly he accepted in 1792 may be more complicated.

[58] KPM, XVII 12, fol. 50 [44].

roth had discovered uranium just two years before, in September 1789.[59] Immediately after his discovery, he examined the possibility of using "uranium calx" for coloring glass and painting on porcelain.[60] Apparently Bergling continued Klaproth's earlier experiments. In March 1792 he reported the results of his experiments, which had been "proposed by professor Klaproth." In each case he added samples of the colored porcelain that he had produced or provided information about failure.[61] In two additional reports from September and December 1792, he presented additional experimental results along with samples of colored material.[62] He apparently succeeded in preparing a new pigment with uranium as well as one with scheelite, which yielded "a nice yellow color."[63]

THE MAKING OF A RELIABLE ARCANIST

On 26 May 1791, the King decreed a "new *Reglement* for the personnel of the Porcelain Manufactory in Berlin."[64] The order effectively made the arcanists responsible for the two parts of the arcanum, that is to say, the making of the paste along with glazing as well as the preparation of pigments and auxiliary materials for painting. This decree was fortuitous for Bergling because in previous months he had started a series of new experiments to become acquainted with the work of arcanists as well.

Bergling had performed wet quantitative analyses of porcelain earths stemming from different deposits, and of feldspar, quartz, and some other kinds of stones. He did this with Klaproth and another savant, the mineralogist and mining official Dietrich Ludwig Gustav Karsten (1768–1810). Karsten had first studied at the Mining Academy of Freiberg from 1782 to 1786 and then continued his studies for one year at the University of Halle. In 1789 he became a mining assistant in Heinitz's Mining and Smelting Department in Berlin and began to teach mineralogy in the Mining Academy of Berlin, and in 1792 he was promoted to mining councillor. Karsten was a well-known mineralogist as well as a technical expert in mining.

At this time the Royal Prussian Porcelain Manufactory got its "porcelain earth" (a mixture of kaolin and quartz) from deposits near the town of Halle at Brachwitz, Beidersee, Morl, and Sennewitz. Other sources included Silesia (mines around Ströbel) and the town of Passau (which had previously been its main supplier). Its feldspar stemmed from Silesian deposits (Krumhübel, Lomnitz, Schreibershau).[65] The use of raw minerals coming from different natural deposits, or from different parts of the same deposit, presented a problem because their composition was never

[59] Klaproth reported his discovery to the Royal Prussian Academy of Sciences on 24 September 1789; see Registres de l'Académie, ABBAW I-IV-32, fol. 81. His first publications on the discovery are: Martin Heinrich Klaproth, "Mémoire chimique et mineralogique sur l'Urane," *Mémoires de L'Académie Royale des Sciences et Belles Lettres* 1786–7: 160–74 (the issue was backdated); Klaproth, "Chemische Untersuchung des Uranits, einer neuentdeckten metallischen Substanz," *Annalen der Chemie* 1789 (II), 387-403. Klaproth's "uranium" presumably was a lower uranium oxide. See also Klein, "Klaproth's Discovery" (cit. n. 33).

[60] Klaproth, "Mémoire chimique" (cit. n. 59).

[61] KPM, XVII 12, fols. 54 [48]–56 [50].

[62] Ibid., fols. 57 [51]–58 [52] and 59 [53].

[63] Ibid., fol. 58 [52]. "Uranium yellow" [*Urangelb*] is mentioned in a table of porcelain colors from 1838 by the manufactory's director, Georg Friedrich C. Frick; see Köllmann and Jarchow, *Berliner Porzellan* (cit. n. 23), 323 (color number 28 in the table).

[64] Geheimes Staatsarchiv Preußischer Kulturbesitz (hereafter cited as GStAPK), I. HA Rep. 151 Finanzministerium, Abt. IC, Nr. 9469, fols. 64–71.

[65] Siebeneicker, *Offizianten und Ouvriers* (cit. n. 23), 99.

exactly the same. Yet knowledge about the chemical composition of the ingredients of porcelain paste was crucial for the success of manufacture. The wet quantitative chemical analysis of minerals—one of the most recent chemical methods—yielded the most exact knowledge in this respect. Around 1790 at the very latest, it became a significant technique involved in the manufactory's quality control of materials. Thus, one of the most advanced parts of chemical science became useful in manufacture.[66]

On 6 January 1791, Bergling finished his first report on the wet chemical analysis of four samples of porcelain earth stemming from the deposits at Morl, Beidersee, Ströbel, and Passau.[67] He also performed wet quantitative analyses with feldspar, quartz, and a few additional stones, which he reported a couple of weeks afterward.[68] This shows that Bergling, the would-be arcanist, was learning by doing. But he was receiving additional instructions from Karsten and Klaproth. Karsten first read his report and then commented on it in written form.[69] His comment criticized several of Bergling's analytical results, but it ended with the remark that the substances involved in this case caused "the most difficulties in their analysis." He added that even "very good chemists, who are much more experienced, often made multifarious errors in this case."[70] Thus Karsten spelled out clearly that Bergling did chemistry here.

On 3 March 1791, Bergling submitted his written response to Karsten. It was partly a defense and partly an acceptance of Karsten's objections. He had contacted his mentor Klaproth for this purpose and had "talked to him in detail and also showed the earths to him."[71] The direct personal contact between Bergling and Klaproth apparently yielded support for the former. Bergling pointed out, for example, that Klaproth "had examined the calcareous earth in his presence and found that his results were correct and that the earth was pure calcareous earth."[72] But Bergling had yet another, even stronger argument in his favor: "I took the vitriolic acid from Professor Klaproth," he stated, "and I assumed it was certain that it is entirely pure, since it was destined for medical use."[73] Likewise, he had taken "the acetic acid from Professor Klaproth, who had confirmed that it was entirely pure."[74] On the other hand, Bergling also conceded several times that he had erred and "that Herr Assessor Karsten was right."[75] He ended by stating, "it will always be the greatest encouragement for my service to be supported in my experiments by experienced chemists."[76] Again, we encounter here a case of interaction between a laboratory worker and a savant official that displays features of collaboration rather than simple subordination of the former under the latter.

[66] The question of whether the manufactory's elderly arcanists, Schopp and Manitius, performed some kind of chemical analysis must remain open, since they did not report their activities, due to secrecy.

[67] KPM, XVII 12, fols. 66 [60]–67 [61].

[68] Ibid., fol. 68 [62].

[69] For more details concerning the interaction of Bergling and Karsten, see Klein, "Chemical Experts" (cit. n. 22).

[70] KPM, XVII 12, fol. 65 [59]. For more details, see Klein, "Chemical Experts" (cit. n. 22).

[71] Ibid., fol. 70 [63].

[72] Ibid.

[73] Ibid.

[74] Ibid., fol. 71 [64]. The apothecary-chemist Klaproth prepared his acids for medical use or for his chemical experiments.

[75] Ibid., fols. 71 [64], 72 [65].

[76] Ibid., fol. 73 [66].

In June 1793, after the elderly arcanist, Manitius, had retired, Bergling was formally promoted to the position of arcanist. From that time forward he was the manufactory's most important chemical expert, apart from the older chemist arcanist Dr. Schopp. In the spring of 1795 he compiled a comprehensive inventory, in the form of a table, of all pigments and substances contained in the manufactory's laboratory. This reveals that the laboratory held its pigments in store for at least six months, as Bergling and the inspection committee had recommended.[77] In the same year, the manufactory began to produce a new good: the so-called sanitary dishes [*Gesundheitsgeschirr*].[78] The paste for this new kind of porcelain was made with the same ingredients as the manufactory's true porcelain ("porcelain earth," feldspar, and a bit of clay) but contained a considerably larger proportion of clay than true porcelain. Because it was fired at lower temperatures than true porcelain, it was possible to use hitherto unused space in the furnaces, which meant a more economic use of fuel. Thus the sanitary dishes were cheaper than true porcelain and attracted new groups of consumers, which led to an increase in the manufactory's revenue between 1795 and 1805.[79]

The new paste was a coinvention of Bergling and a newly employed laboratory worker named Johann Georg Roesch (1767–1821), whose career was similar to Bergling's.[80] Roesch was first an apprenticed painter at the manufactory and then learned chemistry at the Mining Academy in Berlin. His teachers were Klaproth, Karsten, and the chemist Sigismund Friedrich Hermbstaedt.[81] In 1791 he became a laboratory assistant, and he was promoted to laboratory worker in 1793. In 1795, after his return from travels to porcelain manufactories and potteries in Saxony, Thuringia, and Austria, Roesch was promoted to the position of a vice-arcanist. This example further demonstrates the success of Heinitz's strategy to establish a refined form of chemical expertise alongside new forms of chemical education. The culminating point in this evolution was the establishment of a distinct research and teaching institution (the "*chemisch-technische Versuchsanstalt*") in the manufactory in 1877.[82]

CONCLUSION

Robert K. Merton, Edgar Zilsel, and some other historians argued in the 1930s and 1940s that the history of the early modern sciences must be studied in the context of technology and economy.[83] Mining, colonial trade, navigation, fortification and artillery, and the machines and transportation devices involved in these technological systems, these historians pointed out, influenced learned men's choice of objects of study. They further argued that scholars began to converse with craftsmen and

[77] Ibid., fols. 76 [68]–79 [71].

[78] The name referred to the fact that it was not glazed with lead compounds, which were poisonous. A similar product had been manufactured earlier in France. See Siebeneicker, *Offizianten und Ouvriers* (cit. n. 23), 36–7.

[79] Ibid., 37.

[80] Ibid., 149, 486.

[81] Referring presumably to Roesch, the chemist Sigismund Friedrich Hermbstaedt wrote in his teaching report from 13 October 1796 to Minister von Heinitz that "one of the young *Arcanisten* of the Porcelain Manufactory" had attended his course; GStAPK, I. HA, Rep. 121, Nr. 7959 (Acta Gen. den wissenschaftlichen Unterricht der Berg-, Hütten- und Salinen-Aspiranten betreffed), fol. 96.

[82] Siebeneicker, *Offizianten und Ouvriers* (cit. n. 23), 81–2, 161–2.

[83] Merton, *Science, Technology and Society*; Zilsel, *Social Origins of Modern Science* (both cit. n. 2).

engineers, adopted their empirical methods, and combined them with erudite knowledge and logical reasoning. These early studies and arguments, which were largely ignored during the Cold War, stimulate present historiography but also have some shortcomings. Almost all of these earlier studies were preoccupied with the question of how technology conditioned the emergence of the early modern sciences, rather than analyzing interconnections in both directions.[84] Furthermore, they drew heavily on the ancient distinction between the scholar and the craftsman.[85] While Zilsel, for example, demonstrated in several of his empirical case studies that it was actually difficult to distinguish engineers' knowledge from the knowledge of certain empirically minded scholars, he kept the simple distinction between craftsmen and scholars in his general conclusions and theoretical arguments. Zilsel did not introduce new categories that would have done justice to the broad spectrum of practitioners evolving from the late Middle Ages, such as engineers and architects, practical mathematicians and instrument makers, gunners and bombardiers, assayers and alchemical projectors, apothecaries and chemical physicians, technical officials and savant officials.

In this essay I have studied a group of chemical experts that was new in the eighteenth century: laboratory workers, arcanists, and savant chemists involved in the manufacture of European hard-paste porcelain. The arcanists and laboratory workers of the royal porcelain manufactories in Continental Europe were not individual practitioners but rather "expert officials" who held key positions within a well-developed system of division of labor organized by the state. The ideal expert for a minister like Heinitz was no longer the itinerant individual projector or alchemist, who boldly advertised his secret personal knowledge, but rather a more humble, disciplined figure. Hence, the terms "projector" and "alchemist" took on a definite negative meaning in the eighteenth-century context discussed here. He (not she!) would be knowledgeable, reliable, and ready to perform long-term service for the state.

I would further argue that the manufactory's external chemical experts—most notably M. H. Klaproth—were not just enlightened savants who would have mobilized rhetorical means in order to control the *ouvriers* and rationalize labor on the basis of preexisting knowledge. The Enlightenment and cameralist discourse certainly contributed to the events described above, but it is not easy to clearly point out what

---

[84] In this essay I have focused on the way in which chemistry entered porcelain manufacture, and not on the inverse question of how manufacture affected chemistry. But it should be mentioned here briefly that the inverse relation existed as well. Klaproth's chemistry was conditioned significantly by his work in the Berlin Porcelain Manufactory and his connections with the central Prussian Mining and Smelting Department. The material objects that were at the center of porcelain manufacture and mining—earths, stones, metal oxides, and other minerals—conditioned his choice of objects of study. Beginning in 1786, his scientific interests, which were previously wavering between pharmaceutical chemistry and analyses of all kinds of substances, decidedly turned to minerals. He became an expert on mineral analysis, which was a precondition for his discovery of new substances. Manufacture also provided some more concrete resources for his chemical analytical program. For example, beginning in 1786, Klaproth used the furnaces of the Porcelain Manufactory, which yielded the temperature of 1450°C, for his mineral analysis, and he also performed experiments in the Berlin Mint. See Klein, "Klaproth's Discovery" (cit. n. 33).

[85] On this issue, see also Long, *Artisan/Practitioners* (cit. n. 4), 92. Merton was more sensitive to the new types of practitioners. He also pointed out that "the inventor and the scientist were often one" in the early modern period (Merton, *Science, Technology and Society* [cit. n. 2], 146). This "scholar-and-craftsman thesis," which is not included in Zilsel's work, was heavily attacked by Rupert Hall ("The Scholar and the Craftsman" [cit. n. 3], 17).

exactly their contribution was. Instead, I have analyzed the pragmatic reasons for combining natural and technical knowledge along with the concrete ways in which this was actually performed.

Klaproth was Bergling's teacher, but the two men also cooperated in the laboratory, producing new forms of combined knowledge from the bottom up. Much of what laboratory workers, apprenticed arcanists, chemist arcanists, and external savant chemists involved in porcelain manufacture knew and did consisted of locally bound knowledge and methods learned on the spot. For example, knowledge about porcelain earths of different origins and their varying composition did not belong to chemists' typical canon of empirical knowledge. On the contrary, it belonged to the manufactory's *arcanum,* as did knowledge about pigments and fluxes for painting on porcelain. By means of state-organized collaboration, laboratory workers, arcanists, and chemists like Klaproth, who acquired his own expertise on porcelain earths and pigments during his tenure on the manufactory's committee, combined their local knowledge about materials with more general knowledge about chemical species and methods of chemical analysis. In this way they refined their knowledge and made first steps toward the standardization of materials.

The old argument that science had no impact on technology prior to the second half of the nineteenth century, or even the twentieth century, relies on two major presuppositions that are inappropriate in the case studied here: the equation of science with theory in the style of early modern theoretical mechanics, and of technology with simple workshop production. In the big state-organized industry of the time there was a high degree of division of labor along with a broad spectrum of different types of workers and officials, which ranged from largely untrained workers to very well trained, experimenting officials who are difficult to distinguish from savants experimenting in academic contexts. I argue that the technological and organizational complexity of this kind of industry created a demand for experts who possessed combined technical and scientific chemical knowledge as well as new values. These experts were not mere administrators and supervisors but also fulfilled distinct technical functions in the system of division of labor.

The Royal Prussian Porcelain Manufactory in Berlin was a site of production that implemented chemical science in a systematic fashion, organized by the state. Chemistry played a role in the production of porcelain paste, glazing, the preparation of pigments, and auxiliary materials for ornamenting porcelain, and to some extent for the construction of furnaces and choice of fuel as well. In addition, chemistry helped to establish new values such as cleanliness and order in the laboratory, precision of quality control of materials, standardization of materials, disciplined observation, and the willingness to write experimental reports. The most important contribution of chemistry to porcelain manufacture was chemical analysis. Wet quantitative analysis of minerals, in particular, performed at the manufactory after its reorganization in 1786–7, belonged to the most advanced chemical methods of the late eighteenth century. Unknown in any single art and craft of the time, it was a distinct part of chemistry. It contributed to the quality control and standardization of porcelain paste and thus to the manufacture of high-quality porcelain. Furthermore, chemical purity of materials was of great importance for the preparation of pigments, as minute variations in their composition caused great differences in the aesthetic quality of colors.

Pharmacy and Chemistry in the Eighteenth Century:

What Lessons for the History of Science?

*by Jonathan Simon**

ABSTRACT

This essay questions the continuity of chemistry across the eighteenth century based on an analysis of its relationship to pharmacy in France. Comparing a text by Nicolas Lémery (1675) with one by Antoine Baumé (1773), the article argues for a key transformation in chemistry across this period. The elimination of the practical side of pharmacy (indications and dosages) from chemistry texts is symptomatic of a reorientation of chemistry toward more theoretical or philosophical concerns. The essay considers several possible explanations for this change in orientation, including developments within pharmacy, but in the end privileges an approach in terms of the changing publics for chemistry in eighteenth-century France.

If we consider chemistry as a practice in its broadest sense rather than as a "modern science," with all that this latter denomination entails in terms of institutional, professional, and ideological alignments, this reflection can, even today, take us a long way from the academic discipline as taught in schools and universities. The British use of the term "chemist" to refer to a drugstore is just one example of the wider sense of the word and reminds us that it is not limited to academic or industrial researchers.[1] But what about the period before there was a recognizable modern academic discipline of chemistry?[2] The word "chemistry" is by no means new; we find texts on chemistry and references to chemists stretching back several centuries, and yet the nature of chemistry and the chemists who practiced it have changed a great deal over time. The name provides a link across the centuries, and a basis for a respectable subdiscipline, the history of chemistry. This would probably not be the case had the authors of the 1787 nomenclature reform taken the opportunity to rebaptize the science itself.[3] But

---

* S2HEP (EA 4148), Université Lyon 1, 69622 Villeurbanne Cédex, France; jsimon@univ-lyon1.fr.

[1] The most famous is, of course, Boots the Chemist. Jesse Boots was the name of the entrepreneurial apothecary who founded the company, and "chemist" was a term commonly used to refer to apothecaries in England—by metonymic shift it has come to denote the pharmacies themselves in Britain.

[2] Rudolf Stichweh identifies the second half of the eighteenth century as the origin of the modern scientific discipline that would come to replace the academic discipline as a structuring element across the nineteenth and twentieth centuries; Stichweh, "The Sociology of Scientific Disciplines: On the Genesis and Stability of the Disciplinary Structure of Modern Science," *Sci. Context* 5 (1992): 3–15.

[3] The reform of 1787 introduced a significant rupture in the history of chemistry, but while supporting the oxygen theory, the new nomenclature did not entirely abandon traditional terms. See

---

this did not happen; the history of chemistry did not lose this nominal continuity, and so thinking about what we mean by chemistry remains, for the moment at least, a relevant task.[4]

Here, I want to explore this issue of the identity of chemistry for the eighteenth century in France with particular reference to pharmacy and pharmacists. To do this, I will look at a successful chemistry book from the seventeenth century and compare it with another from the eighteenth, both written by pharmacists. The aim is to see what these texts can tell us about the changing nature of chemistry and in particular its relation to pharmacy or pharmaceutical chemistry. While there is a certain "genidentity" or continuity in chemistry between the seventeenth and eighteenth centuries, we can also identify significant shifts that signal the rise of a form of chemistry that is more familiar to us today.[5] I take this shift in chemistry, which, I argue, was characterized by the rise of philosophical chemistry at the expense of the chemical arts, to be the most significant element in the history of chemistry during this period, and I want to put this shift in relation to reforms in pharmacy as well as much wider changes with respect to the place and diffusion of chemistry in French society during this period.

There is a double methodological interest to this approach. The first idea is that we can see more clearly what constituted (and constitutes) chemistry by avoiding defining it from the inside out, as it were, and thinking of it instead in terms of its neighbors. The justification for this method is that describing an object of study in terms of what it is not opens up interesting perspectives for analysis that remain closed when one limits the question to what it is.

The second is now a classic methodological position in the history of science that invites us to valorize the embodiment of chemistry in chemists. Posing the question of who is doing chemistry how and why keeps us from the temptation of treating the science as a free-floating entity independent of the practical and professional lives of those who are taken to represent it. Thus, rather than asking simply what is chemistry, we will reorient the question by asking what is chemistry for, or what does chemistry do? Already turning the question in this sense leads us to think more in terms of what do chemists do when they are doing chemistry? This line of argument leads us quite naturally to the question of identification at the level of the historical actors: who counts as a chemist?

To open this reflection, which also contributes to the wider historical debate over continuity and rupture, or similarities and differences with respect to chemistry and pharmacy, I want to consider a well-known and often cited treatise on chemistry from the second half of the eighteenth century that clearly promotes the values of rupture and renewal.

In his entry on chemistry that appeared in the *Encyclopédie* in 1753 (vol. 3), Gabriel François Venel complained that too many people "limit the idea of chemistry to

Louis-Bernard Guyton de Morveau, Antoine-Laurent de Lavoisier, Claude-Louis Berthollet, and Antoine-François de Fourcroy, *Méthode de nomenclature chimique* (Paris, 1787).

[4] For reflections on the future of chemistry in the age of nanoscience, see the final chapter of Bernadette Bensaude-Vincent and Jonathan Simon, *Chemistry: The Impure Science*, 2nd ed. (London, 2012).

[5] The rapid success of the new chemical nomenclature proposed by Guyton de Morveau, Lavoisier, Fourcroy, and Berthollet in 1787 did much to make chemistry before this time seem more foreign than it might otherwise have appeared. For more on the effects of the nomenclature reform, see Bernadette Bensaude-Vincent and Ferdinando Abbri, eds., *Lavoisier in European Context: Negotiating a New Language for Chemistry* (Canton, Mass., 1995).

its medicinal uses." He elaborated this complaint in the following terms: "they ask of the product of an operation, what does it cure? They do not know *Chemistry* except through the remedies owed to it by practical Medicine, or at the very best by this route and through the hypotheses with which it has furnished the theoretical Medicine of the schools."[6]

To put this argument in its context, we should point out that this is the third (and last) in a list of misconceptions enumerated by Venel. After making the tendentious but highly significant claim that "Chemists still form a distinct people, very small in number, with its own language, laws and mysteries,"[7] Venel bemoans their confusion with alchemists and then, more generally, the incapacity of the uneducated to distinguish between chemists and artisans.

> Among these ill-educated people, there are those who think that to be a chemist is to have a laboratory, to prepare perfumes, phosphorus, colors, enamels, to be familiar with a large chemistry manual and the most unusual and least known procedures, in a word, to be a workman [*ouvrier*] of operations and possessor of arcana.[8]

Venel's terms of reference for these activities that the ignorant wrongly take to be "chemistry" are revealing: perfumery, forms of phosphorus (of interest due to their dramatic inflammable properties), dyeing, and enameling. That all these activities require a knowledge of chemicals and a certain skill in their manipulation is beyond question, but for Venel they do not count as chemistry. He also takes this opportunity to dissociate chemistry and chemists from the professional guild traditions that dictated the rules of practice in these areas, denouncing the secrecy surrounding many of these artisanal processes. The reference to instruction manuals is more intriguing, as learning chemistry using such a manual contrasts with the guild tradition of apprenticeship and the passing down of skills from generation to generation by word of mouth and practical work under the watchful eye of the master. Nevertheless, Venel includes these instruction manuals in his condemnation because of their lack of chemical theory and their tendency simply to list instructions. This descriptive instructional approach—still to be found in many laboratory manuals—did not fit with Venel's theory-oriented vision of the authentic science. Be that as it may, his attacks on these activities that people confused with chemistry serve to remind us how many occupations in the eighteenth century demanded a more or less extensive knowledge of chemistry—whether theoretical or (merely?) empirical: metallurgy, mineralogy, perfumery, dyeing, explosives manufacture, and so forth. According to Venel, when asked, "What is chemistry?" most people get the answer wrong. He, of course, believes he has got the answer right, but what is interesting from my perspective is that there existed a variety of answers to the question, many more, I would suggest, than today. Furthermore, Venel was just one of a number of chemists at the time trying to impose his own particular vision of what chemistry should be.

Let us return to the third of Venel's denunciations, that of the widespread identification of the term chemistry with medicine, both in the context of theories about the

---

[6] Gabriel François Venel, "Chymie," in Denis Diderot and Jean D'Alembert, *Encyclopédie ou dictionnaire raisonné des sciences, des arts et des métiers par une société de gens de lettres* (Paris, 1751–65), 3:408.

[7] Ibid.

[8] Ibid.

functioning and dysfunctioning of the human body and, more importantly for me here, "its medicinal uses." The question asked of the product of a chemical operation, "What does it cure?" is a pertinent one for patients, doctors, and pharmacists alike, but is it relevant for the chemist who performs the operation? Indeed, is it a chemist who performs the operation (and, if not, can the operation be considered chemistry)? We need to bear in mind that the preparation and sale of medicinal products that were the result of chemical operations was the legal monopoly of the pharmacist, a monopoly that has never been lost in France (although what constitutes a medicinal product, as well as the training and work of a pharmacist have changed almost beyond recognition over the past two centuries). Well into the twentieth century, pharmacists were judged by their peers and the public according to the quality of the medicines they produced, and many of the preparations required the execution of more or less complex chemical operations. As we have seen, for Venel this kind of pharmaceutical chemistry did not count as chemistry, but—by his own admission— many thought that it did. While Venel wanted to see chemistry renewed through a theoretical refoundation, citing as ideals the Cartesian, corpuscular, and Newtonian systems,[9] others, it seems, were content to consider chemistry as the science of transforming raw materials into medicines. Looking at some books on chemistry will help us better to understand why.

Thus, I want to consider two classic works on chemistry, the first by Nicolas Lémery (from 1675) and the second by Antoine Baumé (from 1773). I do not mean to hold these texts up as authoritative sources for understanding chemistry—even restricted to French chemistry—at this time but rather consider them to be exemplary texts that help us better appreciate the dynamics of the relationship between pharmacy and chemistry. This being said, the texts have not been chosen at random, as before being authors of chemistry texts both Lémery and Baumé were successful pharmacists.

NICOLAS LÉMERY

Let me start then with the *Course of Chymistry* by Nicolas Lémery, first published in 1675. This book serves as a particularly good reference on one hand because of its remarkably long publication record and on the other because of its well-established place in the history of chemistry. Thus, we still find new editions of this work being published in the eighteenth century, albeit in a highly annotated version in which the comments by Théodore Baron d'Hénouville are almost as long as the original text (see the edition of 1756).[10] The full title of the book already gives a good idea of its orientation: *A Course of Chymistry. Containing the Manner of performing those Operations that are in Use in Medicine by means of an easy method. With reflections on each operation, for informing those who want to apply themselves to this science.*[11]

---

[9] Ibid., 409.

[10] For more biographical information about Lémery, see Bernard Joly's essay, "Etienne-François Geoffroy (1672–1731), a Chemist on the Frontiers," in this volume.

[11] Nicolas Lémery, *A Course of Chymistry* (London, 1677). The original French title was *Cours de chymie, Contenant La Manière de Faire les Opérations qui sont en usage dans la Médecine, par une Méthode facile. Avec des Raisonnements sur chaque Opération, pour l'instruction de ceux qui veulent s'appliquer à cette Science*, which was translated for Walter Harris's English translation of 1677 as *A Course of Chymistry. Containing The Easiest Manner of performing those Operations that are in Use in Physick. Illustrated With many curious Remarks and Useful Discourses upon each Operation.*

Nevertheless, while Lémery's professed goal is to help those who want to perfect their skills in preparing medicines, the book has assumed a high profile in the history of chemistry because of the author's use of corpuscular theory to explain chemical properties and the course of reactions. Although this corpuscular theorizing may answer Venel's demands for a theoretical chemistry, it is not representative of the book's contents. The vast majority of the *Course of Chymistry* is composed of more or less detailed recipes for preparing chemical compounds as well as information concerning their medicinal use. Despite his corpuscular approach, which remains interesting—I do not want to deny this—Lémery's chemistry was resolutely oriented toward the question, "What does it cure?" Lémery is exemplary of a tradition of pharmaceutical chemistry that assumed an important place in French chemistry well into the eighteenth century. Indeed, this orientation was probably the principal reason for the book going into so many editions. Lémery's *Course of Chymistry* is a useful book, published in a small format that must have seen much action in pharmacists' *officines* and other chemical laboratories across Europe.

To illustrate what I mean, I will take an example from Lémery that can be compared with the same operation when we look at Baumé. Thus, I will consider the preparation of martial saffron (*Saffran de Mars* or red iron oxide). Here is the simple procedure described by Lémery for preparing this substance:

> Wash a number of iron filings well and expose them to the dew for a long time. They will rust and you should collect this rust. Put the same filings in the dew once more and gather the rust as before. Continue in this manner until you have enough.[12]

As we can see, this is not a particularly sophisticated manipulation, although it illustrates one of the basic operations that interested chemists of the time, calcination. After describing this excessively slow method using dew, Lémery presents two alternative techniques that evidently save time, although, according to Leméry, they result in an inferior product. But his entry on martial saffron does not end with the instructions on how to prepare it. He goes on to give information concerning the indications for diseases and the appropriate doses.

> It is used with success for pale complexions and the retention of menses, hydropsy (edema), and other diseases arising from obstructions: The dose is from 10 grains to two scruples in the form of tablets or pills.
> Several would have Mars taken along with purgatives, which is a very good practice.[13]

Iron is one of many substances that have the property of unblocking obstructions, be it within an organ or more generally, and Lémery provides an envelope of doses and recommendations concerning the best use of this product. While this is the aspect of Lémery's chemistry course that I want to emphasize, this is not the end of his entry on martial saffron. In the comments following the information on its preparation and use, we find the corpuscular thinking that has made Lémery such an important figure in the history of chemistry. Here, he evokes the form of the iron and steel corpuscles in order to explain the difference between the efficacy of the products obtained by the

[12] Nicolas Lémery, *Cours de chymie* (Paris, 1675), 111.
[13] Ibid., 112.

calcination of one or the other.[14] He concludes that iron is to be preferred over steel in the mixt because the pores in the iron are not as "firm" as those in steel, which means that the iron is more easily liberated and so can act more readily on the organism via its "salt." Furthermore, while steel might be purer, Lémery is concerned that the purification process drives off the most salutary parts of the iron.[15]

I do not have the space here to cite other examples, but the reader can easily verify for him- or herself that this entry for martial saffron is quite typical in its form. The entries in the book follow the same general pattern: instructions for preparing the substance, information concerning its therapeutic use, and—as an option—a theoretical discussion. The core of the work in terms of practical instruction is a list of techniques for preparing substances followed by indications for their medicinal use. If this is the standard format for the contents of a "chemistry course," what exactly does Lémery see as chemical in all this? Is it just the manipulation, the theorizing about how and why the matter transforms, or the utility of the substances produced? Lémery's definition of chemistry fails to answer these questions. He proposes a concise version of a commonplace conception, where chemistry is seen to turn around analysis:

> Chemistry is an art that teaches us how to separate the different substances that come together in a mixt.[16]

That chemistry teaches how to separate suggests that the techniques take precedence over questions of utility. But if the only place where this art is used is in the preparation of medicines, we can still legitimately ask whether it is not a subfield of pharmacy, a pharmaceutical art.

Lémery does not provide any explicit discussion of the relationship between pharmacy and chemistry, but we can nevertheless observe that chemistry seems to be something distinct from his global practice of the preparation of medicaments. Thus, one significant feature of Lémery's *Course of Chymistry* is its marked orientation toward mineral chemistry. In a 534-page book, gold, silver, tin, bismuth, lead, copper, iron, mercury, antinomy, and arsenic already take us to page 230, while chalk, coral, salt, saltpeter, ammoniac, vitriol, sulfur, and amber close out the mineral section and its twenty-one chapters. The exploration of the vegetable kingdom begins only on page 367, and the nineteen chapters on vegetable products, 140 pages in all, represent fewer than half the number of pages dedicated to the mineral kingdom. The final four chapters on animal products cover only eighty-two pages. This distribution contrasts with the pharmacopoeia, where the vast majority of the medicinal preparations were drawn from the vegetable kingdom, as we shall see in what follows.

When Lémery explicitly treats pharmacy and not chemistry, as in his *Universal Pharmacopoeia* from 1697, the distribution between the kingdoms of nature is quite different.[17] The subtitle claims that this book contains "all the pharmaceutical com-

---

[14] Lémery considered steel to be a purified form of iron.

[15] For more on this question of purity, see Jonathan Simon, "The Production of Purity as the Production of Knowledge," *Found. Chem.* 14 (2012): 83–96.

[16] Lémery, *Cours de chymie* (cit. n. 12), 2–3.

[17] Lémery, *Pharmacopée Universelle contenant toutes les compositions de pharmacie qui sont en usage dans la Médecine, tant en France que par toute l'Europe; leurs Vertus, leurs Doses, les manières d'opérer les plus simples & les meilleures* (Paris, 1697).

pounds used in Medicine, in Europe as in France; their virtues, doses and the best and simplest methods of proceeding." In order to cover this vast but very practical ground, this book stays very close to the pharmacopoeia, with entries literally constructed around ones drawn from these texts. If we look at the entries in the index under "S" (where we find martial saffron), they are dominated by the syrups—a category that does not even appear in his *Course of Chymistry*. Out of some 250 syrups, the overwhelming majority are prepared using plants. What this difference between Lémery's pharmacy text and his chemistry course suggests is that the orientation toward the mineral kingdom is characteristic of chemistry within pharmacy, a point I will return to later. At the beginning of his *Universal Pharmacopoeia*, Lémery logically enough provides a definition not of chemistry, but of pharmacy.

> We define pharmacy as an art or science that teaches us how to select, prepare and mix medicaments. It is a part of therapy or curative medicine. We divide it into two parts, Galenic and Chemical. Galenic pharmacy remains at the level of a simple mixing together, without making the effort to seek out the substances that naturally compose each of the drugs. Chemical pharmacy is concerned with the analysis of natural bodies in order to separate out the useless substances and to make more exalted, more essential remedies out of them.[18]

Galenic pharmacy concerns the simple drugs and their combinations, but if analysis is involved, the pharmacist is practicing chemistry. Thus, although Lémery invokes the alchemical terminology of exaltation and essences, he remains true to his idea of chemistry separating the components of a mixt, albeit here with a specifically practical goal. In this context, Lémery is explicitly limiting himself to a definition of chemical pharmacy or pharmaceutical chemistry—an application of chemistry to pharmacy within pharmacy. It is not, however, the only definition we find of chemistry in his *Universal Pharmacopoeia*. In his glossary we can read the following definition: "Chymia is the part of pharmacy that teaches us how to analyze mixts."[19] There are two ways to read this phrase, either an independent chemistry has a role to play in pharmacy, or chemistry is, properly speaking, a component of pharmacy. This pithy definition, in all its ambiguity, provides a good place to leave Lémery and turn to the eighteenth century.

ANTOINE BAUMÉ

In 1773, Antoine Baumé published a book called *Experimental and Rational Chemistry*, which is a more or less direct descendant of Lémery's *Course of Chymistry*. Indeed, Baumé regularly cites Lémery as an authoritative source, and yet the style and content of the book are significantly different, as we shall see. It is important to bear in mind that while he has not enjoyed the same recognition as Lémery by historians, when this book came out, Baumé was a very well established chemist. Thus, in 1772 he entered the chemistry section of the Académie des Sciences in Paris, one of a small number of pharmacists to attain this summit of official recognition in France.[20] And yet, on the rare occasions that Baumé is mentioned in histories of chemistry, it is,

---

[18] Ibid., 1.
[19] Ibid., 29.
[20] He replaced Lavoisier as *membre adjoint* when Lavoisier was promoted to the status of *membre associé*, and Baumé received the same promotion when Lavoisier was promoted *pensionnaire* in 1778.

along with Sage and Priestley, as an opponent of Lavoisier's oxygen theory. Indeed, he was probably the most prominent pharmacist (or chemist) to speak out against Lavoisier's new chemistry in France, and in return historians have not been kind to him. This being said, the fact that his own attempts at establishing a "modern" chemistry have been eclipsed by his status as an opponent in the canonical Chemical Revolution makes Baumé an interesting figure.

A qualified pharmacist, for over a decade Baumé served as the experimenter's hands for Pierre-Joseph Macquer in his Parisian chemistry courses.[21] These chemistry courses were, according to Baumé's own admission, the principal source for his *Experimental and Rational Chemistry* and, as he explains in his introduction, provided the basis for his chemical authority:[22]

> The work which I present here in the form of a body of fundamental operations in chemistry is the fruit of more than twenty-five years of labor. During this time, I was demonstrating chemistry with Monsieur Macquer, and we conducted sixteen chemistry courses together, with each course featuring more than two thousand experiments. Beyond this, I have conducted more than 10,000 experiments as a supplement to these courses, which have been the object of many memoirs, with several having been read at the Academy.[23]

This argument by sheer numbers is a reminder, if any is needed, that chemistry was considered an experimental science premised on its constitutive operations.

Appearing in three volumes, for a total of over 2,000 pages, *Experimental and Rational Chemistry* is a substantial work, and, while it made serious theoretical contributions to chemistry, it was also meant to be of practical use. Thus Baumé starts with fifty pages on the instruments needed to constitute a good laboratory, reducing the apparatus to the minimum necessary for performing chemical investigations. This kind of logic is aimed at an audience of "amateurs" who want to pursue chemistry on their own, a point I will return to shortly.[24]

Despite the very practical orientation of the opening sections, Baumé keeps to his purpose of presenting theoretical reflections in parallel with circumstantial descriptions of the chemical operations—that would doubtless be useful to student doctors and pharmacists—but these descriptions are stripped of the information concerning the medical uses or other applications of these products. This gives the book quite a different look and feel from Lémery's *Course of Chymistry*. To illustrate this point, one need only look at Baumé's presentation of martial saffron and compare it with Lémery's treatment of the same substance cited earlier.

> Place several pounds of iron filings in a large, flat, clay dish, spreading them out thinly: expose this to the dew: the surface of each filing turns into rust. Once a certain quantity has been formed, crush the filings in an iron mortar: a yellowish dust is produced that can be separated out using a sieve made of silk; that which passes through is *martial saffron*.[25]

[21] See Christine Lehman, "Pierre-Joseph Macquer: Chemistry in the French Enlightenment," in this volume.

[22] Jonathan Simon, "Authority and Authorship in the Method of Chemical Nomenclature," *Ambix* 49 (2002): 207–27.

[23] Antoine Baumé, *Chymie expérimentale et raisonnée*, 3 vols. (Paris, 1773), 1:ii.

[24] In 1778, Baumé published a book exclusively dedicated to describing the best instruments for distilling alcohol; Antoine Baumé, *Mémoire sur la meilleure manière de construire les alambics et fourneaux propres à la distillation des vins pour en tirer les eaux-de-vie* (Paris, 1778).

[25] Baumé, *Chymie expérimentale* (cit. n. 23), 2:546.

While Baumé's description of the technique provides more detail, the operation is evidently the same as the one from a century earlier. Furthermore, Baumé, like Lémery, rounds out the entry with a theoretical analysis of the reaction. The air and the dew water act like fire to remove the phlogiston from the metal, leaving a calx—the rust—an explanation of calcination that, thanks to the promotion of the "phlogiston theory" by Thomas Kuhn among others, is familiar to the majority of historians of science today. Baumé's discussion of this iron product illustrates how a number of empirical investigations had changed chemists' views on iron, and he cites the experiments performed by Réaumur to show that steel and iron are not two different metals. But what is particularly significant for the argument I am developing here is that Baumé offers no information on the therapeutic use of the substances in question—neither medical indications nor dosages. Indeed, for this kind of practical pharmaceutical information, the reader would have to turn to Baumé's *Elements of Pharmacy*, which first appeared in 1762.[26] Thus, while Baumé reproduces Lémery's approach, he has eliminated the detailed information about the medical usage of his preparations. The entry on martial saffron is not exceptional in this respect, as this information is absent from all the chapters of his *Experimental and Rational Chemistry*. While Baumé provides plenty of details concerning industrial processes, with, for example, a long section on desalination,[27] he does not have anything to say about the uses of chemical products, whether in medicine or elsewhere. As was characteristic of the entries on the arts in the *Encyclopédie* in the 1750s, Baumé is providing the information that might be of interest to a generalist educated public rather than to specialist professionals such as metallurgists, dyers, or doctors. While interested in the manufacturing processes used in France, the potential bourgeois reader would have no direct professional reason to master them. Finally, I should point out that the three volumes of *Experimental and Rational Chemistry* are entirely dedicated to the mineral kingdom and its products, a significant choice that I will return to below.

What changes might have prompted this shift in Baumé's presentation of the science? An examination of his definition of chemistry does not provide us with very many clues. Chemistry "is a science based on experience; its object is the analysis or decomposition of all bodies found in nature, and the combination of all these bodies, or their principles, among themselves to form new compounds."[28] Although it opens out onto the issue of chemical synthesis, the conception of chemistry is not very different from that proposed by Lémery a century earlier. Furthermore, Baumé insists heavily on the utility of chemistry, citing glass- and metalworking as examples of the arts that can benefit from a command of the science. Thus, he explains how iron will receive particularly close attention because of its "utility in the arts and medicine."[29] Nevertheless, it is not this orientation toward the utility of chemistry that comes across as the central message of his book. What interests Baumé much more is the chemical system he is arguing for, based on a sophisticated version of the ancient

[26] Antoine Baumé, *Élémens de pharmacie théorique et pratique . . . avec une table des vertus et doses des médicamens* (Paris, 1762); on p. 210 of the 1818 edition, iron is ascribed the paradoxical properties of being able to both provoke menstruation in women and to counter excessive "evacuation."

[27] Baumé, *Chymie expérimentale* (cit. n. 23), 3:568–89.

[28] Ibid., 1:1. For Baumé, it is also "chymie" spelled with a "y," but his image is far removed from the new historiographical category of "chymistry" used to describe alchemical techniques and knowledge.

[29] Ibid., 1:xxxi.

four-element theory. His idea is that all of chemistry can be understood in terms of an increasingly complex combination of these elements. Thus, while he might justify the treatment of each individual item by its use in the arts, his real goal seems to be promoting his combinatorial theory.

Baumé's text is symptomatic of a distinctive form of chemistry that was emerging in the eighteenth century in response to various developments that did not flow directly from chemical investigations. Globally, I would describe this change as a reorientation of chemistry toward more theoretical concerns or the rise of "philosophical chemistry," as this kind of theorizing was known at the time. Over time, this reorientation would lead to a change in the nature of the practitioners of chemistry, with new chemists espousing a new chemistry. Paradoxically, while pharmacists like Baumé were fully implicated in this transformation and embraced the more philosophical chemistry, in the end this evolution would disqualify pharmacists as the proponents of chemistry, casting them as the more or less passive beneficiaries of a science that was conducted and taught by chemists. But why was chemistry changing in this sense? While I cannot pretend to provide a complete answer to this question, which, I believe is *the* central question for the history of eighteenth-century chemistry, I can suggest two different types of explanation, one that looks to the changing status of French pharmacy for clues and another that refers to the changing audience for chemistry courses in France.

THE EVOLUTION OF PHARMACY

While it is common to think of pharmacy before the twentieth century as conservative and essentially unchanging, this caricature does not stand up to closer scrutiny. Starting with the pharmacopoeia we can see that the practice of pharmacy was evolving in the seventeenth and eighteenth centuries. It was the pharmacopoeia (albeit a formally unofficial constraint before the reforms at the beginning of the nineteenth century) that circumscribed the day-to-day work of pharmacists in producing medicines, and thus the form, content, and legal status of these pharmacopoeias influenced the extent and nature of the pharmacists' chemistry. These quasi-official (and subsequently official) lists of medicines were enriched in the seventeenth century with a growing number of preparations. Based on a study by J. Bergounioux, a historian of pharmacy from the 1920s, we can clearly see the major developments from the seventeenth into the eighteenth century. Looking at the Codex Medicamentarius of 1638 and comparing its content with the one from 1732, he noted more than twice as many medicaments (see table 1). That means twice as many products that the pharmacist had to learn to prepare. Admittedly, these two pharmacopoeias are separated by more than a century (and successive editions after 1732 suggest that this rate of growth in the materia medica was not maintained), but still, it is clear that the apprentice apothecary had an increasing repertoire to master.

Despite this growth in the total numbers of medicines, the proportions of approximately 80 percent plant-based medicines and 10 percent each from the other two kingdoms of nature remained constant between 1638 and 1758. In sum, vegetable products dominated the materia medica. Furthermore, this is just a list of the accepted medicines; it does not provide any information concerning the frequency of their prescription. It would be reasonable to suppose that vegetable preparations accounted for well over 90 percent of the apothecary's business. Nevertheless, while the majority

Table 1. Number of Preparations in Successive Editions of the Codex Medicamentarius Organized by Kingdom of Nature (Percentages in Parentheses)[a]

Date	Vegetable	Mineral	Animal	Total
1638	353 (80.8)	46 (10.5)	38 (8.7)	437
1732	730 (78)	104 (11.1)	102 (10.9)	936
1748	709 (76.1)	117 (12.6)	105 (11.3)	931
1758	707 (79.4)	97 (10.9)	86 (9.7)	890

[a] Table drawn up by J. Bergounioux; see Bergounioux, "Les éditions du Codex Medicamentarius de l'Ancienne Faculté de Médecine de Paris," *Bulletin de la Société d'histoire de la pharmacie* 58 (1928): 70–9, on 75. This article is available online at http://www.persee.fr/web/revues/home/prescript /article/pharm_0995-838x_1927_num_15_54_1668 (accessed 12 May 2014).

of the pharmacist's commerce involved vegetable products, the unchanging proportions of the three kingdoms should not blind us to the rise in the number of mineral preparations demanded of the pharmacists. This growth cannot, as in the vegetable kingdom, be attributed to the introduction of hitherto unknown plants brought back to Europe from overseas exploration but is due instead to the increasing sophistication of the medicines prepared by the transformation of known mineral substances.

In addition, not all of the plant preparations required chemical manipulation; many were simples or combinations of simples—that is to say, the plant or a part of the plant as it had been collected (its leaves, roots, bark, etc.). Furthermore, the techniques used in the preparation of other plant-based medicines were generally repetitive— maceration, distillation, etc.—similar operations leading to similar products, albeit with different pharmaceutical properties.[30] The means for preparing those medicines that did demand chemical transformation, be it extraction or reaction, were described in the handbooks that served as supplements to the central educational practice of apprenticeship. Nevertheless, the point is that despite the numerical superiority of plant products in the pharmacopoeia, the range of manipulations aimed at bringing about chemical transformations was much greater in the mineral kingdom, and the rise in pharmaceutical chemistry is tied to an increasing interest in these transformations of nonorganic materials. The proportions of the different sections of Lémery's book already attest to this phenomenon, and this orientation is only confirmed by Baumé's choice to exclude the vegetable and animal kingdoms entirely and dedicate all three volumes to the chemistry of the mineral kingdom.

Against this background of a more demanding professional practice of pharmacy that we can trace back to the seventeenth century, and which doubtless contributed to the success of Lémery's textbook and its successors, the eighteenth century was also a time of significant institutional reform in French pharmacy. Such reforms are particularly striking in a guild-based profession that even today projects a markedly conservative image. While it is impossible to generalize about reforms across France because of the essentially local nature of the functioning of guilds, pharmacists in

---

[30] To illustrate this point, one need only think of the hundreds of different herbal infusions, each prepared in the same way but with different medicinal properties depending on what leaves, roots, or bark go into its concoction.

Paris attained a new level of recognition and social promotion in the eighteenth century, culminating in the creation of the College of Pharmacy in 1777. Under royal protection, this new guild separated the pharmacists from a centuries-long union with the spicers and promised a bright future for this subaltern medical profession—a future in which pharmacists could potentially rival the status of the Paris surgeons if not the elite caste of medical doctors that dominated French medicine.[31]

Chemistry features in this institutional reconfiguration, and one sign of the rise of pharmacists was the official recognition of their right to teach chemistry—among other subjects—at the college. Starting from 1780, this right was extended to public lectures, although chemistry was only included in this public teaching program in 1797. In this period, however, the French Revolution and its aftermath radically altered the institutional situation, if not the practice of French pharmacists. Indeed, while probably unimaginable in 1777, the changes in the twenty-five years that followed would be even more dramatic for pharmacists than the creation of the college. These changes must have at first appeared catastrophic, as they started with the legal dismantling of the guild infrastructure by a pair of laws promulgated in 1791. The first established the freedom to exercise any profession (the *Allarde* decree), while the second abolished the guilds altogether (the *Le Chapelier* law). Nevertheless, this same wave of liberalization would ultimately lead to the refounding of French pharmaceutical institutions in 1803, harmonizing regulations across the country and placing pharmacy in a very strong position with respect to a substantially weakened medical corps.[32]

The law of Germinal (April 1803) cast the state in the role of guarantor of the pharmacists' monopoly over the production and sale of medicines while granting the profession considerable autonomy. This law also inaugurated a system of state-sponsored education in pharmacy that would gradually displace the apprenticeship tradition. Chemistry was included in the teaching program of the new schools, and the chemistry courses taught at the new School of Pharmacy in Paris by Vauquelin and Bouillon-Lagrange consisted in large part of a presentation of Lavoisier's new chemistry, with relatively little time spent on properly pharmaceutical chemistry.[33] Indeed, as I have argued in detail elsewhere, the beginning of the nineteenth century was a key period in the separation of chemistry from pharmacy in France.[34] All this can be included in an overall narrative of the rise of the French pharmacist at the end of the eighteenth century, and Baumé's book is, I believe, symptomatic of certain aspects of this history. His willingness to subordinate all the directly pharmaceutical applications (displacing them from his chemistry book to a pharmacy text) to chemical theory is related to pharmacists' changing self-perception. Baumé saw himself as a legitimate proponent of chemical theory who no longer needed the alibi of the pharmaceutical utility of chemical products to put forward his philosophical opinions.

[31] On the evolving structure of French medicine in the eighteenth century, see Toby Gelfand, *Professionalizing Modern Medicine: Paris Surgeons and Medical Science and Institutions in the 18th Century* (Westport, Conn., 1980).

[32] For an overview of the changing fortunes of the medical profession after the Revolution, and notably the integration of surgeons and doctors, see the first chapters of Russell Charles Maulitz, *Morbid Appearances: The Anatomy of Pathology in the Early Nineteenth Century* (Cambridge, 2002).

[33] The notes from this course, taken by a student named Moutillard, are in the archives of the Bibliothèque Interuniversité de Pharmacie, Paris, MS 26.

[34] Jonathan Simon, *Chemistry, Pharmacy and Revolution in France 1777–1809* (Aldershot, 2005).

Over the longer term, the promotion of philosophical chemistry by pharmacists like Baumé, while initially a way of raising their general status, would ultimately undermine their status as scientists. Thus, the promotion of this new kind of chemistry contributed to its clear separation from pharmacy and the loss of control that pharmacists previously exercised over the field. It was chemistry, after all, that brought Baumé into partnership with the doctor Pierre-Joseph Macquer and eventually allowed him to transcend this relationship by publishing his own theoretically oriented chemistry text.

I would argue, therefore, that this institutional history concerning the changing status of pharmacists is not only relevant for understanding the evolving relationship between chemists and pharmacists but also a pertinent factor in the constitution of the new chemistry at the end of the eighteenth century. Nevertheless, one rarely finds any reference to such trends or developments in pharmacy in histories of chemistry, while historians of pharmacy, sadly, would never dare to make this kind of argument.

CHEMISTRY'S PUBLICS

The second element I want to develop in trying to understand why philosophical chemistry came to the fore in the eighteenth century also concerns teaching but is not limited to the chemical education of pharmacists. A number of studies have recently brought to light an important phenomenon that influenced the status and trajectory of chemistry in the eighteenth century—the rise of the public chemistry lecture. The popularity of such lectures by certain chemistry lecturers, like Rouelle and Macquer, given in celebrated venues like the Jardin du Roi (later the Jardin des plantes) has long been known, but they appear to be only the tip of an iceberg of public and private chemistry courses that attracted a large, socially diverse audience.[35] While apprentice pharmacists and medical students had traditionally made up the overwhelming majority of the audience for these kinds of chemistry courses, starting from the middle of the eighteenth century they drew in a much wider public.[36] Pharmacists and doctors were still well represented, of course, but chemistry was attracting an increasing number of adepts who did not fit this pattern of professional training. Thus, to consider only the best-known cases, we can count among this new public for chemistry courses the political philosopher Jean-Jacques Rousseau and the two most famous lawyers ever to have practiced chemistry: Louis-Bernard Guyton de Morveau and Antoine-Laurent Lavoisier.[37]

This new public had different exigencies for chemistry. While the explosions, noxious smells, and spectacular effects were doubtless an important part of the appeal, the demand was also for philosophy and not just experiments. The medical uses of martial saffron might well have been of some interest to the generalist public, but these auditors were not going to be examined on how to prepare and administer such drugs. Thus, we can hypothesize a virtuous circle of reciprocal influence; the chang-

[35] John Perkins, "Creating Chemistry in Provincial France before the Revolution: The Examples of Nancy and Metz. Part 1. Nancy," *Ambix* 50 (2003): 145–81.

[36] Bernadette Bensaude-Vincent and Christine Lehman, "Public Lectures of Chemistry in Mid-Eighteenth-Century France," in *New Narratives in Eighteenth-Century Chemistry: Contributions from the First Francis Bacon Workshop*, ed. Lawrence M. Principe (Dordrecht, 2007), 77–96.

[37] Bernadette Bensaude-Vincent, "L'originalité de Rousseau parmi les élèves de Rouelle," *Corpus* 36 (1999): 81–101.

ing public for chemistry made different demands on the science, pushing it toward theory, while in return the more theoretically sophisticated chemistry that resulted was fashioning its audiences to have a different vision or at least different expectations of the science. What emerged was the group of "true" chemists that Venel referred to, and a new chemistry that would find its ultimate expression in Lavoisier's work, where there was absolutely no place for medical uses and dosages.[38]

A question remains, however; what if anything is the connection between these two histories I have described above: the institutional history of pharmacy and the place of chemistry in the cultural or social history of enlightenment France? We could consider them to be essentially unrelated, but I want to argue that there are significant connections. Thus, the chemistry courses I have been talking about not only drew a new public into chemistry but also offered new opportunities for pharmacists (and others) who were teaching them. Someone like Baumé, although a qualified pharmacist, spent the majority of his professional life teaching as well as engaged in entrepreneurial chemical manufacturing schemes. The interest that pharmacists like Baumé took in philosophical chemistry in turn served to bolster the image of pharmacy as something more than the routine preparation of drugs using timeworn recipes. As chemistry (taught almost exclusively by pharmacists and doctors) changed orientation, new opportunities opened up for teachers, and chemistry teaching was no longer so closely tied to the materia medica. The paradox is that this development led to the rise of a new generation of full-time chemists liberated from this pharmaceutical association and promoting a vision of a quite independent chemistry. Indeed, after a long gestation period in which chemistry continued to be taught almost exclusively in medical faculties in France,[39] the success of the German model eventually led to the introduction of chemistry into the universities as an independent science,[40] replete with increasingly sophisticated theory, although admittedly of great practical use, particularly in its industrial applications.

HISTORIOGRAPHICAL LESSONS

What are the wider lessons that can be drawn from this approach to the history of eighteenth-century chemistry? Here, as elsewhere, I have been emphasizing the evolution of the relationship between chemistry and pharmacy and arguing that it contributed to the very nature of chemistry as it developed through the nineteenth century and well into the twentieth. Reflecting on how different chemistry might have looked had it remained the exclusive domain of pharmacists (or of metallurgists, for that matter) during this period is just one way to assess the pertinence of this approach. Thus, a step to one side of a classic perspective in the history of science offers valuable insights into how and why a discipline developed the way it did. The switch from an internal logic of disciplinary development to a complementary view from the perspective of a field that is generally considered to lie outside reveals more

[38] For more on the attitude Lavoisier and his colleagues adopted toward pharmacy, see Jonathan Simon, "The Chemical Revolution and Pharmacy: A Disciplinary Perspective," *Ambix* 45 (1998): 1–13.

[39] Organic chemistry, in particular, found its institutional home in the French medical faculties. We see this quite clearly in Adolphe Wurtz's institutional trajectory; see Alan J. Rocke, *Nationalizing Science: Adolphe Wurtz and the Battle for French Chemistry* (Cambridge, Mass., 2001).

[40] The classic paper on the Giessen model is J. B. Morrell, "The Chemist Breeders: The Research Schools of Liebig and Thomas Thomson," *Ambix* 19 (1972): 1–46; William Brock has returned to this subject more recently: William H. Brock, "Breeding Chemists in Giessen," *Ambix* 50 (2003): 25–70.

clearly the work, both conscious and unconscious, that goes into establishing and defending a certain vision of the way the science should be conceived. While it may, paradoxically, reinforce the apparent coherence of a discipline's trajectory, the view from without nevertheless allows a critical analysis of this development. This lesson is clearly not limited to the case of pharmacy and chemistry but is applicable to any domains that lie close to one another.

An important element in the relationship between chemistry and pharmacy, one that is implicated in the development of public chemistry courses as well as the institutional evolution of pharmacy, is the status of the two fields. Status, be it social or scientific, is often at the center of the conflicts or at least tensions between neighboring disciplines. In this case, the question of the status of chemistry and pharmacy—art versus science or commercial activity versus philosophical knowledge—was never far from the surface during the realignment of chemistry that I have limned above.

Finally, for the skeptic, if this history is as important as I have suggested, why is it not more present in the history of chemistry we read in our introductory textbooks? This question opens up the wider issue of how to write the history of science. The theoretical orientation of mainstream history of eighteenth-century chemistry from Berthelot's classic rendition of the Chemical Revolution in a book written in 1890 to a very similar vision promoted by the more philosophical treatment in Thomas Kuhn's *Structure of Scientific Revolutions* seventy years later has left no place in the story for pharmacists (as a professional group).[41] Making space for pharmacy in the history of chemistry requires rethinking this approach. There have, of course, always been other ways of doing the history of chemistry, with a rich tradition notably in Germany, open to a number of different influences and not confined to recounting the evolution of chemical theory.[42] Indeed, Archibald Clow and Nan L. Clow's classic book on the Chemical Revolution represents a vision of a different series of events where theory had a limited role to play.[43] The challenge is to find a version of the history that lies between these extremes, one that makes space for the social, the useful, and the theoretical. One way to arrive at this point is to take a different starting point, such as asking what chemistry was for, and paying special attention to the questions of who was doing chemistry and what they were doing.

---

[41] Marcellin Berthelot, *La révolution chimique: Lavoisier, ouvrage suivi de notices et extraits des registres inédits de laboratoire de Lavoisier* (Paris, 1890). This version was substantially the same as that used by J. B. Conant for his course at Harvard published in 1948; James Bryant Conant, *The Overthrow of the Phlogiston Theory: The Chemical Revolution of 1775–1789* (Cambridge, Mass., 1956); and it was reused by his best-known student, Thomas Kuhn, *The Structure of Scientific Revolutions* (Chicago, 1962).

[42] For an overview of these approaches to the history of chemistry, see Marco Beretta, "The Changing Role of the Historiography of Chemistry in Continental Europe since 1800," *Ambix* 58 (2011): 257–76.

[43] Archibald Clow and Nan L. Clow, *The Chemical Revolution: A Contribution to Social Technology* (Philadelphia, 1992). This book was originally published in 1952.

Concluding Remarks:
A View of the Past through the Lens of the Present

by Bernadette Bensaude-Vincent*

ABSTRACT

Reflecting on the upsurge of interest among historians of chemistry in the material, artisanal, and commercial aspects of early modern chemistry, this essay argues that they are attracting attention because of a number of similarities between the style of chemistry cultivated in this period and the new cultures of chemistry being developed today. The close interactions between knowing and making, academic knowledge and practical applications, the social value and prestige attached to chemistry, the public engagement in chemical culture, the concern with recycling, and even a specific relational ontology instantiated in the term "rapport" are characteristic features of the current technoscientific culture. However, these analogies between early modern chemistry and the technoscientific paradigm may turn into obstacles if they end up in hasty rapprochements and whiggish interpretations of the past. In keeping with the attempts displayed in many articles in this volume to identify and understand the meaning of the actors' categories, this essay emphasizes the contrast between the visions of the past and the future developed by eighteenth-century chemists and the concept of time that prevails nowadays. The concept of "regime of historicity" provides a useful conceptual tool to take a view of chemistry as embedded in a culture and integral part of the horizon of expectation of an epoch. On the basis of this contrast between the regimes of historicity, the essay recommends the pluralism of concepts of time (polychronism) as an antidote to anachronisms.

Following a decade of intense scholarship on the Chemical Revolution, over the past few years we have witnessed a striking revival of interest in early eighteenth-century chemistry. As Frederic L. Holmes pointed out in 1989, for many decades historians of chemistry have been mainly concerned with the emergence of modern science. As they were commenting on and dissecting Lavoisier's published works and manuscripts, they tended to consider the eighteenth century as the "stage on which the drama of the chemical revolution was performed."[1] For them, early modern chemistry was nothing but pre-Lavoisieran chemistry.

This volume, which provides a sample of the most recent historiography of chemistry, testifies to the radical change of view that has occurred over the past three decades. Early modern chemistry is interesting in itself. By scrutinizing the actual prac-

* Department of Philosophy, Université Paris 1-Panthéon-Sorbonne, 17 Rue de la Sorbonne, 75231 Paris Cedex 05, France; bensaudevincent@gmail.com.
[1] Frederic L. Holmes, *Eighteenth-Century Chemistry as an Investigative Enterprise* (Berkeley, Calif., 1989), 3.

tices of chemists and paying attention to the multiple sites where chemical operations and demonstrations were performed in the early modern period, this volume gives an idea of what it was to be a chemist in the early modern period. On the one hand, the essays identify stable features beneath the changes of denomination from "alchemy" and "chymistry" to "chemistry." On the other hand, they try to highlight and understand the categories used by the actors themselves. This dual agenda triggers historiographical reflections on the broad question, How does one write the history of science?

The concern with materials strikes me as one of the most obvious and permanent features in the ensemble of practices described in this volume. Whether they deal with metals, glass, acids, or plants, the chemists depicted here are negotiating with the properties and behaviors of individual substances. In all cases, their manipulations of materials—whether meant to transmute them or to use them for specific purposes—were intimately linked with cognitive goals. Chrysopoeia was a quest to understand the natural world and to make use of its powers (Principe, Roos). The success of Stahlianism in the mid-eighteenth century relied not only on the theory of combustion but also on the study of acids and salts, which offered both cognitive and practical advantages (Chang). While making useful products for medicine or for public welfare, most chemists at the same time tried to understand and predict the properties of the particular substances that they manipulated. This dual agenda was summarized by Herman Boerhaave (quoted by Powers) in the early eighteenth century: "When the Chemist explains to you the nature of glass, he at the same time teaches you a sure way of making it."[2] Most essays suggest that the combination of crafts and science, of concerns with utility and truth claims, was a major feature of early modern chemistry.

On the other hand, a number of essays gathered in this volume follow Hélène Metzger's injunction to historians of science: they should position themselves as contemporaries of the works under study.[3] Together with Alexandre Koyré, she insisted that historians of early modern science should not project their own categories onto early modern science but instead should forget about the present in order to uncover the meaning of the actors' categories. Many essays in this volume try to identify the criteria for being recognized as a chemist in a given context: secrecy for Dr Plot (Roos), a combination of eloquence and material practices in the case of Libavius (Moran), academic membership in the case of Geoffroy (Joly), collaborative academic work in the case of Macquer and Lavoisier (Lehman), the combination of manual skills with theoretical ambitions for the "philosophical chemist" of Diderot's *Encyclopédie* (Simon), combining botanical and chemical knowledge in the case of Olmedo (Crawford). Together they provide a clear view of a science that had been overthrown, actively forgotten, and rendered obscure, unreadable, and unintelligible to us by the focus on the origins of modern chemistry.

In this essay I would like to offer some historiographical reflections on the revival of interest in this period of chemistry. How are we to understand that a style of chemistry usually dismissed as being premodern or prescientific appeals to so many histo-

---

[2] Herman Boerhaave, *Elements of Chemistry*, trans. Timothy Dallowe (London, 1735), 1:51.

[3] Hélène Metzger, "L'historien des sciences doit-il se faire le contemporain des savants dont il parle?" *Archeion* 15 (1933): 34–44. Reprinted in Metzger, *La méthode philosophique en histoire des sciences*, Corpus des oeuvres philosophiques en langue française (Paris, 1987), 9–21. See also Gad Freudenthal, ed., *Studies on Hélène Metzger* (Leiden, 1990).

rians? What makes it so interesting in the early twenty-first century? I will argue that the early modern period regains our attention because the values attached to current science are changing and the patterns of science in society are less and less alien to those of early modern chemistry. As Lucien Febvre, the founder of the *École des Annales*, noted, "Man does not remember the past; rather he always reconstructs it. . . . He does not keep the past in his memory, as the Northern ices keep millenary mammoths frozen. He starts from the present and it is through the present that he knows the past."[4] Since looking back to the past from the present is a "normal" attitude for historians, it is unavoidable to be sensitive to similarities between past and present situations. If early modern chemistry makes sense to us, it may be because it resonates with some specific features of the present episteme. By the mid-twentieth century, theoretical physics appeared as the model science and "pure science" was the most valued. By contrast, in the early decades of the twenty-first century, practical applications seem to be the main driver of research efforts, and partnerships between academia and industry are encouraged. Science policy is driven by a report significantly entitled "Society, the Endless Frontier," as an echo to the famous 1945 report, "Science, the Endless Frontier."[5] Chemistry as an "impure science," a science that has always been engaged in productive practice, a science proud of being socially useful rather than confined to arcane theory, has a chance to be more highly valued than it used to be.[6]

I will first emphasize the striking similarities between the style of chemistry cultivated in the context of mid-eighteenth-century France and the new cultures of chemistry being developed today. I will point not only to the close interactions between knowing and making, academic knowledge and practical applications, but also to various commitments that go with this epistemic attitude: the social value and prestige attached to chemistry, the public engagement in chemical culture, the concern with recycling, and even a specific relational ontology instantiated in the term "rapport." While the analogies between eighteenth-century chemistry and the technoscience[7] of the turn of the twenty-first century drive the attention of historians, they nevertheless may turn into obstacles when they end up in hasty rapprochements and whiggish interpretations of the past. In keeping with the attempts displayed in many articles of this issue to identify and understand the meaning of the actors' categories, I will contrast the visions of the past and the future developed by eighteenth-century chemists

---

[4] Lucien Febvre, *Combats pour l'histoire* (1952; Paris, 1992), 14.

[5] Parakskevas Caracostas and Ugur Muldur, *Society, the Endless Frontier*, European Commission/DG/XII R&D, Etudes (Luxembourg, 1997).

[6] Bernadette Bensaude-Vincent and Jonathan Simon, *Chemistry: The Impure Science* (London, 2008).

[7] The Oxford English Dictionary defines "technoscience" as "Technology and science viewed as mutually interacting disciplines, or as two components of a single discipline; reliance on science for solving technical problems; the application of technological knowledge to solve scientific problems." This standard definition only provides the earliest meaning of the notion coined by the Belgian philosopher Gilbert Hottois in *Le signe et la technique. La philosophie à l'épreuve de la technique* (Paris, 1984), 60–1. However, in the 1980s, a number of philosophers and social scientists who fought against the idealized view of a "pure science" detached from social and economic interests took up the term to refer to the true face of science constructed by a multitude of heterogeneous actors (Bruno Latour, *Science in Action* [Milton Keynes, 1987], 174). More broadly, the term is currently used to denote a specific regime of research determined by science policy and oriented toward societal or economic demands (B. Bensaude-Vincent, *Les vertiges de la technoscience* [Paris, 2009]). Thus, the notion of technoscience denotes much more than the actual hybridization of science and technology. It is a new ideal of scientific practice, which no longer claims to be disinterested or neutral or amoral.

and the concept of time that prevails nowadays. Early modern chemists did not have the same experiences of time as current scientists; they lived in a different landscape, more precisely in a different "timescape." The concept of "regime of historicity," introduced by French historian François Hartog, captures the special way of articulating the past, the present, and the future that characterizes each society.[8] It is a useful conceptual tool to take a view of chemistry as embedded in a culture and an integral part of the horizon of expectation of an epoch. On the basis of this contrast between the regimes of historicity, I emphasize in conclusion the significance of assuming a historical relativity of our concepts of time (polychronism)—as an antidote to anachronisms. While acknowledging the influence of the present on our interpretation of the past, it is important to keep the distance of the past, to experience the estrangement of historicity by historicizing the notion of time itself.

SCIENCE, ARTS, ET MÉTIERS

In 1699, chemistry was officially established as one of the classes of the Paris Academy of Sciences along with geometry, astronomy, mechanics, anatomy, and botany. The utility of science was the main feature set out by the first perpetual secretary of the Academy, Louis Bernard Le Bovier de Fontenelle, to legitimate the patronage of scientific research by the king. In the preliminary discourse to the *Histoire et mémoires de l'Académie* he stated that chemistry—then viewed as a branch of physics, the science of nature—was useful to "sustain life."

> The utility of mathematics and physics, although in verity not obvious, is no less real for that. Considering man in a state of nature, nothing is more useful to him than that which can preserve his life, and that which can produce the arts. These arts are both a great aid to him and a great ornament to society. That which can preserve life belongs in particular to physics, and from this perspective, it has been divided into three branches in the Academy—making three different species of academician—Anatomy, Chymistry, and Botany. It is easy to see how important it is to know the human body with exactitude, and the remedies one can draw from Minerals and Plants.[9]

While in the late seventeenth century the culture of science was essentially oriented toward curiosities, in the eighteenth century utility became the prime mover.[10] From this perspective, chemistry enjoyed a privileged status, because of its close connection to the manufacture of drugs and many other products indispensable in everyday life such as ceramics, glass, glue, and metals. This utility, Fontenelle suggests, was more directly evident than that of astronomy, for instance. The prevailing view of science for public utility undoubtedly contributed to the prestige of chemistry in the French Enlightenment.

The close interaction between making and knowing that Ursula Klein has emphasized in a number of publications justifies the title of one of them: "Technoscience

---

[8] François Hartog, *Régimes d'historicité—Présentisme et expériences du temps* (Paris, 2003).

[9] Louis Bernard Le Bovier de Fontenelle, "Discours préliminaire sur l'utilité des mathématiques et de la physique," in *Éloges des académiciens* (Paris, 1699). See also Roger Hahn, *The Anatomy of a Scientific Institution: The Paris Academy of Science, 1666–1803* (Berkeley and Los Angeles, 1971).

[10] W. Clark, J. Golinski, and S. Schaffer, eds., *The Sciences in Enlightened Europe* (Chicago, 1999); L. Stewart, *The Rise of Public Science* (Cambridge, 1992). On the culture of curiosities, see Horst Bredekamp, *The Lure of Antiquity and the Cult of the Machine: The Kunstkammer and the Evolution of Nature, Art and Technology* (Princeton, N.J., 1995).

avant la lettre."[11] The use of the fairly recent notion of "technoscience" points to striking similarities between the style of chemistry that flourished in the eighteenth century and the new cultures of chemistry that developed in the late twentieth century. Both periods are marked by close interactions between knowing and making, between academic knowledge and practical applications, as well as by various commitments that go along with this epistemic attitude: the social value and prestige attached to research for society and public welfare and a public engagement in science.

However, the interplay between cognitive and practical aims, which characterized eighteenth-century chemistry as well as current technoscience, was designated by more appropriate phrases. In a time when scholars would never have dared forge a compound term mixing Greek and Latin roots (as happened with "technoscience"), the terms "theory" and "practice" were commonly used to characterize the Janus face of chemistry. Indeed, Johann Gottschalk Wallerius, who held the chair of chemistry at the University of Uppsala, coined the phrases *chymia pura* and *chymia applicata*. Christoph Meinel argued that the above distinction based on goals played a strategic role in the legitimation of chemistry on the academic stage.[12] However, in France few chemists adopted the distinction between pure and applied sciences before the nineteenth century.[13] Instead in D'Alembert and Diderot's *Encyclopédie*, the authors operated with a threefold distinction between *science*, *arts*, and *métiers*. Chemistry had a place in all three categories. The entry "chymie," authored by Gabriel François Venel, celebrated the dual nature of chemistry as science and art with general chemistry as the "shared trunk" of a large spectrum of chemical arts. Venel insisted less on the conventional theme of "science improving the arts," preferring to emphasize the reciprocal contribution of the arts to the advancement of science.[14] In his view, the savant chemist could neither perform experiments nor draw general conclusions and predictions unless he had mastered the skills that the artisans (he termed them "artists") acquired by handling chemicals in the workshop. Venel advocated a sensorial empiricism[15] based on the *coup d'œil*, the intimate knowledge of colors, smells, textures, and temperatures that resulted from the *habitus* forged over a long period of practical experience.

> In short, one needs to be an artist, a trained and experienced artist, if for no other reason than to be able to execute or direct different processes with the same facility, the same knowledge of resources, the same rapidity which make these operations a game, a pleasure, and a captivating spectacle instead of a long, painful, discouraging exercise during which disheartening obstacles interfere and success is uncertain. All such isolated phenomena, these so-called bizarre operations, this variety of products and singularity of the results of experiments that half-chemists blame on the techniques of chemistry or

[11] Ursula Klein, "The Laboratory Challenge: Some Revisions of the Standard View of Early Modern Experimentation," *Isis* 99 (2008): 769–82; Klein, "Blending Technical Innovation and Learned Knowledge: The Making of Ethers," in *Materials and Expertise in Early Modern Europe: Between Market and Laboratory*, ed. U. Klein and E. C. Spary (Chicago, 2010), 125–57; Klein, "Technoscience *avant la lettre*," *Perspect. Sci.* 13 (2005): 226–66.

[12] Christoph Meinel, "Theory or Practice? The Eighteenth-Century Debate on the Scientific Status of Chemistry," *Ambix* 30 (1983): 121–32.

[13] The distinction between pure and applied chemistry only prevailed in the postrevolutionary era with Jean Antoine Chaptal, *Chimie appliquée aux arts* (Paris, 1807). Prior to this famous treatise, the term occurred in the title of a single chemical treatise: Dhervillez and Lapostolle, *Plan d'un cours de chymie expérimentale, raisonnée et appliquée aux arts* (Amiens, 1777).

[14] Gabriel François Venel, "Chymie," in *Encyclopédie* (Paris, 1753), 3:420.

[15] On sensorial empiricism, see Jessica Riskin, *Science in the Age of Sensibility* (Chicago, 2002).

on unknown properties of the materials they are working with can all be attributed to the artist's inexperience. They rarely occur with experienced Chemists. It is very rare, and perhaps it never happens, that a true chemist is unable to replicate a product using the same materials.[16]

While the impact of science on the arts consists in incremental progress, Venel claimed that the arts were the source of science: no chemical knowledge without knowhow, no broad view of chemical phenomena without paying attention to the details and circumstances of chemical operations. There is a striking asymmetry in his view of the interaction between science and the arts in favor of the arts.

To what extent, however, was this promotion of the value of the arts—typical of the *Encyclopédie*—more pronounced in the presentation of the practices of chemistry than the other sciences? A detailed examination of the narratives of experiments at the Paris Royal Academy of Sciences in the eighteenth century by Christian Licoppe suggests a close link between the pressure to improve experimental practices and the concern with utility. In particular, the reproducibility of results became a major requirement and criterion of validity at the Paris Royal Academy. Whereas at the turn of the century the culture of curiosities fostered the production of unexpected and spectacular effects and wonders, utility required the stabilization and robustness of experimental data. In particular, Licoppe's analysis of Charles C. Dufay's memoirs on phosphorus shows how utility could influence the practice of academic research. The purpose of this series of experiments conducted in the 1720s was to establish that the light emitted by phosphorus was neither a wonder of nature nor a magic property of any particular phosphorus, such as the one from Bologna. He tried to demonstrate that all substances could be considered as phosphorescent provided they are prepared according to a simple and robust protocol.[17] And he promised that this common property would prove extremely useful.

The promises of utility could well have been just a rhetorical ornament, an excuse for conducting curiosity-driven research in an academic context, as is often the case for current technoscientific research. However, the French government encouraged the links between science and the arts throughout the century. The members of the Paris Royal Academy of Sciences were regularly commissioned to find solutions to practical issues related to public health, urban life, or agriculture.[18] The unification of laboratories with workshops was also institutionalized with the creation of academic positions in the royal manufactures. For instance, the royal manufacture of tapestries—Les Gobelins—created in 1731, was successively "inspected" by Charles-François de Cisternay du Fay, Jean Hellot, Macquer, and Claude-Louis Berthollet. All of these inspectors attempted to provide theoretical accounts of the operations performed by craftsmen in dyeing with the aim of improving or rationalizing their practices.[19]

---

[16] Ibid.; English translation by Lauren Yoder, University of Wisconsin, http://quod.lib.umich.edu/d/did/ (accessed 25 April 2014).

[17] Dufay, "Mémoire sur un grand nombre de phosphores," *Mémoires de l'Académie royale des sciences* (1730), 522–35; Christian Licoppe, *La formation de la pratique scientifique: Le discours de l'expérience en France et en Angleterre (1630–1820)* (Paris, 1996).

[18] Bernard Joly's essay, "Etienne-François Geoffroy (1672–1731), a Chemist on the Frontiers," in this volume, mentions Geoffroy's expertise with regard to technological issues.

[19] Hellot published *L'Art de la teinture des laines et des étoffes de laine en grand et en petit teint* in 1750; Macquer published *L'Art de la teinture en soie* in 1763; and Berthollet published *Éléments de*

In addition, public shows, museums, shops, and workshops displayed chemical products and instruments for people to enjoy or to buy.[20] Recent studies of the chemistry courses delivered in Paris provide a new perspective on the interactions between science and the arts. Chemistry was so fashionable that fifty public and private courses were delivered in Paris in the 1750s and dozens in provincial cities.[21] The training of pharmacists and doctors was the primary motivation for the creation of public and private chemistry courses. In many cases the courses were delivered by a duo made up of a pharmacist and a physician, as medical doctors who were required to teach in full costume were not allowed to perform experiments.[22] The number of public and private courses increased over the century as the utility of chemistry and experimental demonstrations attracted a wider audience. According to John Perkins, about 45,000 people studied chemical science circa 1740 in France.[23] One of the most famous public courses was given at the Jardin du Roy by Guillaume François Rouelle from 1742 to 1764. It was attended by crowds of people from all social categories including ladies, the philosophes such as Rousseau, Diderot, Turgot, Condorcet, and leading figures of academic chemistry, such as Macquer and Lavoisier. Rouelle's public demonstrations are famous among historians of chemistry for being the vehicle for the propagation of Stahl's phlogiston theory.[24] However, it was certainly not Stahl's theory that attracted hundreds of people to the auditorium at the Jardin du Roy and dozens to Rouelle's private course on rue Jacob. His public demonstrations taught the art of making chemical preparations and how to use the famous affinity tables that he had himself enriched to glassmakers, dyers, metallurgists, and other artisans.[25]

Nevertheless, not all metiers of chemistry were enlightened by theoretical science. A large number of anonymous amateurs also performed experiments for fun or for entertainment. The *Gazette*, an early newspaper created in 1631 by the physician Théophraste Renaudot, which advertised chemistry courses, mentioned that Paris hosted practitioners of chemical operations who did not combine knowing and making. "Our paper would not suffice to make known in detail all those who burn char-

---

l'art de la teinture in 1791. See Agustí Nieto-Galan, *Colouring Textiles: A History of Natural Dyestuffs and Industrial Europe* (Dordrecht, 2001).

[20] Liliane Pérez, "Technology, Curiosity and Utility in France and in England in the Eighteenth Century," in *Science and Spectacle in the European Enlightenment*, ed. B. Bensaude-Vincent and C. Blondel (Aldershot, 2008), 25–42.

[21] Bernadette Bensaude-Vincent and Christine Lehman, "Public Lectures of Chemistry in Mid-Eighteenth-Century France," in *New Narratives in Eighteenth-Century Chemistry: Contributions from the First Francis Bacon Workshop*, ed. Lawrence M. Principe (Dordrecht, 2007), 77–96; John Perkins, "Creating Chemistry in Provincial France before the Revolution: The Examples of Nancy and Metz, Part 1: Nancy," *Ambix* 50 (2003): 145–81, "Part 2: Metz," 51 (2004), 43–75.

[22] Thus, in Paris, the physician Pierre-Joseph Macquer taught with the apothecary Antoine Baumé; in Montpellier, Gabriel François Venel, who was a physician, paired up with Jacques Montet; in the 1780s, the apothecary de La Planche started a new course with Jean-Baptiste Bucquet, a medical doctor.

[23] John Perkins, "Chemical Paris: Laboratories and Other Sites, 1750–90" (paper presented at "Sites of Chemistry in the Eighteenth Century," Oxford, 4–5 July 2011).

[24] Rhoda Rappaport, "G. F. Rouelle, His *Cours de Chimie* and Their Significance for Eighteenth-Century Chemistry" (master's thesis, Cornell Univ., 1958); Rappaport, "G. F. Rouelle: An Eighteenth-Century Chemist and Teacher," *Chymia* 6 (1960): 68–101; Rappaport, "Rouelle and Stahl—the Phlogistic Revolution in France," *Chymia* 7 (1961): 73–102.

[25] Mi Gyong Kim, *Affinity, That Elusive Dream* (Cambridge, Mass., 2003); Lissa Roberts, "Chemistry on Stage: G. F. Rouelle and the Theatricality of Eighteenth-Century Chemistry," in Bensaude-Vincent and Blondel, *Science and Spectacle* (cit. n. 20), 129–40.

coal in Paris in order to illuminate such-and-such a chemical truth, and a full folio would hardly suffice just to name those who burn charcoal without knowing why they do it."[26]

Paris also hosted crowds of obscure practitioners who performed routine chemical operations in more or less hazardous and unhealthy conditions. According to André Guillerme, a historian of chemical crafts in France, a huge number of workshops in Paris employed thousands of people with no connection whatsoever to the academic milieu.[27] This rogue proletariat, although indispensable for the chemical arts, is omitted from the detailed descriptions of craftsmen such as metallurgist, glassmaker, hat-maker, dyer, or pharmacist to be found in Diderot's *Encyclopédie*. One merit of Guillerme's description of the chemical industries in Paris in the second half of the eighteenth century is to bring out the background role of collectors of raw materials: women, children, and workers of the underclass wandered through the streets of Paris to collect ashes, rugs, urine, excrement, and hair from houses as well as bones, skins, and horns from the slaughterhouses. The more that the population of Paris ate and consumed, the more the capital produced food and commodities. Thanks to the abundance and quality of water in Paris, an intense activity developed based on the recycling of all kinds of waste: urine was used to make volatile alkali; rugs were recycled to make paper, cardboard, and felt; decaying houses were the main source of niter and were used for the preparation of saltpeter; the blood of cows was calcinated with the addition of potash residues of Prussian blue to prepare animal black;[28] glue was made out of debris from sheep and veal skin, from claws, and from used gloves or clogs. The quality of the final product depended on the nature of the animal debris used. Ironically this premodern chemical industry based on biochemical operations such as maceration, fermentation, putrefaction, at room temperature and ordinary pressure corresponds better to the model of soft chemistry that has been developed over the past decades than the chemical industry of the past two centuries.[29] However, this archaic biochemical economy would not meet all the requirements of today's chemical industries since it was neither clean nor healthy. The smell and the pollution in certain districts of Paris were so unbearable that after repeated complaints from neighbors the municipality banned all chemical workshops from the center of the city and most of them relocated to the suburbs.

FAMILY RESEMBLANCES IN DIFFERENT REGIMES OF HISTORICITY

The search for resemblances between the chemistry of today and that of the eighteenth century could be pushed further. In particular, parallels could be drawn at a more abstract level of epistemology and ontology. The systematic campaigns of experiments of displacement reactions performed by eighteenth-century chemists to set out the tables of affinity could be compared to the systematic exploration of the

---

[26] *Gazette de Médecine*, 1761, 199–200.

[27] André Guillerme, *La naissance de l'industrie à Paris, entre vapeurs et sueurs* (1780–1830), (Seyssel, 2007). See also Sabine Barles, *La ville délétère: médecins et ingénieurs dans l'espace urbain* (Seyssel, 1999).

[28] A black pigment made by carbonizing animal bones.

[29] On soft chemistry (*chimie douce*), see J. Livage, "Vers une chimie écologique. Quand l'air et l'eau remplacent le pétrole," *Le Monde*, 26 October 1977; and C. Sanchez et al., "'Chimie douce': A Land of Opportunities for the Designed Construction of Functional Inorganic and Hybrid Organic-Inorganic Nanomaterials," *Comptes Rendus Chimie* 13 (2010): 3, doi:10.1016/j.crci.2009.06.001.

combinatorial practices used in current chemical and pharmaceutical industries. The ontological status of the substances included in the *tables de rapports*—defined by their interrelations rather than by their compositions—could serve as a model for improving the definition of nanoparticles since their properties and behaviors seem to depend almost exclusively on their surface area and consequently on their reactivity.[30] Just as eighteenth-century chemists focused on the dispositions of chemicals and their interactions, nanotechnologists could well refocus their investigations on the dispositions and interactions of nanoparticles rather than on their technological performance.[31] However, the purpose here is neither to draw lessons from history nor to provide guidance for active nanoscientists and science policy makers.

Instead, I would like to develop some historiographical reflections about this parallel between eighteenth-century and current chemistry. Generally, when we can identify a family resemblance between two individuals belonging to distinct generations, we can conclude that they might share at least a few genes. No such inference can be made about patterns of research unless we assume the existence of genes or their equivalent in scientific disciplines. Instead of essentializing disciplines such as chemistry in this way, I suggest comparing the cultural heritage transmitted from generation to generation to a sort of "language game," which is part of a broader "form of life." Thus, I would argue that Ludwig Wittgenstein's concept of "family resemblance"—denoting loose analogies connecting particular uses of the same word without assuming any common essence—is more appropriate in this context than talk of genes.[32] The cultural heritage of chemistry built around the space of the laboratory, with recipes, protocols, methods, concepts, and so on, undoubtedly shaped a recognizable identity of chemistry. However, the practices inherited by means of cultural heritage are shaped by the historical context where they are enacted. In particular, the identity of chemistry is continuously reshaped according to the way its practitioners envisage their position in history.

In his essay, Jonathan Simon suggests that it might be more fruitful to describe an object of study in terms of what it is not rather than in terms of what it is. Here I would like to take a similar approach. Emphasizing differences rather than similarities between eighteenth-century and current experiences is a methodological approach aimed at preventing tacit projections of our concepts and concerns onto past science. Recent historiographical research shows that the vision of time is a cultural construction. Each society has a special way of articulating the past, the present, and the future. The notion of regime of historicity forged by François Hartog is a tool for characterizing the order of time created in each culture. Time is not a universal metaphysical entity but rather the experience that people have in a context, how they envision the relations between the past, the present, and the future. "A regime of historicity is nothing more than a way of engaging with the past, the present and the future or combining the three categories in some manner."[33] In the early eighteenth

[30] Lissa Roberts, "Setting the Table: The Disciplinary Development of Eighteenth-Century Chemistry as Read through the Changing Structure of Its Tables," in *The Literary Structure of Scientific Argument*, ed. Peter Dear (Philadelphia, 1991), 99–132. On the ontological status of chemicals, see Bensaude-Vincent and Simon, *Chemistry: The Impure Science* (cit. n. 6), 201–29.

[31] B. Bensaude-Vincent and Sacha Loeve, "Metaphors in Nanomedicine: The Case of Targeted Drug Delivery," *Nanoethics* (2013), http://dx.doi.org/10.1007/s11569-013-0183-5 (accessed 25 April 2014).

[32] Ludwig Wittgenstein, *Philosophical Investigations* (Oxford, 1953), secs. 66–7, p. 32e.

[33] François Hartog, *Régimes d'historicité: Présentisme et expérience du temps*, 2d ed. (Paris, 2012), 14.

century, time was still experienced as the cyclic repetition of heroic moments, or exemplars. The famous image of "dwarfs on the shoulders of giants" gives a sense of the regime of historicity that prevailed before the emergence of the modern notion of revolution. According to Reinhart Koselleck, it is around the late eighteenth century that the modern notion of history (*die Geschichte*) emerged in Prussia.[34] The past started to be seen as a vector made of unique events and oriented toward the future. In France, the French Revolution of 1789 is considered as the turning point when the notion of revolution ceased to be understood as a cyclic movement and came to denote a radical break with the past. History then became an experimental field open to the future rather than the narratives of heroic moments that provided guidelines for the future. Bernard Joly's portrayal of Geoffroy in this volume instantiates the premodern regime of historicity. In his exploration of the links between the alchemical tradition and the more recent tradition of mechanical chemistry cultivated at the Paris Academy of Sciences, Geoffroy never assumed a radical break with the alchemical past. Rather, he tried to promote innovations as the continuation of a longstanding quest that required the continuous efforts of many generations. He was not unique in considering the present and the future as the continuation of the past. In the article "Chemistry" in Diderot's *Encyclopédie*, Gabriel François Venel created a horizon of expectation by expressing the wish for "a New Paracelsus" who would bring about a revolution in chemistry.[35] However, his view of revolution was nothing like a break with the past or the present. Instead, Venel called for a hero capable of restoring the high profile of chemistry. His vision of the future was shaped by the heroic past, and his concept of revolution suggested the repetition of an epochal moment, a landmark providing guidelines for the future. It thus becomes clear that most eighteenth-century chemists did not need grand narratives of change built on broad revolutionary claims.[36] It is only in the course of Lavoisier's career that the term "revolution" came to refer to a radical break with the past. In his "Réflexions sur le phlogistique" of 1785, Lavoisier explicitly invited the reader to forget about the past and to assume that phlogiston had never existed.[37] In this way, the future became the reference point, providing guidelines for the present and a teleological framework for rewriting the history of the past.

There is a striking contrast between these two regimes of historicity and the current experience of time in today's technosciences, which subordinates the future to the present, and frames the future according to today's prevailing values. The past no longer sheds a light on the present or the future. It is viewed only as a document, a trace, or a monument. The past is divorced from the present, completely detached from what is going on now. Traces of the past are certainly kept as sanctuaries of a defunct world, but there is no lesson to be drawn from the past. Each emerging technology claims to bring about a new revolution, the nth Industrial Revolution. However, neither is the future the goal determining the present, as it used to be in the twentieth century. According to Hartog, in the modern regime of historicity that prevailed between 1789 and 1989 (the symbolic date of the fall of the Berlin Wall), the future shed a light on the present and on the past. Now the present sheds a light

---

[34] Reinhart Koselleck, *Future Pasts: On the Semantics of Historical Time* (New York, 1985).
[35] Venel, "Chymie" (cit. n. 14).
[36] I. Bernard Cohen, *Revolution in Science* (Cambridge, Mass., 1985).
[37] Bernadette Bensaude-Vincent, *Lavoisier: Mémoires d'une révolution* (Paris, 1993).

on the past and on the future. The faith in progress, the belief that tomorrow would be better than yesterday and even today, has given way to a kind of anxiety about the future. Indeed, research and development in nanotechnology and biotechnology are still officially made in the name of the future with such promises as food for all, energy, new diagnostic techniques and therapies, longevity, and so on. However, this future is no longer an *avenir radieux*, a bright future lighting up the temporal horizon. Rather, biochemical technologies are promoted as attempting to remediate the problems (pollution, scarcity of resources) caused by the development of synthetic chemistry and a frantic rhythm of innovation and productivity. The main concern is to maintain the present state of affairs, to continue business as usual. Even nature is treated as a heritage that needs to be preserved. In this regime, where the present is indefinitely extended, inflated, and omnipresent, the meaning of chemistry has radically changed. Whereas in the eighteenth century chemistry was a popular and fashionable science in harmony with the prevailing values of the Enlightenment (public good and perfectibility), today chemistry is perceived as a major cause of pollution and a threat to the future of atmospheric air, the climate, and human health. In the early twenty-first century, there are no longer any chemical plants in Paris. Chemical factories have largely been delocalized to emerging countries, and the consumption of chemicals and plastics continues to increase. Chemistry has to be made invisible because it is at odds with the prevailing concern to preserve nature instead of celebrating the transforming power of technology.

POLYCHRONISM VERSUS ANACHRONISMS

The purpose of this parallel between early modern chemistry and the technosciences in which today's chemists are exercising their professional skills is to disentangle a methodological puzzle in the practice of history of science. At first glance, looking at the past through the lenses of the present seems to challenge the movement of anti-whiggism, which has eroded the biased view of early modern as pre-Lavoisieran chemistry over the past decades.[38] If alchemy is no longer an obscure mystical set of beliefs and images, if eighteenth-century chemistry is more than a stage for the drama of the Chemical Revolution, it is undeniably because two generations of professional historians got rid of the narratives forged by the scientific winners. It was impossible to capture the meaning and restore the consistency of early modern practices of chemistry as long as the Chemical Revolution was viewed as the necessary outcome of the truth and legitimacy of Lavoisier's chemistry.

However, to gain an insight in obsolete science, to restore its significance and consistency, historians do not have to retreat to an ivory tower and work in isolation from their time. One can follow Metzger's efforts to make herself the contemporary of the works under study without forgetting about current science and without adopting a puritan attitude of abstention and abstraction from the outside world. It is illusory to think that we can exclude all forms of presentism. The patterns and the values of

---

[38] The notion of whig history derived from British political history has been introduced in the history of science by Herbert Butterfield, *The Whig Interpretation of History* (London, 1931). For later discussions, see John Schuster, *The Scientific Revolution: An Introduction to the History and Philosophy of Science* (Wollongong, 1995), chap. 3, "The Problem of Whig History in the History of Science"; and Nicholas Jardine, "Whigs and Stories: Herbert Butterfield and the Historiography of Science," *Hist. Sci.* 41 (2003): 125–40.

present-day science certainly influence the topics and periods that we elect to investigate as well as the terms that we use to describe the past. In history, as in natural sciences, there is no objective perspective from nowhere or no-time. Far from being neutral, disinterested, or dispassionate, historians of early modern science need to engage with their subjects as they have to actively counteract the discrediting of the losers and to come out with a more positive view of their achievements.[39] Whatever the personal empathy of historians for their subjects, this valuation is made possible by present-day concepts and controversies. The valuation of the role of craftsmen and of social utility in early modern chemistry resonates with the values promoted in current technosciences. And the controversies raised by this regime of knowledge suggest that the old values of pure science are still alive. Therefore, the most powerful antidote against whiggism is not to forget about present-day science but to be aware of the influence of the present on historical practices.

To avoid the unconscious imposition of current categories on the works of past chemists who ignored them (anachronism), we have to go further and also be aware of the historical multiplicity of historical concepts of time (polychronism). In the essay in which Metzger discussed the question, "Should the historian of science make herself the contemporary of the scientists under scrutiny?" she herself acknowledged that it is an impossible task. As she pointed out the various obstacles to a full understanding of the works of early modern chemists, she anticipated the idea of incommensurability later introduced independently by Thomas Kuhn and Paul Feyerabend. Basically it is impossible to fully grasp their concepts and to compare paradigms because they are based on different assumptions, values, and worldviews. The concept of incommensurability has to be extended to our experiences of time. To erode the whiggish idea of the accumulation of scientific knowledge with the passage of time, it is vital to question the very notion of "the passage of time." Although the image of the time arrow and the progress of mankind is deeply engrained in our mentality it is by no means an impermanent reality. It is a historical construction submitted to revisions that encourages a critical attitude to the present as much as to the past. The identity of chemistry has been framed—and is still being reconfigured—through a variety of timescapes.

---

[39] The famous principle of symmetry in the analysis of controversies formulated by STS scholars—paying equal attention and respect to the losers and the winners—cannot overlook that the winners' views have gained legitimacy and shaped the future. See Latour, *Science in Action* (cit. n. 7); Isabelle Stengers, *Cosmopolitics 1* (Minneapolis, 2010).

Notes on Contributors

Bernadette Bensaude-Vincent is Philosopher and Historian of Science and Professor at Université Paris I Panthéon-Sorbonne. Her research topics span from the history and philosophy of chemistry to materials science and nanotechnology. Her publications include *History of Chemistry* (with Isabelle Stengers; Cambridge, Mass., 1997); *Chemistry: The Impure Science* (with Jonathan Simon; London, 2008); and *Fabriquer la vie: Où va la biologie de synthèse* (with Dorothée Benoit-Browaeys; Paris, 2011).

Marco Beretta is Professor of History of Science at the University of Bologna. He is editor of *Nuncius: Journal of the Material and Visual History of Science*. His recent books include *The Alchemy of Glass* (Sagamore Beach, Mass., 2009) and *F for Fakes: Hoaxes, Counterfeits and Deception in Early Modern Science* (edited with Maria Conforti; Sagamore Beach, Mass., 2014). He is currently working on a monograph about Lavoisier's laboratories.

Ku-ming (Kevin) Chang is Associate Research Fellow at the Institute of History and Philology of Academia Sinica, Taipei, Taiwan. He is a coeditor of a collected volume, *World Philology* (Cambridge, Mass., 2014), that compared major philological traditions, and is revising a manuscript on the history of the dissertation as a genre of academic writing and publication. He has also published articles, and is preparing a manuscript, on Georg Ernst Stahl's chemistry and medicine.

Matthew James Crawford is Assistant Professor of History at Kent State University. He received his PhD in History and Science Studies from the University of California, San Diego, and is currently finishing a book manuscript that examines the relationship between science and empire in the eighteenth-century Spanish Atlantic World.

Matthew Daniel Eddy is Senior Lecturer in the History and Philosophy of Science at Durham University, England. He researches how knowledge is valued, systematized, and visualized by the academy, the public, and industry. He is currently pursuing research projects that address the history of visual culture, environmental thought, and the relationship between science and governance. His larger field of expertise covers seventeenth- to twentieth-century forms of scientific representation and argumentation, including historicized conceptions of mind, memory, matter, time, language, visuality, informatics, human origins, and religion.

Hjalmar Fors is a researcher and teacher at Uppsala University. His first book is *Mutual Favours: The Social and Scientific Practice of Eighteenth-Century Swedish Chemistry* (Uppsala, 2003). His second book, *The Limits of Matter: Chemistry, Mining, and Enlightenment*, will be published by the University of Chicago Press in 2014.

Bernard Joly is Emeritus Professor of Philosophy and History of Science at the University of Lille (France). His main publications are *La rationalité de l'alchimie au XVIIe siècle* (Paris, 1992), *Descartes et la chimie* (Paris, 2011), and *Histoire de l'alchimie* (Paris, 2013).

Ursula Klein is a senior research scholar at the Max Planck Institute for the History of Science and Professor at the University of Konstanz. She is the author of *Experiments, Models, Paper Tools: Cultures of Organic Chemistry in the Nineteenth Century* (Stanford, Calif., 2003) and *Materials in Eighteenth-Century Science: A Historical Ontology* (with Wolfgang Lefèvre; Cambridge, Mass., 2007). Her current book project is entitled "Alexander v. Humboldt's Prussia: Science, Technology and the State."

Christine Lehman formerly taught physics and chemistry. She is the author of a PhD dissertation entitled "Gabriel-François Venel (1723–1775): sa place dans la chimie française du XVIIIe siècle," which she defended at Paris X-Nanterre University in 2006. She published the transcription of Venel's chemistry course taught in the year 1761 (Dijon, 2010). Her main research interest is eighteenth-century chemistry.

Seymour H. Mauskopf is Professor Emeritus of History at Duke University. His fields of research interest are the history of chemistry (*Crystals and Compounds* [1976], *Chemical Sciences in the Modern World* [Philadelphia, 1993]); the history of marginal science (parapsychology; *The Elusive Science* [with Michael R. McVaugh; Baltimore, 1980]); and the relation between history and philosophy of science (*Integrating History and Philosophy of Science: Problems and Prospects* [with Tad Schmaltz; Dordrecht, 2012]). His recent research has focused on the history of explosives and munitions, and he is currently writing a book on Alfred Nobel's interactions with British munitions scientists in the late nineteenth century. In 1998, he received the Dexter Award for Outstanding Contributions to the History of Chemistry from the American Chemical Society.

Bruce T. Moran is Professor of History at the University of Nevada, Reno, where he teaches courses in the history of science and early medicine. He is the author of *Distilling Knowledge: Alchemy, Chemistry, and the Scientific Revolution* (Cambridge, Mass., 2005) and *Andreas Libavius and the Transformation of Alchemy: Separating Chemical Cultures with Polemical Fire* (Sagamore Beach, Mass., 2007). He is currently preparing a study titled "Experiencing Alchemy: Private Life, Local Meaning, and Alchemical Agendas in Early Modern Europe."

William R. Newman is Professor of History and Philosophy of Science at Indiana University and General Editor of The Chymistry of Isaac Newton (www.chymistry.org). He has published extensively on the history of alchemy and chymistry: recent works include *Atoms and Alchemy: Chymistry and the Experimental Origins of the Scientific Revolution* (Chicago, 2006) and *Promethean Ambitions: Alchemy and the Quest to Perfect Nature* (Chicago, 2004).

John A. Norris studied History of Science at Charles University in Prague (PhD, 2006) and geochemistry at the University of Georgia (MS, 1996). He is a professional rare book seller specializing in the history of science and remains active in bibliographic research and in studying mineral theories in early modern mining and alchemical literature.

John C. Powers is Associate Professor of History and Assistant Director of the Science, Technology, and Society Program at Virginia Commonwealth University. His first book was *Inventing Chemistry: Herman Boerhaave and the Reform of the Chemical Arts* (Chicago, 2012). His current project examines the introduction of thermometry into chemistry.

Lawrence M. Principe is Drew Professor of Humanities at Johns Hopkins University in History of Science and Technology and in Chemistry. His recent books include *The Scientific Revolution: A Very Short Introduction* (Oxford, 2011) and *The Secrets of Alchemy* (Chicago, 2012). He is completing a study of chymistry at the early Académie Royale des Sciences.

Jennifer M. Rampling is Assistant Professor of History at Princeton University; she was formerly a Wellcome Trust Postdoctoral Research Fellow at the University of Cambridge. She is currently completing her first book, *The Making of English Alchemy*.

Anna Marie Roos is Senior Lecturer in History at the University of Lincoln and Associate Faculty at the University of Oxford. Recent books include *Web of Nature: Martin Lister (1639–1712): The First Arachnologist* (Leiden, 2011) and *Salt of the Earth: Natural Philosophy, Medicine, and Chymistry in England, 1650–1750* (Leiden, 2007).

Jonathan Simon is *Maître de conférences* in History of Science at the Université Lyon 1. He is currently working on the history of serotherapy in France. His books include *Chemistry, Pharmacy and Revolution* (Aldershot, 2005) and *Chemistry: The Impure Science* (with Bernadette Bensaude-Vincent; London, 2008).

Index

SUGGESTIONS FOR CONTRIBUTORS TO OSIRIS

Osiris is devoted to thematic issues, conceived and compiled by guest editors who submit volume proposals for review by the Osiris Editorial Board in advance of the annual meeting of the History of Science Society in November. For information on proposal submission, please write to the Editor at osiris@etal.uri.edu.

1. Manuscripts should be submitted electronically in Rich Text Format using Times New Roman font, 12 point, and double-spaced throughout, including quotations and notes. Notes should be in the form of footnotes, also in 12 point and double-spaced. The manuscript style should follow *The Chicago Manual of Style*, 16th ed.

2. Bibliographic information should be given in the footnotes (not parenthetically in the text), numbered using Arabic numerals. The footnote number should appear as superscript. "Pp." and "p." are not used for page references.

   a. References to books should include the author's full name; complete title of book in *italics*; place of publication; date of publication, including the original date when a reprint is being cited; and, if required, number of the particular page cited (if a direct quote is used, the word "on" should precede the page number). *Example*:

      [1] Mary Lindemann, *Medicine and Society in Early Modern Europe* (Cambridge, 1999), 119.

   b. References to articles in periodicals or edited volumes should include the author's name; title of article in quotes; title of periodical or volume in *italics*; volume number in Arabic numerals; year in parentheses; page numbers of article; and, if required, number of the particular page cited. Journal titles are spelled out in full on the first citation and abbreviated subsequently according to the journal abbreviations listed in *Isis Current Bibliography*. *Example*:

      [2] Lynn K. Nyhart, "Civic and Economic Zoology in Nineteenth-Century Germany: The 'Living Communities' of Karl Möbius," *Isis* 89 (1999): 605–30, on 611.

   c. All citations are given in full in the first reference. For succeeding citations, use an abbreviated version of the title with the author's last name. *Example*:

      [3] Nyhart, "Civic and Economic Zoology" (cit. n. 2), 612.

3. Special characters and mathematical and scientific symbols should be entered electronically.

4. A small number of illustrations, including graphs and tables, may be used in each volume. Hard copies should accompany electronic images. Images must meet the specifications of The University of Chicago Press "Artwork General Guidelines" available from the Editor.

5. Manuscripts are submitted to Osiris with the understanding that upon publication copyright will be transferred to the History of Science Society. That understanding precludes consideration of material that has been previously published or submitted or accepted for publication elsewhere, in whole or in part. Osiris is a journal of first publication.

Osiris (ISSN 0369-7827) is published once a year.

Single copies are $33.00.

Address subscriptions, single issue orders, claims for missing issues, and advertising inquiries to *Osiris*, The University of Chicago Press, Journals Division, PO Box 37005, Chicago, IL 60637.

Postmaster: Send address changes to *Osiris*, The University of Chicago Press, Journals Division, PO Box 37005, Chicago, IL 60637.

Osiris is indexed in major scientific and historical indexing services, including *Biological Abstracts*, *Current Contexts*, *Historical Abstracts*, and *America: History and Life*.

Paperback edition, ISBN 978-0-226-02939-9

Osiris

A RESEARCH JOURNAL DEVOTED
TO THE HISTORY OF SCIENCE
AND ITS CULTURAL INFLUENCES

A PUBLICATION OF THE
HISTORY OF SCIENCE SOCIETY

EDITORIAL OFFICE
DEPARTMENT OF HISTORY
80 UPPER COLLEGE ROAD, SUITE 3
UNIVERSITY OF RHODE ISLAND
KINGSTON, RI 02881 USA
osiris@etal.uri.edu